GOVERNING AGROBIODIVERSITY

Global Environmental Governance

Series Editors: John J. Kirton, Munk Centre for International Studies, Trinity College, Canada and Miranda Schreurs, Freie Universität Berlin, Germany

Global Environmental Governance addresses the new generation of twenty-first century environmental problems and the challenges they pose for management and governance at the local, national, and global levels. Centred on the relationships among environmental change, economic forces, and political governance, the series explores the role of international institutions and instruments, national and sub-federal governments, private sector firms, scientists, and civil society, and provides a comprehensive body of progressive analyses on one of the world's most contentious international issues.

Titles in the series

The Social Construction of Climate Change
Power, Knowledge, Norms, Discourses
Edited by
Mary E. Pettenger
ISBN 978-0-7546-4802-4

Governing Global Health
Challenge, Response, Innovation
Edited by
Andrew F. Cooper, John J. Kirton and Ted Schrecker
ISBN 978-0-7546-4873-4

Participation for Sustainability in Trade
Edited by
Sophie Thoyer and Benoît Martimort-Asso
ISBN 978-0-7546-4679-2

Bilateral Ecopolitics
Continuity and Change in Canadian-American Environmental Relations
Edited by
Philippe Le Prestre and Peter Stoett
ISBN 978-0-7546-4177-3

Governing Global Desertification
Linking Environmental Degradation, Poverty and Participation
Edited by
Pierre Marc Johnson, Karel Mayrand and Marc Paquin
ISBN 978-0-7546-4359-3

Governing Agrobiodiversity
Plant Genetics and Developing Countries

REGINE ANDERSEN
Fridtjof Nansen Institute, Norway

ASHGATE

Published by
Ashgate Publishing Limited
Gower House
Croft Road
Aldershot
Hampshire GU11 3HR
England

Ashgate Publishing Company
Suite 420
101 Cherry Street
Burlington, VT 05401-4405
USA

www.ashgate.com

British Library Cataloguing in Publication Data
Andersen, Regine, 1963-
Governing agrobiodiversity : plant genetics and developing countries. - (Global environmental governance series)
1. Plant diversity conservation - International cooperation
2. Plant diversity conservation - Developing countries
3. Plant diversity conservation - Philippines 4. Plant diversity conservation - Government policy - Developing countries
I. Title
639.9'9

Library of Congress Cataloging-in-Publication Data
Andersen, Regine, 1963-
Governing agrobiodiversity : plant genetics and developing countries / by Regine Andersen.
p. cm. -- (Global environmental governance series)
Includes bibliographical references and index.
ISBN 978-0-7546-4741-6 (alk. paper)
1. Agrobiodiversity--Developing countries. 2. Plant diversity conservation--Developing countries. 3. Agriculture and state--Developing countries. 4. Plant varieties--Patents. 5. Genetically modified foods--Law and legislation. I. Title. II. Series.

S494.5.A43A485 2008
333.95'316091724--dc22

2008003431

ISBN 978-0-7546-4741-6

Mixed Sources
Product group from well-managed forests and other controlled sources
www.fsc.org Cert no. SA-COC-1565
© 1996 Forest Stewardship Council
FSC

Printed and bound in Great Britain by
MPG Books Ltd, Bodmin, Cornwall.

Contents

PART 4: DOMESTIC RESPONSES TO THE INTERNATIONAL REGIME CONSTELLATION PERTAINING TO PLANT GENETIC RESOURCES FOR FOOD AND AGRICULTURE: CASES FROM THE PHILIPPINES

PART 5: CONCLUSIONS, RELEVANCE AND CHALLENGES

List of Figures

List of Tables

Preface

In recent decades, the world has seen a steady increase in international agreements dealing with environmental issues. And so, we should assume that the international governance of serious environmental problems has improved, even if the precise impact of international regimes may vary. At the very least, we should assume that the international regimes themselves are not making things worse.

The research presented in this book tells a different story. Despite the original intentions, international regimes pertaining to plant genetic resources in agriculture have in fact led to a situation where many actors now have the possibilities to exclude each other from the access to, and use of, these vital resources. Not only is this a threat to the conservation and sustainable use of plant genetic diversity in agriculture, but it may also seriously affect food security and the outlook for combating poverty in the world.

This situation came about largely as a result of interaction between international regimes and the driving forces behind them: the WTO Agreement on Trade Related Aspects of Intellectual Property Rights (TRIPS), the Convention of the Union for the Protection of New Varieties of Plants (UPOV), the Convention on Biological Diversity (CBD), the International Undertaking on Plant Genetic Resources (IU), and the International Treaty on Plant Genetic Resources for Food and Agriculture (ITPGRFA). This book investigates their interaction in depth, showing how the driving forces behind them transformed and channelled their interests and ideas along various patterns – mechanisms of influence – which are traced across scales from the international to the national level: (1) structural power, that is the power to shape political-economic structures; (2) learning and norm diffusion; and (3) institutional capacity. We will see how the driving forces shaped regime interaction regarding the management of agricultural plant genetic resources through these mechanisms of influence at the international and national levels, and across scales.

With the International Treaty on Plant Genetic Resources for Food and Agriculture, which was adopted in 2001 and entered into force in 2004, the international community has an instrument with the potential to change this negative trend. Whether that will happen, however, depends crucially on the political will of the contracting parties to the Treaty. This book highlights how the identification of mechanisms of influence and their patterns can provide entry points for actors to shape governance in agrobiodiversity at the international as well as national levels. Moreover, it indicates possible ways of breaking out of the vicious circle of today's management of agricultural plant genetic resources, so that we may ensure the continued maintenance of these resources so vital to food security and poverty eradication.

In terms of regime studies, the book develops an analytical framework and a methodology for the analysis of relative aggregate regime effects. Many regime studies have focused on the effectiveness of international regimes – where effects are measured against the norms and rules of the regimes themselves. By contrast, in this study, the aggregate effects of international regimes are measured against a set of criteria developed independently of these regimes. These criteria have been formulated on the basis of a problem analysis. The analytical framework provides a method for precisely identifying the overlap of functional scopes between international regimes, from which regime interaction and assumptions as to aggregate effects can be derived. It further offers a method of tracing these effects across scales to the national level, systematically taking other influencing factors into account.

The focus is on the developing countries. Most of our agricultural plants originate from these countries, and today this is where the greatest genetic diversity can still be found in the fields. Genetic diversity is a form of life assurance for traditional small-scale farmers in developing countries, as it helps to spread the risk of crop failure due to pests and diseases or adverse climatic conditions like drought. Thus, the topic is crucial also from the perspective of poverty eradication. And yet, with their weak institutional capacity, the developing countries face special challenges in implementing the wide array of international regimes. This makes it important to understand the conditions for such implementation processes in developing countries, the driving forces behind them, and, not least, the mechanisms of influence through which these driving forces work.

To enable in-depth investigation of the effects at the national level, case studies were carried out in the Philippines – a developing country that, in addition to being representative in terms of various criteria and rich in agricultural biodiversity, was known for its work in pioneering the sustainable management of biological diversity. Thus, I anticipated, it would be possible to highlight some much-needed positive examples of the effects of international regimes.

Even during my first field trip to the Philippines in 2000 I found my assumptions seriously challenged. This became even clearer in the course of the second field trip two years later. Although considerable effort had been undertaken to conserve hot spots of wild biodiversity in the country, scant political attention was paid to the value of plant genetic diversity in agriculture and the necessity of sustainable management. Gene-bank conservation had some support, but otherwise, all incentive structures pointed towards the accelerated use of modern high-yielding, genetically homogeneous varieties of plants – so important for meeting the food demands of a quickly increasing population – but without measures to compensate for the rapid loss of genetic diversity in the fields. There were a few highly promising projects being carried out by non-governmental organizations – but not enough to balance the picture.

Moreover, I found that the legal space for sustainable management of these resources was being steadily reduced, particularly through legislation on bio-prospecting of genetic resources and on intellectual property rights.

How, I wondered, could this situation come about in a country that had worked so hard to pioneer the implementation of the Convention on Biological Diversity to ensure the conservation and sustainable use of its rich biological diversity and equitable sharing of benefits from the use of these resources?

This book charts the history of the advocacy coalition that promoted the bio-prospecting regulation, and how its short political time-horizon combined with the election of a new president in the Philippines undercut the very foundations for implementation of the much-praised regulation. The book also tells how powerful foreign interests intervened in the promising Philippine efforts that sought to produce legislation on intellectual property rights in line with the needs and priorities of the country as well as the CBD – and, through infiltration and incentives, managed to change the direction of the whole process. When this story came out – along with revelations of similar patterns in a number of other legislative processes – politicians and the public raged, and the newspapers were filled with headlines about foreign take-over and imperialism in the Philippines. But it was too late. The legislation had already been adopted – and it was largely in line with precisely those foreign interests.

International regimes played an important role in both cases. The bio-prospecting regulation was derived from parts of the CBD, selected in an effort to counterbalance the anticipated effects of the TRIPS Agreement. The intellectual property regulation was adopted to comply with the TRIPS Agreement by following much of the UPOV Convention. The precise ways in which implementation took shape were much the result of the interplay between these regimes, the driving forces behind this and their mechanisms of influence.

A key challenge in my research project was to see whether the findings from this one case country had relevance for other developing countries as well. After additional data had been collected from a large number of countries, several important parallels emerged. Low political attention paid to agrobiodiversity issues was found in most developing countries. Furthermore, it could be seen that intellectual property rights, access legislation and seed laws were reducing the legal space required for the sustainable management of these resources in more and more countries. And there were clear indications that some of the same mechanisms of influence that I had found in the Philippines were at play in many countries.

The book has been written with a broad target group in mind. Scholars in the field of environmental studies will find information on the scope for implementing environmental regimes at the domestic level in light of other influencing factors, particularly other international regimes. For scholars of development studies, the book provides an in-depth account of how international agreements are dealt with in developing countries, the factors contributing to their implementation in situations with weak institutional state capacity, and the effects of these dynamics. And scholars of regime studies will find a new analytical framework for grasping the relative aggregate effects of international regimes pertaining to an issue area.

For delegates to the Governing Body of the ITPGRFA and other stakeholders, the book offers a detailed historical account of the developments of these regimes and their interaction. Furthermore, it offers central information on the challenges to, as well as options for, promoting implementation of the Treaty. For stakeholders of other international agreements like the CBD, the TRIPS Agreement and the UPOV Convention, the book provides insights into the effects of the interaction between the regimes, and, it is to be hoped, a basis for better balancing the various concerns involved. Decision-makers and practitioners involved in agrobiodiversity issues will find points of orientation in the jungle of international agreements, as well as prospects and options for policy development and action at the national level.

It is my sincere hope that this book can contribute to the many efforts already underway, aimed at improving the management of plant genetic resources for food and agriculture. I also hope that it will advance our understanding of how international regimes can better be employed as instruments for strengthening global governance in environmental issues.

Regine Andersent
Senior Research Fellow
Fridtjof Nansen Institute

Acknowledgements

I would like to express my sincere gratitude for all the support received for the project that led up to this book. The project was generously financed by the Research Council of Norway (RCN grant no. 134770/730) and the Fridtjof Nansen Institute (FNI). I would like to thank all directors of the Institute during the project period – Kåre Willoch, Arild Moe and Peter Johan Schei – for all their moral support.

Deep appreciation goes to Prof. Arild Underdal of the University of Oslo and Prof. Cary Fowler (then at Noragric, Norwegian University of Life Sciences) for inspiration and useful comments. I would also like to thank FNI Senior Research Fellows Dr G. Kristin Rosendal and Morten Walløe Tvedt for helpful comments and considerable inspiration. I am grateful to Associate Prof. Trygve Berg of Noragric for explaining the biological foundations of the topic and for all his moral support. My former colleague Svanhild-Isabelle Batta Torheim has been a most appreciated sparring partner along the way.

In the Philippines a great many people shared their insights and experiences with me and provided the documentation necessary for my fieldwork. Nicanor Perlas, president of the Centre for Alternative Development Initiatives, provided moral support and help with practical fieldwork arrangements. Neth Daño, then director of Southeast Asia Regional Initiatives for Community Empowerment, opened the brilliant SEARICE archives for me, providing core information. I am grateful to her and to her successor Wilhelmina R. Pelegrina for all their help. Prof. Pamela G. Fernandez (University of the Philippines Los Baños) housed me during my stay in Los Baños and supplied core information from an agronomist's point of view.

Let me also thank Prof. Jose T. Esquinas-Alcazar, former Secretary of the FAO Commission on Genetic Resources for Food and Agriculture in Rome, for all his inspiration, and Prof. M.S. Swaminathan (MS Swaminathan Research Foundation, India), for encouragement along the way. Also Senior Advisor Grethe Helene Evjen (Ministry of Agriculture and Food, Norway) and Senior Advisor Jan P. Borring (Ministry of the Environment, Norway) have been important sources of inspiration and provided valuable comments. I have enjoyed close co-operation with Annette von Lossau of the German GTZ, and wish to thank her for support and interesting discussions. I am also grateful to Prof. Stephen B. Brush (University of California, Davis) for valuable exchanges.

In connection with the Farmers' Rights Project,[1] I have been able to work closely together with Dr Anitha Ramanna of the University of Pune; Regassa Feyissa, director of Ethio-Organic Seed Action; and Manuel Ruiz Muller, director of the Peruvian Society for Environmental Law. They all provided indispensable insights into the situations in India, Ethiopia and Peru.

1 <www.farmersrights.org>.

Several colleagues at FNI have commented on parts of the text. Here let me mention in particular Senior Research Fellows Steinar Andresen, Lars H. Gulbrandsen, Dr Gørild Heggelund, Dr Jon Birger Skjærseth, and Olav Schram Stokke. I would also like to thank Bente Herstad (then in the Development Fund, Norway) and Dr Hanne Svarstad (then at the Centre for Development and the Environment, University of Oslo) for inspiration in the initial stages of this work, and Frøydis Kvaløy (then at Noragric) for professional exchanges.

Thanks in particular to Prof. John Kirton, Dr Miranda Shreurs (editors of this series) and Madeline Koch (University of Toronto) for all their encouragement throughout the process of finalizing this book.

Heartfelt gratitude goes to Susan Høivik (Norwegian Institute of International Affairs, NUPI) for excellent language editing and to Maryanne Rygg (FNI) for professional layout assistance. I also wish to thank Kari Lorentzen, Claes Lykke Ragner and Ivar Liseter (all from FNI), for valuable library and technical assistance.

My deepest appreciation goes to my husband Christian Messenbrink and our daughter Mai Elise, who have supported me so lovingly and patiently throughout this whole project. Heartfelt thanks also to my father, Prof. Emeritus Per Sveaas Andersen, and my mother Elise Marie Andersen for invaluable moral support.

In the end, of course, final responsibility for this text and any shortcomings remains with me.

List of Abbreviations and Acronyms

ACF	Advocacy Coalition Framework
AFMA	Agricultural and Fisheries Modernization Act (Philippine)
AGILE	Accelerating Growth, Investment, and Liberalization with Equity – a program under USAID
APAARI	Asia-Pacific Association of Agricultural Research Institutions
ASOMPS VII	Seventh Asian Symposium on Medicinal Plants, Spices and Other Natural Products
ASSINSEL	International Association of Plant Breeders for the Protection of Plant Varieties
BFAR	Bureau of Fisheries and Aquatic Resources (Philippine)
CADI	Centre for Alternative Development Initiatives (Philippines)
CADT	Certificates of Ancestral Domain Titles (Philippines)
CBD	Convention on Biological Diversity
CBDC	Community Biodiversity Development and Conservation Program
CGIAR	Consultative Group on International Agricultural Research
CGRFA	Commission on Genetic Resources for Food and Agriculture
CIAT	Centro Internacional de Agricultura Tropical (International Center for Tropical Agriculture)
CIMMYT	Centro Internacional de Mejoramiento de Maiz y Trigo (International Maize and Wheat Improvement Center)
CIP	Centro Internacional de la Papa (International Potato Center)
CIPR	Commission on Intellectual Property Rights
CONSERVE	Community-based Native Seeds Research Center Inc. (Philippines)
COP	Conference of the Parties (CBD)
CORRA	Council for Partnership on Rice Research in Asia
CPGR	Commission on Plant Genetic Resources (FAO – later named CGRFA)
CPP	Communist Party of the Philippines
CTE	Committee on Trade and Environment (WTO)
DA	Department of Agriculture (Philippine)
DAI	Development Alternatives, Inc., a US-based consultancy firm
DAO	Department Administrative Order (Philippines)
DENR	Department of Environment and Natural Resources (Philippine)
DOH	Department of Health (Philippine)
DOST	Department of Science and Technology (Philippine)

DUS	Genetic **d**istinctness, **u**niformity and **s**tability (UPOV criteria)
EFTA	European Free Trade Association
EO 247	Executive Order 247 on the Prospecting of Biological and Genetic Resources (Philippines)
ERDB	Ecosystems Research and Development Bureau (Philippine)
ETC Group	Action Group on Erosion, Technology and Concentration (formerly RAFI)
EU	European Union
FAO	Food and Agriculture Organization of the United Nations
GATS	General Agreement on Trade in Services
GATT	General Agreement on Tariffs and Trade
GEF	Global Environment Facility
GFAR	Global Forum on Agricultural Research
GMA	*Ginintuang Masaganang Ani* – High Value Commercial Crops Program (Philippines)
GMO	Genetically Modified Organisms
GPA	Global Plan of Action for the Conservation and Sustainable Utilization of Plant Genetic Resources for Food and Agriculture
GRAIN	Genetic Resources Action International, a NGO
GSP	Generalized System of Preferences
GTZ	Gesellschaft für Technische Zusammenarbeit (German donor organization)
GURT	Genetic Use Restriction Technology
IACBGR	Inter-Agency Committee for Biological and Genetic Resources (Philippines)
IARCs	International Agricultural Research Centers
IBPGR	International Board for Plant Genetic Resources
ICRISAT	International Crops Research Institute for the Semi-Arid Tropics
IEC	Integrated Information, Education and Communication system for biodiversity conservation
IGC	Intergovernmental Committee on Intellectual Property and Genetic Resources, Traditional Knowledge and Folklore (WIPO)
IITA	International Institute of Tropical Agriculture
IPGRI	International Plant Genetic Resources Institute (now Biodiversity)
IPM	Integrated Pest Management
IPO	Intellectual Property Office (Philippine)
IPR	Intellectual Property Rights
IPRA	Indigenous Peoples' Rights Act (Philippine)
IR	International Relations
IRR	Implementing Rules and Regulations

IRRI	International Rice Research Institute
ITPGRFA	International Treaty on Plant Genetic Resources for Food and Agriculture
IU	International Undertaking on Plant Genetic Resources
IUCN	World Conservation Union
KMP	*Kilusang Magbubukid ng Pilipinas* (Peasant Movement of the Philippines)
LDCs	Least-Developed Countries
MASIPAG	*Magsasaka at Siyentipiko Para sa Pag-unlad ng Agrikultura* (Farmer-Scientist Partnership for Development, Inc.)
MAT	Mutually Agreed Terms
MSI	Marine Science Institute (Philippines)
MTA	Material Transfer Agreement
NAST	National Academy of Science and Technology (Philippines)
NBSAP	National Biodiversity Strategy and Action Plan (Philippines)
NCIP	National Commission on Indigenous Peoples (Philippine)
NCPGR	National Committee on Plant Genetic Resources (Philippine)
NDUS	Novelty and genetic distinctness, uniformity and stability (UPOV criteria)
NEDA	National Economic and Development Authority (Philippine)
NIPAS Act	National Integrated Protected Areas System Act
NGO	Non-governmental organization
NPA	New People's Army (Philippines)
NPGRL	National Plant Genetic Resources Laboratory (Philippine)
ODA	Official Development Assistance
OECD	Organization for Economic Co-operation and Development
PA 21	Philippine Agenda 21
PAWB	Protected Areas and Wildlife Bureau (Philippine)
PCARRD	Philippine Council for Agriculture, Forestry and Natural Resources Research and Development
PCSD	Philippine Council for Sustainable Development
PGRFA	Plant Genetic Resources for Food and Agriculture
PhilRice	Philippine Rice Research Institute
PIC	Prior Informed Consent
POs	People's Organizations
PSBDC	Philippine Strategy for Biological Diversity Conservation
PSC	CGIAR Private Sector Committee
PSIA	Philippine Seed Industry Association
PVP Act	Plant Variety Protection Act (Philippines)
RAFI	Rural Advancement Foundation International (later renamed ETC Group)
SBSTTA	Subsidiary Body on Scientific, Technical and Technological Advice (CBD)

SCP	Standing Committee on the Law of Patents
SEARICE	Southeast Asia Regional Initiatives for Community Empowerment
SINGER	System-wide Information Network for Genetic Resources
SMTA	Standard Material Transfer Agreement
SPLT	Draft Substantive Patent Law Treaty
TRIPS	Agreement on Trade Related Aspects of Intellectual Property Rights
UNCED	United Nations Conference on Environment and Development
UNDP	United Nations Development Program
UNEP	United Nations Environment Program
UNESCO	United Nations Educational, Scientific, and Cultural Organization
UPOV	Union for the Protection of New Varieties of Plants
US	United States of America
USAID	US Agency for International Development
WIEWS	World Information and Early Warning System on Plant Genetic Resources for Food and Agriculture
WIPO	World Intellectual Property Organization
WTO	World Trade Organization
WWF	Worldwide Fund for Nature

Plant genetic resources – one of the most fundamental and essential of all resources on Earth – are seriously threatened. Their loss will touch each one of us and endanger future generations.
Jacques Diouf[1]

To inherit is an art in itself; for inheriting, in the final analysis, is culture.
Thomas Mann[2]

---

1 Director-General, FAO. From the foreword in the Global Plan of Action for the Conservation and Sustainable Utilization of Plant Genetic Resources for Food and Agriculture (FAO 1996a, 3).

2 (1875–1955). Original text: '*Aber schließlich, zu erben muss man auch verstehen, erben, das ist am Ende Kultur*'. From a letter to Klaus Mann in 1939 (Mann 1963, 107). Official English translation confirmed per mail by Buddenbrookhaus, 6 March 2006.

PART 1
Introduction

When the first hunters and gatherers became farmers some 10,000 years ago, they started out with only a few crops and varieties to grow. Through careful selection of the best seeds and propagating material, and exchange with other farmers, it became possible to develop these few varieties into many. Also new crops were found in the wild that could be cultivated. Through the continuous management and innovation by farmers over thousands of years, the few initial crops and varieties evolved into an unconceivable wealth of crop diversity. An estimated 7000 species of crops have been cultivated or collected by humans for food (Wilson 1992, 275), and the estimated number of distinct varieties of each of these crops varies up to more than 100,000[1] (FAO 1998, 18).

However, the development of crop diversity changed profoundly with the Green Revolution.[2] Whereas the Green Revolution contributed to a substantial increase in food production, it also decimated untold food crop varieties that were vital to small-scale farmers and the future of plant breeding. In an attempt to rectify the loss, domestic and international gene banks were established. Gene banks are now essential to conserving and maintaining the varieties that were – and still are being – saved. However, widespread genetic erosion is also taking place in some, perhaps even many, gene banks, due to poor management and poor maintenance (FAO 1998, 129), which can be seen as the result of scarce financial resources as well as limited institutional capacities. In addition, legal restrictions on access to still-available genetic resources are emerging as an increasing problem for all stakeholders in agriculture. For these reasons, the maintenance of plant genetic diversity in agriculture is becoming increasingly difficult for the traditional farmers who are the custodians of plant genetic diversity in the fields. In light of the 10,000 years of farming, the current crisis in the management of plant genetic resources for food and agriculture (PGRFA)[3] is evolving with dramatic speed.

International regimes are perhaps the most important instruments available to the international community for meeting such a crisis. But how do the current international regimes work? How do they meet the challenges of PGRFA management? How do they actually affect such management? These are the questions addressed in this book, which focuses on the effects in developing countries, where the diversity of PGRFA is particularly important for food security among the rural poor.

---

1 This figure is for rice.

2 The 'Green Revolution' refers to the increase in food production due to the introduction of improved, high-yielding and genetically homogeneous varieties of wheat, rice, maize and other cereals in the 1960s.

3 For a definition of PGRFA and the foundations of the concept, see chapter 2.

This first part of the book explores the relevance of the issue, determines the objective of the analysis, and provides a brief overview over the contents of the book (Chapter 1), before the scientific foundations of the topic are established (Chapter 2) – as the basis and context for the ensuing analysis.

Chapter 1

Relevance and Objectives of the Study

This chapter details why the PGRFA management is so important for food security, poverty eradication and future generations – and it shows how this book aims to contribute to the growing research body on international governance.

Relevance for Food Security and the Fight Against Poverty

Biological diversity has usually been associated with wild animals and plants. There has generally been less political awareness of the crucial importance of genetic diversity in agriculture – and yet, plant genetic diversity is decisive to the breeding of food crops and is thus one of the central preconditions for food security throughout the world.

Access to diverse genetic resources is vital to modern plant breeding, as it provides the genetic traits required to deal with crop pests and diseases, as well as with changing climate conditions. History can offer many examples of the crucial importance of genetic resources for food security and human survival.[1] During the potato blight in Ireland in the 1840s, for instance, more than one million people starved to death and another one to two million migrated to America while the harvests rotted in the soil. The genetic resources that provided the necessary resistance to potato blight were finally found in the Andes and in Mexico and subsequently brought to Ireland, putting an end to the crisis. In 1870, coffee rust wiped out the coffee industry in huge parts of Asia and Africa, thereby eliminating the economic basis for untold numbers involved in production. Cotton was hit by epidemics in the 1890s. Rust struck the US wheat crop in 1904. Brown spot disease devastated India's rice crop, resulting in the disastrous Bengal Famine of the 1940s. Corn blight struck the US in the 1970s. These and other examples have one element in common: the genetic resources needed to produce the necessary resistance against the various pests and diseases were still available and accessible. Without these genetic resources, the battles would have been lost. In order to avoid crises like these, modern plant breeders depend on the diversity of PGRFA.

Plant genetic diversity is also essential for traditional small-scale farming, on which approximately 1.4 billion people worldwide depend for their livelihoods.[2] Traditional small-scale farmers maintain the yields and quality of their crop varieties by saving and exchanging seeds and seedlings, as they have done since the dawn of

1 All examples in this paragraph are from Fowler and Mooney (1990, 42–53).

2 Approximately 1.4 billion people live in farm families that are largely self-reliant and self-provisioning as to seeds and other planting materials, according to Cary Fowler et al. (in Brush 2000, 1).

agriculture. Without genetic renewal, yields will decrease and quality will deteriorate. Plant genetic diversity is vital in marginal environments in order to have plants that can adapt to difficult conditions. Diversity between and among crops is a means of spreading the risk of crop failure due to pests and diseases or adverse climatic conditions like drought. Today, 75 per cent of the world's poorest 1.2 billion people live in rural areas and depend on farming (IFAD 2001). Therefore, plant genetic diversity is an indispensable factor in the fight against poverty, and is thus a central precondition for achieving UN Millennium Development Goal 1 on the eradication of extreme poverty and hunger.

However, the diversity of domesticated plant varieties is disappearing at an alarming rate. For several major crops, losses in variety have reportedly been up to 80–90 per cent over the past century.[3] At the same time, interest in the commercial use of genetic resources has increased with the growing economic stakes of biotechnologies, followed by demands for intellectual property rights. As patent systems are costly institutions, the capacity of developing countries to develop and effectively use such systems is limited. There has been much protest against intellectual property rights from the South, along with demands for protecting the rights of farmers and indigenous peoples and ensuring a fair and equitable sharing of the benefits arising from the use of these resources.

The international community has responded to these developments with several regimes that pertain fully or partly to PGRFA: the Convention on Biological Diversity (CBD), the Agreement on Trade Related Aspects of Intellectual Property Rights (TRIPS) of the World Trade Organization,[4] and the International Treaty on Plant Genetic Resources for Food and Agriculture (ITPGRFA) with its predecessor, the International Undertaking on Plant Genetic Resources (IU). These regimes have emerged from differing rationales and interests, resulting in more or less different functional scopes, goals and emphases. What they all have in common is that they affect the management of PGRFA. Since multilateral agreements are the most important instruments available to the international community for influencing PGRFA management, it is central to understand how they work and interact – and how this affects the situation at the national level.

Given the crucial importance of agro-biodiversity for the rural poor in developing countries, it is particularly important to study the effects of these international regimes in developing countries. The hope is that this book can be a contribution to the ongoing work of improving the management of these vital resources – as well as contributing to the efforts to combat poverty.

3 There are few exact figures on the extent and pace of genetic erosion in agriculture. However, nearly all the 154 countries reporting to the FAO for the Leipzig Conference in 1996 (FAO 1998) maintained that genetic erosion was a serious problem. In China an estimated 90 per cent of the 10,000 wheat varieties grown a century ago have been lost. In Mexico an estimated 80 per cent of the maize varieties grown in the 1930s are gone. These are estimates reported by the countries themselves.

4 Following from the provisions in the TRIPS Agreement on intellectual property rights to plant varieties, also the International Union for the Protection of New Varieties of Plants, UPOV, has become relevant in this context, as will be further explained in this book.

Relevance in an Intergenerational Perspective: Ethics

The food security of future generations depends crucially on agricultural genetic diversity. There are no means known to us today that could compensate for that.[5] We are the beneficiaries of all the plant varieties passed on to us by previous generations – and yet, the present generation has the power to eradicate most of the genetic diversity needed by future generations. The ongoing erosion of PGRFA combined with the mounting regulative hurdles to the management of these vital resources give rise to the moral question of our responsibility towards future generations.

Terence Ball (2001) describes how our sense of responsibility has changed during the past centuries. Every ethical code contains an implicit time-scale. Through the last millennia of deep Christian influence on Western civilization, the norm spread that all individuals should act towards posterity as benefactors. The prevailing ethics was one of stewardship. As Edmund Burke put it in 1790 (referred to in Ball 2001, 93), each generation ought to be caretakers, responsible to their ancestors and their distant descendants alike. The first philosopher to question these values was Thomas Paine, who stated in 1791 that each generation should be autonomous and free to do as its individual members pleased without regard to the wishes of their ancestors or the well-being of their successors (referred to in Ball 2001, 93). Paine claimed the idea of intergenerational partnership to be 'the most ridiculous and insolent of all tyrannies' (ibid.). Ball holds this contribution to be an important part of the root system that shaped our modern mentality. Since the global environmental problems caused by modern civilization came on the agenda, such norms have increasingly been questioned and criticized by many philosophers, other intellectuals and groups in civil society.

The question of intergenerational responsibility is particularly relevant in moral philosophy. One group of contemporary philosophers argues that the moral justification for responsible action is based on the norm of reciprocity (for example, Balakrishnan and Narayan 1996; Pogge 1999 and 2001; Shue 1996). Ball (2001, 95) maintains that such a norm may work well among contemporaries but not between generations. The relationship between generations, he says, is asymmetric in its very nature and can thus not be one of reciprocity. If we have duties only towards those who are in a position to reciprocate by respecting our rights, the idea is impossible to apply for issues that concern other generations than those of our own times.

Another group argues that the norms of humanity form the moral justification for responsible action (Singer 1972; Watson 1977; Goodin 1985; O'Neill 1993). This approach uses the perspective of vulnerability. 'Vulnerability' in this sense is not about the probability of getting harmed, but about the need for help in order for life to be good and not turn bad. It is closely related to dependence. The work of Goodin is especially applicable with regard to the question of intergenerational responsibility for PGRFA management.

Goodin's (1985) point of departure was that we feel particularly strong moral claims towards our families and other close persons, claims that override those we may have to aiding others at large. This reflects a sense of vulnerability. We

5 It should be borne in mind that also modern biotechnology requires genetic diversity.

feel responsible towards those who are vulnerable to us, and the extent of our responsibility is connected to the extent of vulnerability of the persons to us. When vulnerability is the factor that generates responsibilities, we feel the same sort of strong responsibilities towards all those who are vulnerable to our actions and choices. Goodin found that obligations of the present generation towards future generations derive from the actions of past generations. We would not be where we are today without them. A logical way of looking at intergenerational relations could thus be through the concept of 'trusteeship'. As trustees, we would have obligations 'both to the benefactors who created the trust and the intended beneficiaries of the trust' (Goodin 1985, 175). As future generations are completely dependent on us, their vulnerability to our actions and choices would be 'the strongest basis for assigning to present ones strong responsibilities for providing for them' (ibid., 177). Goodin admits that there could be some scope for discounting of the future, due to the possibilities that future generations may solve their problems in other ways. However, he maintains, 'that does not allow for anything like the dramatic discounting that is practised at present' (ibid.).

As future generations are extremely vulnerable to us with regard to plant genetic diversity to ensure their food security, from a moral perspective it is necessary to seek ways and means to stop genetic erosion and ensure the sustainable management of these vital resources. This, it can be argued, is a moral obligation. This book is meant as a contribution towards this end.

Relevance for the Study of International Governance

In a globalizing world with mounting environmental and social problems across borders, international co-operation is fundamental. On the other hand, the proliferation of a growing number of international regimes in recent decades also poses a new challenge for international governance: increasingly more than one international regime pertains to one and the same issue area, and regimes often pull in different directions – as is the case with PGRFA management. There is thus an urgent need to grasp analytically the dynamics of interaction evolving between and among these regimes – and their effects for the issue areas in question.

Interaction between international regimes is a new field of study in political science. The increasing body of work on regime studies addresses regime formation and implementation, including the effectiveness of international regimes (see, among others: Haas et al. 1993; Young 1996 and 1997; Hasenclever et al. 1997; Victor et al. 1998; Wettestad 1999; Miles et al. 2002). However, most implementation studies have been confined to OECD countries, and few have focused on developing countries (see Keohane and Levy 1996; Gupta 1997; Mugabe et al. 1997; Rosendal 2000; Svarstad and Dhillion 2000).

An increasing number of studies address regime interaction (for example, von Moltke 1997; Zhang 1998; Rosendal 2000 and 2006; Stokke 2000, 2001a and 2001b; Gehring and Oberthür 2004). Most of these studies investigate interaction at the international level and/or effects on regime implementation at the national level.

Analyzing the effectiveness of regime implementation is, however, in principle a different task from studying the effects of interacting regimes in a country, and requires a basically different analytical grasp. Regime effectiveness concerns the achievement of regime objectives, with implementation measured against the objectives. The effects of international regimes involve behavioral changes regarding an issue area, changes that can be traced back to the regimes in question – or, with overlapping regimes, to regime constellations. Here, however, they are measured not against the regime objectives, but against a set of criteria developed independently of these with regard to the issues at stake – in our case, PGRFA management. Understanding the combined effects of international regimes for specific issue areas is vital in a world where international governance is becoming increasingly more complex, as it may provide crucial correctives to the course of development.

It has not been possible to find studies that have sought to investigate systematically and in depth the effects of overlapping international regimes at the national level this way. This book thus seeks to contribute to regime studies by developing a framework for analyzing issue-specific effects at the national level of interacting international regimes.

Moreover, it is vital to understand the mechanisms of influence through which overlapping international regimes transpose into national policies and goal achievements at the national level. The more we know about these processes, the better will we be able to design international regimes and determine the action required at the international level to ensure compliance in line with the intentions behind these regimes – and, it is hoped, avoid undesired effects. A further contribution of this book is therefore to offer a theory model and research design for analyzing mechanisms of influence.

Finally, by focusing on developing countries, I seek to grasp analytically the challenges facing countries that have only limited institutional capacity to respond to the rapidly increasing number of international agreements with partly overlapping regulations. In establishing an analytical framework of mechanisms of influence, particular account will be taken of developing countries and their institutional capacities for complying with international regimes.

Objectives and Structure of the Book

On this background, the overall objectives of the book are to contribute to our understanding of how international regimes affect PGRFA management in developing countries – and to develop an analytical framework for studying the issue-specific effects of international regimes at the national level.

Chapter 2 of this first part explores what PGRFA are from a biological angle as well as in a cultural and socio-economic context. It establishes the scientific foundations of the book, which is necessary to understand and assess international processes and national policies pertaining to this issue area.

Part 2 outlines the research design. In Chapter 3 the research questions are posed, the analytical framework developed and propositions formulated with regard to mechanisms of influence of international regimes on PGRFA management.

An important element of Chapter 3 is the identification of criteria for PGRFA management, against which international regimes, regime constellations and national policies will later be assessed. Chapter 4 details a research strategy for analyzing relative aggregate regime effects and explaining them.

Part 3 is devoted to the analysis of the international level. Here we begin by studying the development of each of the international regimes in question – with a particular view to relevance for PGRFA management – and identify norms and rules in that regard in Chapters 5–7. This forms the basis for the analysis in Chapter 8 of the overlaps of functional scopes between the regimes, their interaction in these regards and resultant regime constellations for developing countries. On this basis aggregate effects for developing countries are proposed – taking into account an analysis of the driving forces behind the regime developments and their mechanisms of influence.

These assumptions are in turn tested in case studies from the Philippines in Part IV, where the relative effects of the international regime constellations on PGRFA management are analyzed. Here a bottom-up approach is applied, putting the propositions aside for a while, in order to assess the state of PGRFA management in terms of policy decisions and goal achievements independently from our assumptions. We begin, in Chapter 9, with a brief overview over contextual factors before identifying and presenting all relevant policies and legislation. On this background we analyze the state of policy decisions and goal achievements according to the criteria developed for PGRFA management (in Chapter 3). Taking these findings as points of departure, we identify the extent to which international regime constellations have affected this management – and, not least, how they have affected PGRFA management. In Chapter 10, we analyze in depth two Philippine policies of particular importance to the issue area, in order to identify the mechanisms of influence of international regimes. The findings are explained in the light of a theory framework on mechanisms of influence of international regimes (developed in Chapter 3), taking the involved actors as points of departure.

Part 5 sums up the findings, discusses their relevance for other developing countries, and presents the conclusions. In Chapter 11, we draw together the findings across scales and discuss their general relevance for other developing countries. In Chapter 12, conclusions are derived and implications discussed with regard to international governance, PGRFA management, and future research.

Chapter 2

Plant Genetic Resources for Food and Agriculture: Foundations of the Topic

Political decisions, structures, processes and options cannot be adequately analyzed unless there is a basic understanding of the empirical context in which they are embedded. Whereas the management of PGRFA is a political issue, PGRFA as such is basically a biological topic that is closely interconnected with culture and socio-economics. This chapter presents a brief introduction to these foundations.

Plant Genetic Resources for Food and Agriculture

Plant genetic resources for food and agriculture (PGRFA) encompass the diversity of genetic material in traditional varieties and modern cultivars, as well as crop wild relatives and other wild plant species used as food, fodder for domestic animals, fiber, clothing and shelter, according to the Food and Agriculture Organization of the United Nations (FAO 1998, 1). To understand PGRFA, we need to start out with the ecosystems where they originated and the agro-ecosystems in which they are cultivated. We will then proceed into the parts and explain the relevance of this knowledge for PGRFA management along the way.

Ecosystems and Agro-Ecosystems

An ecosystem comprises the organisms living in a particular environment, and the physical part of the environment that impinges on them (Wilson 1992, 396). The term 'ecosystem' was introduced by Sir Arthur Tansley in 1935, as a way to describe an organized unit of the environment. The concept was 'the idea of progress towards equilibrium, which is never, perhaps, completely attained, but to which approximation is made whenever the factors at work are constant and stable for a long enough period of time' (Tansley 1935, quoted in Odum 1997, 43). An ecosystem has all the components necessary for its organisms to function, and to survive in the long term.

All domesticated agricultural plant varieties have their origin in terrestrial ecosystems, and as long as the related ecosystems are intact, it is often possible to identify their wild relatives there. However, with the rapid extinction of terrestrial ecosystems around the globe, the natural habitats of wild relatives to our agricultural plant varieties are disappearing at a fast pace. This represents a serious challenge to PGRFA management.

There are great differences between agro-ecosystems with regard to their development towards ecological equilibrium. Industrial agro-systems tend to have a high input of energy in the form of fertilizers and of pesticides. Instead of striving towards equilibrium with the environment, such systems tend to combat their environments in order to achieve the highest yields in the shortest possible time (Kellman and Tackaberry 1997, 234–67). By contrast, traditional farming systems are usually better adapted to the environment. Although they may not boast the same maximum yields as industrial agrosystems, their adaptability to the local environments tends to make them more stable in the long run. One reason is that traditional farming systems are usually more diverse with regard to PGRFA than industrial agrosystems, and thus more genetically flexible (see section on varieties and cultivars below). As a reaction to the environmentally negative effects of industrial agrosystems, the concept of integrated pest management (IPM) has been introduced in many parts of the world (Robinson 1996, 109). It still involves only a fraction of the total area of industrial agriculture, but is widely recognized as successful and with potentials for wider diffusion. For these reasons, traditional agriculture is of particular value in PGRFA management.

Species

For a non-biologist, the meaning of the term 'species' might seem obvious at first glance. There is an evident difference between apples and plums and between roses and lilies, so what is the problem? However, as Mayr (1982, 870) notes, the number of species definitions in the biological literature is virtually unlimited. This is due to the enormous amount of variation within and between what we may label as species, and the difficulty of drawing clear demarcations between higher and lower taxa (ibid., 252). A basic question is how and to what extent groups of individuals must differ from each other for them to be labeled as separate species.

A definition that has found wide recognition is that a 'species is a reproductive community of populations (reproductively isolated from others) that occupies a specific niche in nature' (Mayr 1982, 273). It is not unproblematic to apply this definition on agricultural diversity. For the sake of simplicity we may assume that a 'species' comprises such agricultural varieties and their wild relatives as can be interbred by using traditional methods – and this may be sufficient for a range of plant types. However, for the world's oldest and most widespread cultivated plants, the definition does not help us to understand their distinctions and interrelations. The wheat genus is a classical example. As Darwin remarked, there is doubt as to the number of wheat species, as they tend to be ranked as 'varieties' (Darwin 1868, Vol. 1, 334). In the 1920s, Vavilov collected massive amounts of wheat samples from Europe, Asia and Africa. He found no less than six different species with altogether twenty so-called subspecies among them – in addition to all the varieties (Vavilov 1949/50, 175–6). Harlan and deWet (1971, in Harlan 1992, 106–113) sought to solve such problems by recommending that all formal taxonomy be abandoned in order to permit the use of an informal system targeted at the genetic potentials of cultivated plants, to be used by those who work professionally with cultivated plants.

This discussion is relevant for the political aspects of PGRFA management because it illustrates the difficulties in assessing the diversity of PGRFA as a basis for policy development and for measuring impact.

Varieties and Cultivars

Varieties and cultivars are the sub-populations of species and sub-species. But taxonomy does not get any easier when it comes down to the parts. For wild plant populations, ecologists use the term 'ecotype'. An ecotype possesses special characteristics suited to its own particular locality within the ecosystem, and is genetically diverse as well as genetically flexible (Robinson 1996, 58). 'Genetic flexibility' means that a population responds to selection pressures in that its genetic composition changes and it adapts. The flexibility is dependent on the variability of the genetic diversity within the varieties, because the different individuals need to have the reproductive advantage of the others in order to respond (ibid., 57–8).

For domesticated plants the picture is more complicated. Several different terms are used, without clarity. The terms 'race', 'variety' and 'cultivar' are often understood as more or less synonymous. For that reason, Harlan even recommended avoiding the term 'variety', and substituting the term 'race' instead (1992, 110). However, there are major differences in genetic variability between the most ecotype-like varieties or races used by subsistence farmers and the distinct, stable and uniform ones produced by professional plant breeders – and there are many nuances among them. We need to have good concepts to highlight these differences and their interrelations if we are to understand the dynamics of PGRFA erosion and their effects, and be able to identify entry points for good management.

According to one standard dictionary (Hornby 1974, 969), 'variety' in biological terms means 'subdivision of a species'. We may thus use 'variety' as a common denominator for different categories pertaining to genetic characteristics. The categorization set out in Figure 2.1 is an approach to subdividing groups of varieties according to criteria of genetic uniformity, stability and flexibility which together provide an indication of genetic diversity (based on Esquinas-Alcázar 1993, 37; Zeven 1998; Fowler and Mooney 1990, xv; and Robinson 1996).

CATEGORIES OF AGRICULTURAL PLANT VARIETIES				
Ecotypes (wild and weedy forms)	Land races	Farmers' varieties	Creole varieties	Cultivars (+ breeding lines & genetic stocks)
LEAST ⇐		**Degree of:** Genetic inflexibility Genetic uniformity and stability Commercialisation		⇒ **MOST**

Figure 2.1 Categories of agricultural plant varieties

The term '*ecotypes*' has already been explained. *Land races* are traditional local varieties that have evolved over centuries or millennia mainly due to natural selection, being shaped by the land. They are specifically adapted to local conditions, tend to be named after the place they are grown, and possess considerable diversity within and between them (Robinson 1996, 182; Esquinas-Alcázar 1993). Zeven (1998) describes them as varieties with a high capacity to tolerate biotic and a-biotic stress, resulting in high yield stability and an intermediate yield level under a low input agricultural system. *Farmers' varieties* are varieties that have been developed by farmers – by exchange and careful selection, in the search for desirable traits, in addition to natural selection. They tend to carry names that describe characteristic traits and qualities. *Creole varieties* is a term originating from Latin America, describing varieties that are mixes between farmers' varieties and cultivars. Farmers introduce modern cultivars in their fields to increase yields or for other purposes, re-selecting the best combinations from the old and new varieties. These developments are difficult to capture, but there is reason to believe that such practices are diffusing in developing countries.[1]

A *cultivar* (*culti*vated *var*iety) is usually bred after pure line principles (as discovered by Johannsen 1903) and the pedigree method as a clone or a hybrid variety, and is genetically uniform (Robinson 1996, 414). In this way a cultivar normally meets the 'DUS' criteria – genetic distinctness, uniformity and stability – as defined by the International Union for the Protection of New Varieties of Plants (UPOV). Usually these criteria are applied for the national certification of seeds as well. If a plant variety is DUS and satisfies the requirements of novelty, it can be subject to intellectual property rights protection according to UPOV criteria. *Breeding lines* refers to the material developed by plant breeders as intermediate products in the process of producing a cultivar. They usually possess a narrow genetic base, as they are often developed on the basis of a small number of varieties or populations. *Genetic stocks* refer to other genetic combinations – such as genic, chromosomal or genomic mutants that also result from the process of developing a cultivar. This material cannot be defined as varieties, but is included in this category due to its ties to the cultivar. Breeding lines and genetic stocks can be valuable for later breeding efforts, and are thus part of the genetic diversity that can be conserved *ex situ*.

There is a gradual change between the different categories of varieties. However, the distinctions are important for understanding the level of diversity within the varieties. Although a cultivar may have a relatively broad genetic base, it is uniform and thus also genetically inflexible (Robinson 1996, 178). Cultivars are normally genetically disposed to be higher yielding, but not to respond to selection pressures. The greatest genetic diversity and flexibility is found among varieties at the other end of the spectrum. These varieties are genetically disposed to be lower yielding, but they have high response to selection pressures.

As Figure 2.1 indicates, cultivars are the most interesting varieties in terms of commercialization. Due to the rapid expansion of modern agriculture, cultivars are increasingly dominant in the fields. This results not only in genetic erosion in the fields where traditional crops are replaced, but – due to the massive replacement

1 See for example Salazar et al. (2007), documenting such processes in the Philippines.

of traditional varieties – also in reduced access to such varieties for those farmers who continue to grow crops the traditional way. It is not unusual for farmers to cross their traditional varieties with cultivars, thus paving the way for 'creole' varieties. Although 'creole' varieties may provide good alternatives for farmers, the consequence of expanding such practices is the further erosion of genetic diversity among and between agricultural plant varieties in the fields.

Genes

Genes are understood as units of inheritance, carried on chromosomes. They have information pertaining to the heritability of particular traits, whereby an inherited character may be controlled by a single gene, or by many. Genes have been described as sequences of base pairs in DNA molecules that contain information for the construction of protein molecules.

The molecular revolution has rapidly advanced our understanding of heritability, but the exact meaning and function of 'gene' is still not fully clarified. For example, the way in which proteins are built is a complex process in which the role of genes has not been fully determined. Furthermore, many of the genes in the DNA molecule never get translated into proteins (Dawkins 1976, 47 in Dawkins 1982, 156), and a gene can be turned on and off whenever its product is needed (Mayr 1997, 167) but the driving force behind this dynamic has not been identified. For the sake of simplicity, we may say that a gene contains information on heredity, and that the combination of genes is one important determinant for the development of traits of an organism or – more specifically for our purposes – in a plant variety.

Gene technology is particularly relevant with regard to PGRFA management because of its ability to modify organisms so that their genetic traits change. As an organism is a biological entity that is viable and capable of reproducing or transferring genetic material, genetic traits may not only be passed on to the direct offspring of genetically modified organisms (GMO) – thus creating new varieties of plants – but may also spread to previously existing varieties, thereby altering their genetic composition and traits. This is a serious concern for PGRFA management. A further development of gene technology is Genetic Use Restriction Technology (GURT), often referred to as 'terminator technology', which enables the production of plants with sterile seeds. This makes it impossible to use farm-saved seeds, as these will not be capable of reproduction. If spread to other varieties of the same species in the fields, their reproductive vigor may be seriously damaged. Therefore GURT is also a serious concern for PGRFA management. Whereas the use of GMOs is rapidly increasing, GURTs have not yet been released for commercial production.

Although it is important to be aware of the implications of GMO and GURT for PGRFA management, we will not include these topics in this analysis. For one thing, that would require comprehensive additional investigations not feasible within the limits of this project. Second, considerably more attention is being paid to these topics worldwide than to issues of 'traditional' PGRFA management. Thus the focus of this book has been deliberately restricted to 'traditional' PGRFA management.

Genotype, Phenotype and Genetic Base of Plant Varieties

The genotype is the total genetic constitution of an organism, its genetic makeup, as opposed to its physical appearance (based on Mayr 1982, 958–9; Rosenfield and Ziff 1983, 9). A phenotype is the totality of the characteristics of an individual, its observable properties. The genotype and the environment are important determinants for the development of the phenotype. Plant breeders distinguish between crops with a narrow genetic base (such as many vegetables) and those that have a broad genetic base (like many cereals). Such knowledge is important when developing strategies for the conservation and maintenance of PGRFA (see Cooper et al. 2001).

Wild and Cultivated Plants

Wild plants normally reproduce as long as their ecosystems remain intact. In other words, their continued existence depends on protection from serious interference by human beings. For cultivated plants it is the opposite: Their continued existence depends on farmers and breeders and continuous propagation – they cannot exist without the interference of human beings. These basic distinctions have implications for the management of these resources. Wild plants are dependent on conservation and sustainable use for their further existence. For the further existence of cultivated plants, access to diverse genetic resources for farmers and breeders is the crucial factor. The difference between wild and cultivated plants is summed up in Table 2.1.

Table 2.1 Difference between wild and cultivated plants

	Wild Plants	**Cultivated Plants**
Key characteristics with regard to regeneration	Plants regenerate without human intervention	Plants regenerate due to human intervention
Key challenges with regard to management	Maintenance depends on eco-systems conservation	Maintenance depends on access to propagating material

The main focus of this book is on domesticated plant varieties/cultivars.[2] The reason is that it delimits the scope for this book in a way that enables feasible analysis. If wild relatives were to be included, we would have to deal with such topics as

2 *Domestication* comes from the Latin *domus* for house, dwelling, household, and means bringing into the household. According to Harlan (1992, 64), the term implies genetic changes that adapt the plant to the 'domus'. There are intermediate steps between wild and domesticated varieties. Full domestication would mean that the populations could not survive without man. In contrast, *cultivation* is associated with human activities in caring for a plant, again according to Harlan. We may thus see domestication as the result of cultivation

policies on nature conservation and forest management, which would by far exceed the limits of the analysis.

Implications for PGRFA Management

In this section, we have seen how PGRFA comprises diversity between and among species, varieties and genotypes. Awareness of the distinctions between and among all these levels of diversity and of their contents is necessary for designing targeted policies for PGRFA management, and it is likewise a prerequisite for analyzing international agreements and domestic policies pertaining to PGRFA management. We have also seen that the difference between wild and cultivated plants gives rise to different management challenges, and these must be taken into account when analyzing overlapping international regimes pertaining to biodiversity in general and agrobiodiversity in particular.

Cultural and Socio-Economic Aspects of PGRFA Management

Ever since the dawn of agriculture, the cultivation of plants has been closely linked with culture and socio-economic development (Brush 2004; Diamond 1999; Kloppenburg 1988). These topics are wide-ranging, so we will highlight only a few central aspects of particular relevance to PGRFA management here.

Agrobiodiversity, Cultural Identity and Traditional Knowledge

According to Stephen B. Brush (2004, 259), cultural diversity has not been well studied as a functional variable contributing to agrobiodiversity. Nevertheless, it is recognized that culture and crop varieties are associated through folk taxonomies, ritual uses and cuisine. Agricultural plant varieties have diversified not only due to environmental conditions, but also due to the preferences of the people who grow them (ibid., xiv) – as to taste, texture, nutrition value and uses. Culture is thus one of several factors leading to diversity (ibid., 66).

From this it follows that the cultivation and use of crop diversity are intimately linked with traditional knowledge built up throughout the history of cultivation. This knowledge is indispensable for further maintenance of these resources, as it contains the information necessary for growing and using them. It remains a central challenge of PGRFA management to find ways and means to support farmers in their maintenance and further enhancement of traditional knowledge pertaining to crops, and to conserve such knowledge along with PGRFA for the future.

combined with the interaction between the cultivated plant population and the environment, resulting in genetic adaptation.

Socio-Economic Patterns of Crop Dispersal

Since the beginnings of agriculture, seeds and crops have followed farmers, who have exchanged them over short and long distances. Seeds and crops then spread until they met their environmental limits or were ousted by rival crops (Fowler and Mooney 1990, 38). Long-distance dispersal followed traders and explorers over land and sea. There has always been an interest in new crops. Around 2500 BC Sumerians sent collectors to Asia Minor in search of vines, figs and roses; and Queen Hatshepsut of Egypt sent an expedition to East Africa for collecting incense trees in 1482 BC (Fowler 1994, 4). These are only some of the many examples of the wide distribution of crops in ancient times.

With the rise of the Columbian Exchange,[3] the movement of germplasm attained a global dimension and a pace never seen before. Ascendant capitalism became a further driving force in shaping this process (Kloppenburg 1988, 153). Europeans collected seeds and crops from the colonies extensively for selection, duplication and growing in other colonies where conditions were suitable. The Royal Botanical Gardens at Kew became the most important center in a network of more than sixteen hundred botanical gardens throughout Europe (Fowler 1994, 6), and in the colonies new botanical gardens were established. In the second half of the 19th century, plantations expanded rapidly in the colonies, introducing new crops in areas where the old were displaced. However, this also created conditions for new diversity to emerge (Fowler and Mooney 1990, 41). In the 20th century these actions slowed down, concentration on selected crops intensified, and plant explorers searched instead in the Vavilov centers of diversity (see below) for new genes and traits for breeding programs in order to improve the varieties in use. With new breeding technologies and monocultures, the excessive dispersal of germplasm changed into dramatic erosion of genetic resources.

A central question, on which there has been hefty international debate, is whether it is possible to identify the *origin* of plant varieties, that is, exactly where they have developed their characteristic traits. Due to the extensive exchange of seeds and propagating material throughout history, it would be hard – if not impossible – with regard to most agricultural plants to identify the point in time when a particular variety came into existence and to determine the place or country where that happened (Fowler 2001; Andersen 2001). Let us take a brief look at theories of geographic patterns of crop variation.

---

3 This term was introduced by A.W. Crosby in 1972 to describe the trade emerging after Christopher Columbus brought the first seeds of the maize plant from the New World back to Europe. In the following year Columbus brought wheat, olives, chickpeas, onions, radishes, sugar cane, and citrus fruits to support a colony. After further expeditions and the emergence of colonization, germplasm was exchanged between the two 'worlds'. Maize, common bean, potatoes, squash, sweet potatoes, cassava and peanuts went east, whereas wheat, rye, oats and many vegetables from the Old World went west (Kloppenburg 1988, 155).

Geographic Patterns of Crop Variation

Charles Darwin is known as the first scientist to address the origin of species (Darwin 1859). It is less recognized that Alphonse de Candolle was the first to address the question of the geographic origin of cultivated plants. He published his initial work on geographical botany in 1855. However, his most famous contribution is *Origin of Cultivated Plants*, from 1882. Candolle introduced an interdisciplinary methodological approach that included botany, geography, history, archaeology and philology. On this basis, he presented the first comprehensive overview of species with regard to their anticipated origin, early cultivation, and principal facts about their diffusion. With the evolutionary concepts of Darwin and the findings of Candolle as points of departure, the Russian scientist Nicolay Ivanovich Vavilov developed his hypotheses on the centers of origin of cultivated plants in the early 1920s. His method was based on taxonomical phyto-geography, as developed by himself, and on genetics and cytology.

One of Vavilov's first findings was that it was possible to distinguish between *primary* and *secondary* groups of cultivated plants (Vavilov 1992, 103). Primary crops were the basic ancient cultivated plants known to humanity only in their cultivated state (for example wheat, barley, rice, soybeans, flax and cotton). Secondary crops comprised all the plants derived from what had been weeds that infested the primary crop fields and were found to be useful in their own right (for example rye, oats, false flax). Another important finding was that closely related species often belonged to the same ecological groups. Vavilov compared the ecological peculiarities of a number of primary plants and the wild species known to be close to them, and found that the more closely the wild species or varieties were related to cultivated plants, the more often did they belong to similar ecological groups (ibid., 118). These results, among others, were to lead to the inference that closely related species would often have the same distribution area, and that it would be possible to trace their centers of origin.

The region of maximum variation, usually including endemic types and characteristics, could also be the center of origin. Centers of origin would, as a rule, be characterized by many endemic variable traits and could comprise characteristics of entire genera (Vavilov 1992, 321). Within the centers of origin, Vavilov determined the so-called foci of type formation of the most important cultivated plants (ibid., 325), that is hearts of the centers with regard to genetic diversification and type formation. Vavilov's principles were strengthened throughout the period of empirical investigations, but the numbers and borders of the centers of origin of cultivated plants kept changing. In the end, he suggested the seven centers of origin of cultivated plants presented in Table 2.2 (Vavilov 1992, 429–33 and 1951, 31).

Table 2.2 Vavilov's centers of origin of cultivated plants

Centers of origin	Foci of type formation	Some important crops
1. **South Asiatic tropical center** (Most of South and Southeast Asia)	Indian focus Indo-Chinese focus Island focus, comprising islands in Southeast Asia	Rice, sugar cane, and a large number of tropical fruits and vegetable crops
2. **East Asiatic center** (East China, Taiwan, Korea and Japan)		Soybeans, various species of millet, most vegetable crops and many fruits
3. **Southwestern Asiatic center** (From Turkey to Kashmir)	Caucasian focus Asia Minor focus Northwestern Indian focus	Most endemic species of wheat, rye and fruits, peas, chickpeas and lentils
4. **Mediterranean center** (Mediterranean countries)		Olives, carob trees, many vegetable and forage plants
5. **Abyssinian center** (Highlands of Ethiopia, Eritrea, southwestern parts of Arabia/Yemen)		Teff, Arabica coffee, sorghum, endemic species of wheat, barley and the enset banana
6. **Central American center** (Northern parts and islands of Central America)	Mountainous South Mexican focus Central American focus West Indian focus	Maize, several species of cotton, beans, gourds, cocoa, and probably sweet potatoes, yams, peppers and fruits
7. **Andean center** (Surroundings of the three foci in South America)	Andean focus proper Chilean focus (southern parts) Bogotan focus (eastern Colombia)	Potatoes and other tuber-producing species, the quinine tree, the coca bush

In addition to the theory of the centers of origin of cultivated plants, Vavilov introduced the concept of 'secondary centers of origin' for crops that had originated in primary centers but had developed comprehensive genetic variants in other regions, perhaps far distant from the primary one (Harlan 1975, 183). For example, he found that the center of the original common wheat species (with 42 chromosomes) must have been in the Hindu Kush and the western Himalayas, and also in Transcaucasia, whereas species with 28 chromosomes were concentrated in Abyssinia and in the Near East (Transcaucasia, Turkey and northwestern Iran) (Vavilov 1951, 45).

Vavilov's work brought crop science a huge step forward, and represents a pioneering effort and a major contribution to our understanding of how today's agricultural plants have developed from their ancient origins. He was in a continuous process of improving the map of the centers of origin up to his imprisonment and death. No one can say what the map would have looked like if Vavilov could have continued his work as he had planned.[4] Whereas the basic principles of his theories seem valid for many crops, the location and borders of the centers may still be subject to discussion. The centers, as Vavilov determined them, appeared to be very large, with the main foci of diversity for different plants often widely distributed within these borders. This could indicate a need for greater precision. Also, empirical investigations after Vavilov's death have shown that it may be necessary to correct the determination of centers of origin for single species (as noted in Harlan 1975), and perhaps even question whether all crops covered by his theories do have a center of origin (see next section). Finally, comparison of his travels (Vavilov 1997) with his centers of diversity shows that these areas largely coincide. All the areas he visited – except for some parts of the US, the coast of Brazil, parts of Central Europe and Denmark, and a few areas in the (then) USSR that he also explored – were included in the map as centers of origin. Vavilov's expeditions were comprehensive, and his team members contributed significantly to the work. But there were also vast areas not included in the investigations, such as Africa south and west of Abyssinia, and Australia.

Since Vavilov, many scholars have brought greater detail to the picture with more precise determination of centers of origin for single crops. They have also questioned whether there can be said to be a center of origin for all crops (for example Sinskaja, Fowler, Zhukovsky, Brücher, all referred to in Harlan 1975, 188–9).

Jack R. Harlan has probably delivered the most comprehensive and fundamental contribution in this context. Through a detailed study of a range of crops from different parts of the world, he found that gene centers exist for some crops, but not for all (Harlan 1975, 189). Several crops originated in the centers determined by Vavilov, but many originated elsewhere (Harlan 1992, xii). Furthermore, he found that centers of diversity are not the same as centers of origin, and that there could be microcenters of diversity for different crops within the broader centers (Harlan 1992, 137, 147). Thus, Harlan described the emerging picture as 'one of mosaics over broad fronts rather than centers out of which all things come' (ibid., xiii). Each crop would be a case of its own, albeit with a certain consistency in evolutionary behavior. There would be a basic mode of interaction between space, time, and genetic variability, with different crops showing different geographic patterns of variation. On the basis of his empirical findings, Harlan proposed a classification of crop species according to the interaction of space, time and genetic variation as shown in Table 2.3 (Harlan 1975 and 1992, 137–40):

4 Vavilov was imprisoned in 1940, following Stalin's persecution of leaders of agriculture and agricultural science. According to his son, he was sentenced to death because he would not publicly denounce his work. The execution was commuted to 20 years' imprisonment. In prison he was starved to death. Vavilov died in 1943 (Vavilov 1997, xxvii).

Table 2.3 Harlan's classification of geographic patterns of crop variation

Endemic crops	Species that originated and have been cultivated in only one geographic area, without spreading appreciably (e.g. Guinea millet)
Semi-endemic crops	Species that have originated and been cultivated in definable geographic areas, with limited dispersal to other areas, possibly with some local nodes of high variability there (e.g. African rice)
Mono-centric crops	Species that originated in one identifiable geographic area, were cultivated to crops in this area, and then spread widely. Their centers of origin and their centers of diversity are identical and well defined (e.g. Arabica coffee)
Oligo-centric crops	Species with a definable center of origin, wide dispersal and with several or many centers of diversity. They have developed nodes with high variability in different geographic areas (e.g. common wheat)
Non-centric crops	Species cultivated over wide areas, with variations, but where no particular center or centers of origin or diversity can be identified (e.g. sorghum)

Endemic and semi-endemic crops would be relatively simple to distinguish, because they never spread very far and are *geographically restricted.* Mono-centric crops would also be relatively simple to identify, because they are relatively new, that is *restricted in time*, and may thus be traced to their origins by historical data. More difficult, according to Harlan, were the oligo-centric and the non-centric crops. Oligo-centric crops would be difficult, as they are ancient and widely dispersed, increasing in complexity through swathes of time and space. Non-centric crops would be difficult, because they derive from wild or primitively domesticated progenitors with a wide dispersal that would frequently involve multiple domestication (Harlan 1975, 189). Harlan proposed pearl millet, sorghum and Asian rice as some examples of non-centric crops.

Harlan (ibid., 184) maintains that his classification has intermediate states and conditions, and thus limitations. Although such a categorization is useful and important for illustrating the many patterns of geographic variation, it may be difficult to apply in terms of practical classification of crops. Domestication can take place wherever the environmental conditions are favorable, and the differences among the various categories may be difficult to identify. After a lifetime of studying the origin of crops, Harlan states in his latest writings (1995, 54) that due to the ambiguity of geographic patterns it is best to treat each crop separately and to consider origin on an ecological basis. These well-founded doubts may serve as a reminder to negotiators of international agreements who may think it a simple matter to determine the country of origin of crops.

These theories are important for PGRFA management, for two reasons:

- By indicating where the greatest diversity of PGRFA is to be found today, they show where conservation activities are particularly important. Since resources for conservation activities are scarce, such indications are valuable to ensure the highest possible impact of such efforts (see also Brush 2004, 284).
- In the discussion of how the benefits from the use of PGRFA can be shared fairly and equitably, the question of legitimate parties in benefit-sharing arrangements has been much debated. Our discussion here shows that it is generally very difficult, if not impossible in many cases, to identify such parties on the provider side based on the notion of 'country of origin'. These insights are of particular relevance for the debates on access and benefit sharing under the CBD and the ITPGRFA.

Implications for PGRFA Management

We have seen that the cultivation and use of crop diversity are intimately linked with traditional knowledge. As this knowledge is indispensable for further maintenance of these resources, it is vital to find ways and means to support farmers in their maintenance and further enhancement of traditional knowledge pertaining to crops, and to conserve such knowledge along with PGRFA for the future.

It is also clear that the origin of cultivated plant varieties is often hard – if not impossible – to identify, a finding that has implications for access and benefit sharing regulation, particularly if based on the notion that the country of origin should decide over the resources, as set out in the CBD.

Finally, we found that knowledge on centers of diversity is important for enabling targeted measures for conservation and sustainable use of PGRFA with high impact.

PART 2
Research Design

As we have seen, our generation risks depriving all future generations of vital means to ensure their food security – means that have been developed and handed over to us from many previous generations of farmers. Furthermore, the livelihoods of more than one billion economically poor farmers in developing countries are at stake. Grasping the dynamics behind these developments is therefore a crucial challenge.

Many factors could be relevant in seeking to explain the dynamics, among them technological innovation, trade structures and governance practices. Initial research on the topic showed, however, that central factors can be understood in the context of the implementation of multilateral agreements which address all sides of the management of PGRFA from various perspectives, and which can also be seen as a response to the factors mentioned above.

Analyzing the effects of international regimes – across scales – on PGRFA management in developing countries is a challenging task, due to the high level of complexity of the topic. It includes a whole range of issue-areas, each of which could form the basis for a separate book:

- The analysis of several comprehensive international regimes as they pertain to the management of these resources – including their formation and their interaction.
- PGRFA management, covering the topics of conservation, sustainable use, access to the resources, the sharing of benefits arising from their use, intellectual property rights and farmers' rights.
- The management of these resources in developing countries, with a focus on one case country, its environmental, agricultural, societal, and political context.
- Aggregate effects of the international regimes on PGRFA management in the case country, including the development of a methodology to grasp these effects – and their potentials for generalization.
- Mechanisms of influence explaining these effects, including the development of an analytical framework to grasp them – and their potentials for generalization.

It would have been easier to narrow down the focus, and select only one of these issue-areas. That would, however, not enable an understanding of the effects of the various international agreements across scales. In order to grasp these effects on PGRFA management in developing countries – and seek to explain them in the context of other influencing factors – it is necessary to examine all the above topics, and to analyze each of them as thoroughly as possible. How to do that forms the subject of this part of the book, which sets out the research design.

Chapter 3

Research Questions and Analytical Framework

In this chapter, we will start out by formulating the overall research question for the analysis and further limiting the scope for the study. Against this background, operational research questions will be developed and an analytical framework designed.

Overall Research Question and Limitation of Scope

This book is not primarily an implementation study. The purpose is not to find out whether the international regimes in question have been implemented, or to what extent. What matters is to find out whether and how these agreements address the challenges of PGRFA management. Their implementation is a part of that picture, but it is not the focus of the study.

The *overall research question* is: How do international regimes pertaining to PGRFA affect the management of these resources in developing countries? To make the study researchable, it is necessary to limit the scope as a basis for defining operational research questions:

Overlapping international regimes refer to those regimes that fully or partly pertain to the international regulation of PGRFA management. Because of this overlap of functional scopes, they are called 'overlapping international regimes' (see section below, entitled 'Explanatory Variable: International Regimes'). Four international regimes have been selected for analysis:

- The International Treaty on Plant Genetic Resources for Food and Agriculture (ITPGRFA), which was adopted in 2001 and entered into force on 29 June 2004. Its forerunner, the International Undertaking on Plant Genetic Resources for Food and Agriculture (IU), was adopted in 1983 but was not legally binding.
- The Convention on Biological Diversity (CBD), which was adopted in 1992 and entered into force in 1993. The CBD covers all forms of biological diversity worldwide, including PGRFA.
- The Agreement on Trade Related Intellectual Property Rights (TRIPS), which was adopted in 1994 and entered into force in 1995. The TRIPS Agreement

covers the protection of a wide range of intellectual property rights, among them rights pertaining to plant genetic resources (Article 27.3.b).[1]

- The International Convention for the Protection of New Varieties of Plants, establishing the Union for the Protection of New Varieties of Plants (UPOV), which was adopted in 1961, entered into force in 1968 and has been revised three times. UPOV-corresponding legislation is regarded as one way to implement Article 27.3.b of the TRIPS Agreement, so UPOV will be dealt with in connection with the TRIPS Agreement here.

Developing countries is used in the sense employed by the United Nations. However, the term covers a huge diversity of countries, with varying characteristics as to income, production structures, environment, infrastructure, etc. One country – the Philippines – has been selected for in-depth study. On the basis of the findings from this case study, the relevance for other countries with similar conditions will be analyzed, supported by additional data from these other countries. The explanatory power of such an approach is discussed in the next chapter on method, where also the choice of the case country is elaborated.

Management of PGRFA refers to domestic policies in the case country as well as goal achievement regarding these policies. As the most significant effects in the case country that can be related to the international agreements are two specific acts of legislation in the Philippines, these two have been selected for the analysis:

- Executive Order 247: Prescribing Guidelines and Establishing a Regulatory Framework for the Prospecting of Biological and Genetic Resources, Their By-Products and Derivatives, for Scientific and Commercial Purposes and For Other Purposes, 18 May 1995 (hereafter referred to as EO 247).
- Republic Act No. 9168: An Act to Provide Protection to New Plant Varieties, Establishing a National Plant Variety Protection Board and for Other Purposes, 7 June 2002 (hereafter referred to as PVP Act).

Whereas we will seek to analyze all the effects of the international regimes in question on the domestic policies in the case country, the two acts will be studied in-depth to explain how these effects came about.

The analysis of the international regimes covers the whole time-span from their initiation and up to 2004, as this is necessary for understanding their interaction. We will also discuss the developments after 2004 in this light. The analysis of the policies in the Philippines focuses on the period between 1992 and up to 2004 in order to enable feasibility, but also historical factors will be taken into consideration.

1 The focus in this book will be on plant variety protection, also called plant breeders' rights. Patents will be dealt with only marginally, because they have been comparatively less relevant for PGRFA management in developing countries in the period studied, and because it is necessary to delimit the scope to ensure feasibility.

Operational Research Questions

In order to analyze the effects of the international regimes in question, we first need to understand how they relate to PGRFA management, as well as the relations between and among these regimes. We will have to identify the extent to which their functional scopes overlap regarding PGRFA management, because this is the basis for their interaction. Furthermore, their interaction must be analyzed, because that is likely to be decisive for the resultant constellations of regimes with which developing countries are confronted. The operational research questions pertaining to the *international level* are as follows:

1. How did the functional scopes of the IU/ITPGFRA, the CBD, the TRIPS Agreement and UPOV regarding PGRFA management overlap in the period under study?
2. How did interaction between and among these regimes affect their formation and international implementation?
3. What were the resultant regime constellations for PGRFA management?

To see how these regime constellations have affected the management of PGRFA in the Philippines, we first need to analyze the extent to which the respective policies and their goal achievements are conducive to PGRFA management. Then we can examine the extent to which these findings correspond with the identified regime constellations. A further question is whether and how the findings can be attributed to these regime constellations. This question will be answered on the basis of available information and deepened regarding the two acts of legislation in focus. Here it is important to bear in mind that correspondence is not in itself a condition for attribution: A policy may be attributed to a certain international agreement, without corresponding to its norms and rules – or corresponding only to limited components of the agreement – due to misunderstandings or failures, or perhaps in an effort to counterbalance the perceived negative effects of the international agreement. Our two operational research questions pertaining to the case studies, that is effects at the *national level*, are thus:

4. To what extent were Philippine policies and goal achievements conducive to PGRFA management in the period under study?
5. To what extent did they correspond with – and/or could be attributed to – the identified regime constellations pertaining to PGRFA management?

In-depth explanation is needed to establish an understanding of the causality in these dynamics. That is why we will make a thorough study of the two Philippine regulations, the EO 247 and the PVP Act – first identifying the actors who influenced the formulation of the acts, and then analyzing the effects of this influence on the acts that were adopted. On this basis, we then examine which factors influenced the positions of those actors, with a particular view to the influence of international regimes. It is also important to see how the processes that led to the formulation of these regulations have influenced later goal achievement, since that can shed light on the prospects for goal achievement following from different policy processes. Our last three operational research questions pertaining to the case studies are thus:

6. How did actors who were engaged in the field of PGRFA influence the processes leading up to the adoption of the EO 247 and the PVP Act?
7. What main factors influenced the positions of these actors?
8. How has the process leading up to the adoption of the legislation of the EO 247 and the PVP Act affected their goal achievements regarding PGRFA?

Finally, we discuss the *scope for generalizing* from the findings. This is decisive for how this study can contribute to answering the overall research question. The last operational research question is thus:

9. How, and under what conditions, may the findings from the case study of the Philippines have relevance for other developing countries?

Introduction to the Analytical Model

The *dependent variable*[2] of the analytical model – that is what we want to explain – is domestic policies pertaining to the management of PGRFA in the Philippines, in terms of policy formulation as well as goal achievement. The *explanatory variables* – those that may explain the domestic policies – are the selected international regimes. As international regimes do not affect domestic policies through their mere existence, the core challenge with the analytical model is to explain how influence takes place. How did the relevant actors influence the processes, and how can this be explained? We need to identify pertinent mechanisms through which such influence takes place, the *mechanisms of influence*. As will be elaborated below, three mechanisms of influence have been selected as point of departure for the analysis:

- *Structural power:* the ability to shape and determine the structures of the global political economy within which other states, their political institutions, their economic enterprises and their professional people have to operate (Strange 1988).
- *Learning and norm diffusion*, with emphasis on the influence of *advocacy coalitions:* coalitions of actors sharing a set of policy core beliefs and advocating policy decisions based on these beliefs (Jenkins-Smith and Sabatier 1993 and 1994).

2 A variable is an empirical property that appears with two or more values (Frankfort-Nachmias and Nachmias 1992). Thus, *variables* indicate that empirical properties may differ with regard to the extents, strengths or qualities of the empirical property. The *dependent variable* is the factor to be explained and thereby the focus of the study, whereas the *independent variables* are those that may help to explain the observations pertaining to the dependent variable. The term *explanatory variables* is used interchangeably with 'independent variables'.

- *Institutional capacity:* the capacity of the domestic system to respond to international regimes, as well as to pressure and initiatives from various actors in this regard, to balance pressure and initiatives against other interests, to develop policies and to implement them (based on Hanf and Underdal 1998, 163; Jänicke 1995).

There are two basic distinctions between the first two mechanisms and the third one. Whereas the first two represent a continuum from the international level to the domestic, the third is restricted to the domestic level. In addition, the first two mechanisms are expressions of influence on the system, whereas the third indicates, *inter alia*, how the system responds to such influence. Thereby, the first two mechanisms are assumed to influence the third in various ways and *vice versa* – in that the response of the political system provides new points of departure for action by influencing actors. In Figure 3.1, this is shown by placing *institutional capacity* in the middle, between the two other mechanisms of influence, and by way of the arrows.

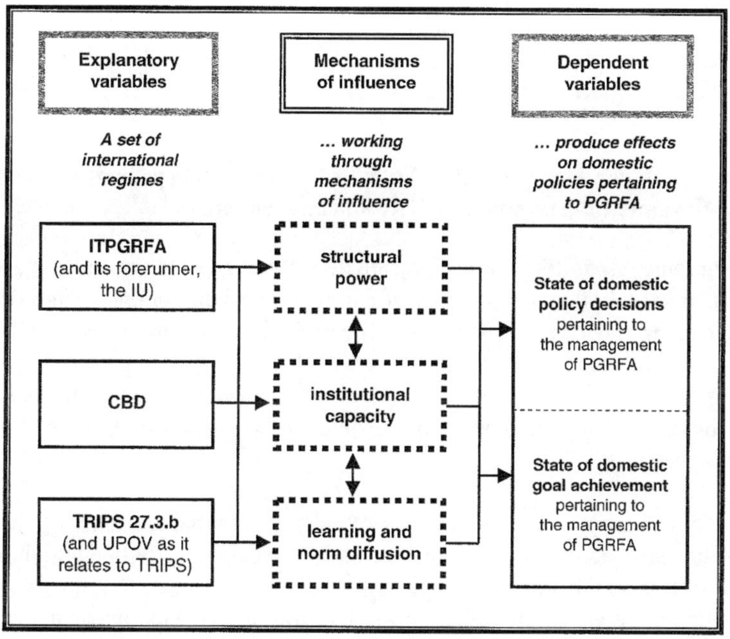

Figure 3.1 Analytical model

The three international regimes are assumed to affect domestic policies pertaining to PGRFA through the three mechanisms of influence. It will be suggested that different mechanisms of influence work through different international regimes – structural power mainly through the TRIPS Agreement, and learning and norm diffusion mainly through the CBD and IU/ITPGRFA. Initially, however, the options should be kept open for all combinations of these factors.[3] There have been heated debates within the TRIPS Council on the legitimacy of the rules pertaining to plant genetic resources, and the events in Seattle and Cancun, with massive anti-WTO demonstrations, have shown that advocacy coalitions are gaining strength. These developments indicate that also learning and norm diffusion may have relevance in explaining the effects of Article 27.3.b of the TRIPS Agreement on domestic policies pertaining to PGRFA. When it comes to the CBD and the IU/ITPGRFA, there have also been traces of structural power in the negotiations. Figure 3.1 reflects these combinations of influence, with arrows from all international agreements through all mechanisms of influence.

Figure 3.1 illustrates the basic analytical framework for this study. What it does not show is the reciprocity between the domestic and the international level – that international regimes are shaped by actors from various countries, and not only *vice versa*. Nor does it identify the mechanisms of influence through which these processes of regime formation take place. The formation of international regimes is not the primary focus of this book – and is therefore not highlighted in the illustration above – but it must be dealt with in the analysis, because it provides a basis for regime interaction. Therefore, we will also seek to identify mechanisms of influence for the international regimes, assuming that there are similarities between the international and at the domestic levels. We will thus explore whether and how mechanisms of influence can be traced across scales.

Defining the Dependent Variables: State of Domestic Policy Decisions and of Goal Achievement Pertaining to PGRFA Management

The operational research questions indicate two interconnected dependent variables. We want to explain the state of domestic policy decisions (output of the decision processes), and of their realization (outcome of the decision processes). This leaves us with two options: either to define the variable as implementation at different levels, or to split them so as to mark the difference between decisions and actual deeds. The first option would be simpler. There are, however, two arguments against using this approach:

- In developing countries there is often a substantial discrepancy between policy decisions and enforcement, and the analytical framework should be able to reflect this circumstance.
- Second, we are interested in more than the implementation of policies, for example initiatives or processes that unintentionally correspond with policy decisions.

3 Moreover, the different mechanisms of influence are not necessarily linked to particular categories of actors. Learning and norm diffusion tend to be associated with civil society actors and to some extent state actors, whereas structural power is usually associated with business corporations and to some extent state actors. However, business actors can also learn, and civil society actors can be co-opted by, for example, business corporations and thereby into a system of structural power.

Therefore, the dependent variable has been split in two. The first dependent variable is defined as the state of domestic policy decisions pertaining to the management of PGRFA in the case country (output). The second one is defined as the state of domestic goal achievement pertaining to PGRFA management (outcome). In Figure 3.1, these two variables are divided by a striped line, indicating that they represent stages along a continuum.

In order to analyze the dependent variables, we will need to establish an understanding of the state of domestic policy decisions as well as of goal achievement pertaining to PGRFA management. In turn, comprehending the state of domestic policy decisions and goal achievement, presupposes a basic understanding of the issue they refer to: the management of PGRFA. We therefore start by defining the contents of *(1) PGRFA management*, as the basis for both dependent variables, and then proceed to *(2) the state of domestic policy decisions,* and finally to *(3) the state of domestic goal achievement.*

The Management of PGRFA: Components and Criteria for Assessments

What constitutes PGRFA management? And what are the indications of conduciveness regarding PGRFA management? The latter is derived from research question four, concerning the extent to which policy decisions and goal achievements are conducive to the management of PGRFA. This indicates a normative stance that is considered necessary to establish fixed points against which policies and goal achievements can be measured, and subsequently against which the relative effects of international regimes can be analyzed. We will not embark on a definition of precise norms (for example, exactly how much genetic diversity must be conserved for *ex situ* management to be deemed conducive), but will focus on directions of the developments, and whether these can be considered conducive. The FAO report *State of the World's Plant Genetic Resources for Food and Agriculture* (FAO 1998) forms a central point of departure for the overview of elements of PGRFA, as well as the indications of conduciveness presented in Table 3.1.

Why are intellectual property rights not included as a central topic in Table 3.1? After all, this is one of the topics pertaining to PGRFA that has caused the most heated debates in negotiations at the international level – and in policy processes at the national level. However, the Table lists only those topics that are considered necessary for PGRFA management – and intellectual property rights are not a precondition for such management. Instead, the topic is highlighted as a factor that affects the management of these resources, particularly access to them and farmers' rights. It is also indirectly relevant to benefit sharing, because this can be linked to royalties from the sales of protected plant varieties. There has been considerable debate concerning this question and how intellectual property systems could be designed to enable the disclosure of the sources of PGRFA for new plant varieties, or even set the documentation of benefit-sharing arrangements as a requirement for certificates of plant variety protection. For these reasons, intellectual property rights will be dealt with in the context of the various PGRFA management components listed in Table 3.1, and not as a separate issue.

Table 3.1 Components of PGRFA management and criteria for assessments

Components	Definitions and explanations	Indications of conduciveness
In situ management of PGRFA	On-farm maintenance and enhancement of crop genetic diversity. *PGRFA also refers to wild relatives of crops found in* in situ *conditions – in other habitats than in agriculture – but this study will focus mainly on domesticated (on-farm) resources.*	• The number of plant varieties per crop grown in the country is stable or increasing. • The amount of land devoted to modern cultivars that are genetically uniform is not increasing without compensating measures. • The number of farmers engaged in *in situ* management of PGRFA is stable or increasing.
Ex situ conservation of PGRFA	Maintenance of seeds and other propagating material outside their natural habitats or on-farm conditions, normally in gene banks. *Ex situ conservation can take place at the local, national and international levels. Co-operation across scales helps to spread the risk of loss.*	• The share of crop varieties conserved *ex situ* is high and/or increasing, as compared to the total amount of crop varieties. • The number of *ex situ* samples per variety enables distribution from active collections. • The samples have passport data, are kept under correct temperatures and conditions, and are regenerated as required.
Access to PGRFA	Legal entitlement or permission to obtain and use available PGRFA, or in countries or cases where this is not required, the free admission to acquire and use such resources. *The two forms of legislation that normally regulate access are laws on bioprospecting and on intellectual property rights.*	• Access is facilitated so as to enable breeding response to pests, diseases and other demands. • Minimal bureaucratic hurdles for farmers and breeders. • Farmers can afford access to the plant varieties they need, and can save, reuse and exchange them according to their customary practices.
Utilization of PGRFA	The utilization of a wide range of crop species and varieties, covering a broad genetic base. *The difference between this element and in situ management is not distinct in practice. In analytical terms, the former is aimed at conservation and the latter at active use, which is considered the best guarantee against genetic erosion.*	• The number of traditional food crops still in use – other than staple crops – is stable or increasing. • The number of varieties of traditional crops – including staple crops – used in food production is stable or increasing. • There are efforts to broaden the genetic bases of crops with narrow genetic bases.

Components	Definitions and explanations	Indications of conduciveness
Benefit sharing	Sharing of benefits from the use of PGRFA with the respective community/ies and the national government by the collector. *Benefit sharing can also be understood in a broader sense, as sharing between beneficiaries at large and populations of custodians of PGRFA, e.g. in the form of development co-operation.*	• Farmers are provided with access on preferential terms to improved varieties of plants by recipients of PGRFA. • Participatory plant breeding takes place in the context of benefit-sharing arrangements. • Monetary transfer takes place in the context of benefit-sharing arrangements. • Development co-operation supports PGRFA management.
Farmers' rights	The rights of farmers to maintain their practices as custodians and innovators of PGRFA, to be recognized and rewarded for their contribution, and to participate in relevant decision-making.	In addition to the above indications on access (customary practices) and benefit sharing (reward to farmers), a further indication is: • Farmers' representation and participation in relevant decision-making bodies.

The State of Domestic Policy Decisions

By 'the state of domestic policy decisions' is meant all policy decisions pertaining to PGRFA management adopted up to the end of 2004. The most important forms of policy decisions, on which this study focuses, are:

- The ratification of international agreements: that the state gives its consent to be bound to an international agreement, on condition that a given number of other countries do likewise.
- The adoption of acts and amendments to acts, thereby referring to acts related to the ratification of international agreements as well as acts resulting from other policy processes.
- Parliamentary budget decisions, referring to the use of financial means and thereby reflecting political preferences.

Domestic policy decisions pertaining to PGRFA management do not exist in a vacuum, but must be seen within the overall political context – in this case, in the Philippines. The political context refers to the Constitution, as well as acts and policy decisions that lie beyond the scope of this study but that may affect goal achievements regarding policy decisions on PGRFA. It also refers to the rules that regulate the policy processes, establishing 'the rules of the game'. These are factors that must be controlled for in the analysis.

The State of Domestic Goal Achievement

A goal can be described as a desired future state or condition that one seeks to achieve or contribute to achieving. The more precisely a goal is defined, in terms of indicators or other criteria, the better is the basis for precision in analyzing goal achievement. Ideally, a goal should be defined in time, space, size and quality, but in this case, we cannot expect the relevant policy goals to be defined so closely. This means we will have to develop assessment criteria based on how the goals were actually formulated.

We may distinguish four main components of goal achievement. The most common way of viewing goal achievement is in terms of the *(1) implementation*[4] of specific policy decisions by national or local authorities (top–down approach). However, implementation is not necessarily only a top–down approach. Also *(2) independent action* may be initiated by target groups or affected groups. Such action may be aimed at policy implementation, at hindering policy implementation, or at shaping it in other directions. Also it may take place independently of implementation, and contribute to goal achievement unintentionally. Nevertheless, goals are not achieved in a vacuum: *(3) other policies* are also being implemented, and whether they are conducive to the specific goals is a question of policy coherence. Not only might other policies support or hamper the achievement of specific policy goals: also *(4) other contextual factors* are important, such as the general attitude of target groups towards the issue, and in our case, the state of the art regarding PGRFA management. These factors are particularly important for identifying any changes that have taken place, which in turn are decisive for determining goal achievement.

Effects of International Regimes

A major challenge in this analysis is to determine the effects of international regimes on domestic PGRFA management. The research strategy for this task is detailed in Chapter 4. The analysis will be instrumental in showing how the country has responded to the international regimes in question, their potentially conflicting provisions pertaining to the same issue-area, as well as how it makes use of potential synergetic overlaps. On that basis, we can start our search for explanations.

Explanatory Variable: International Regimes

The definition of *regime*[5] has been a debated subject among scholars of International Relations for several decades. A milestone was the famous 'consensus definition'

4 This book will focus on implementation outcome (behavioral change) and not on impact (in terms of the real effects on the state of PGRFA). The latter would require considerable natural science expertise and is beyond the boundaries of political science – and the limits of this book.

5 In IR theory, the terms *institution* and *regime* are often used interchangeably. The distinction is marginal: *Institution* is understood as 'sets of rules of the game or codes of conduct that serve to define social practices, assign roles to the participants in these practices,

(see Hansenclever et al. 1996, 178) propounded by Krasner (1982, 185): Here regimes are understood as 'implicit or explicit principles, norms, rules and decision-making procedures around which actors' expectations converge in a given issue-area'. Krasner defined principles as beliefs of fact, causation and rectitude, whereas norms were understood as standards of behavior defined in terms of rights and obligations. Rules were seen as the specific prescriptions for action and decision-making procedures the prevailing practices for making and implementing collective choice.

Critics of this definition have focused on the difficulties of distinguishing its various components (Young 1986; Keohane 1989; Hansenclever et al. 1996). In practice, when analyzing regime issues, it may not be easy to distinguish among principles, norms and rules. Neither is it unproblematic to relate an analysis to converging expectations or to the delimitation of an issue-area. How many converging expectations are necessary, or how narrowly must the issue-area be defined, to allow application of the term 'regime'? An attempt was made to introduce a new definition into the debate that reduced the concept to 'institutions with explicit rules, agreed upon by governments that pertain to particular sets of issues in international relations' (Keohane 1989). However, by excluding the difficult components from the concept, also its richness and usefulness was diminished. Another and probably more fruitful approach is to view the Krasner definition as a working definition that may serve as a point of departure for developing a more comprehensive theory, and subsequently a more precise definition of the concept itself (Hasenclever et al. 1996, 183).

An international agreement is the formal expression of a regime. However, the regime consists of more than the agreement itself – such as interpretation of its rules and regulations, and expectations as to how it is implemented.

Identifying Regimes – And Regime Interrelations

Next, we need to establish an understanding of what constitutes the actual regimes in our analysis. We will analyze how these regimes are interrelated, based on Oran Young's (1996) four-fold categorization of regime interrelations:

- *Embedded regimes* are deeply embedded in overarching institutional arrangements representing a broader context of principles and practices.
- *Nested regimes* are incorporated into broader institutional frameworks on the same issue-area.
- *Clustered regimes* are regimes knit together in more comprehensive institutional packages.
- *Overlapping regimes* are defined as the situation that occurs when individual regimes formed for different purposes intersect on a *de facto* basis, with the functional scope of one regime extending into the functional scope of at least one other regime.

---

and guide the interactions among occupants of these roles' (Young 1994, 3). These rules or codes are both formal and informal, and contribute to the shaping of expectations (Keohane 1989). *Regimes* are a particular kind of institution pertaining to specific issues. International regimes may be found at both the global and the regional levels.

These four are not mutually exclusive. The first three are pertinent to the inner 'taxonomy' of each of the regimes in question, because this may affect regime formation, regime output and regime implementation in various ways. We return to these questions in the next chapter. The fourth category will be central when we seek to grasp regime interaction analytically.

Understanding Regime Overlap and Interaction

As overlapping regimes have often developed from different rationales and actor constellations, they exhibit different characteristics and thereby possibilities for conflicts as well as synergies. An important point is that the many international environmental agreements negotiated so far have rarely involved explicit measures for resolving the frequently conflicting goals of economic and environmental regimes (Rosendal 1999). As damage to the environment typically occurs as a side-effect of legitimate activities undertaken for other purposes (Hanf and Underdal 1998, 152), a successful policy for environmental protection will have to penetrate and modify the activities and policies that cause environmental damage in the first place.

Several scholars have investigated the interaction between overlapping international environmental regimes.[6] Rosendal (2001) suggests a method for identifying whether and how overlapping regimes are conflicting or synergetic. She distinguishes between norms and rules of international regimes, and develops a typology for analyzing the overlap between the functional scopes of norms and rules of such regimes. This typology enables the identification of conflicts and synergies as well as their significance. In this book, Rosendal's method will be used, in a slightly modified version (see Chapter 4).

The foundation for applying this method is the *definitions of norms and rules*. The overall policy objectives and principles of an international regime are often formulated in general terms, as a result of long negotiations between parties with differing interests and often also due to interaction with other regime processes. This may make it difficult to grasp the precise contents of the norms contained in these policy objectives and principles. In our context, *norms* are defined as expectations to the behavior of the Contracting Parties to an International Agreement pertaining to the overall objectives of the agreement, and shared by most of the Contracting Parties. As Rosendal (2001) indicates, we can distinguish between core and secondary norms – referring to whether the norms prescribe core or secondary aspects of the regimes. To grasp core and secondary norms, it is useful to distinguish between different imperative levels. According to Dahrendorf (1964),[7] norms have an imperative dimension, which determines their strength: The expectations converging around a norm may represent a 'must', a 'should' or a 'could'. These three levels of imperative are in fact stages along a gradual continuum between the 'must' and the 'could' options, where the 'should' expectations comprise several levels of imperative. We will operate with four levels of imperative, divided into

6 See, for example, Gehring and Oberthür 2004; Rosendal 2001 and 2000; Stokke 2000, 2001a and 2001b; and Zhang 1998.

7 Referred to in Henecka 1985, 60.

two groups of norms: *Core norms* are defined as expectations concerning what the Contracting Parties 'must' or 'ought to' do, whereas *secondary norms* refer to what the Contracting Parties 'should' or 'could' do, pertaining to an overall objective of an international agreement. The latter category would also apply if there were diverging expectations among the Contracting Parties as to the contents of an objective. If it is difficult to distinguish between levels of imperatives, we may also understand secondary norms as being clearly subordinate to core norms.

Rules are specified as regulations for state behavior in terms of implementation. Rosendal (2001) also distinguishes between programmatic and regulatory rules: regulatory rules prescribe implementation measures, whereas programmatic rules refer to efforts to enhance knowledge about the issue-area. The latter distinction is less relevant in our context, since there are few programmatic rules in the regimes under study, so it will not be applied here. We will distinguish between core and secondary regulatory rules: the core rules are the mandatory rules and the secondary rules are those formulated as more or less optional to the Parties to the respective agreements.

To identify the precise scope and contents of regime overlap, the identified core and secondary norms and rules of the various regimes will be compared with a view to pattern matching, further described in Chapter 4.

An important distinction must be made between the study of regime *overlap* and regime *interaction*. Whereas regime overlap refers to structures, regime interaction refers to processes, and is dynamic. It is about how one international regime affects the directions of regime formation as well as implementation in another international regime – and the converse. Regime interaction can also take place between more than two regimes (as in our case), which increases the level of complexity considerably.

Behind all forms of regime interaction are actors with interests and ideas and differing levels of bargaining power. Regime interaction basically takes place on the initiative of actors and is carried out by actors, which makes it particularly important for us to understand how these actors use international regimes for their purposes.

Forum shopping Forum shopping has emerged as a relevant concept and a promising approach to analyzing regime interaction regarding PGRFA management. In our context, it refers to two interlinked phenomena:

- how countries and actors choose different international forums pertaining to similar issue-areas as their main arena for negotiations, according to their interests and perceived bargaining power;
- the tendency of actors with sufficient bargaining power to move issues from one forum to another, depending on where they consider themselves to have the greatest advantages.

The phenomenon of forum shopping has been investigated in several issue-areas of international politics and law, and has proven its explanatory power (Giellespie

2002; Kellow and Zito 2002; Rothwell 2004).[8] Analyzing forum shopping is closely interlinked with the analysis of mechanisms of influence at the international level – which basically underlie forum shopping. For the sake of analytical clarity we will analyze forum shopping as a means to reveal how regime interaction takes place (explanatory variables),[9] whereas the mechanisms of influence will form the basis for the explanation of why such interaction took place.

In order to understand the conditions under which forum shopping takes place, yet another perspective must be taken into consideration. Regime processes evolve and develop over long periods of time, so their dynamics must be understood in relation to time-related phenomena, that is the constellations of interests and bargaining power and the general developments within different regimes with overlapping functional scopes. This indicates that the time dimension is central in the study of regime interaction (Andersen 2002).

The time dimension in regime interaction Time *per se* cannot explain the interaction between different regimes. It is not the age of a regime that determines whether it will be in a position to influence another regime. For some regimes, only a short time is needed for an agreement to be adopted and implemented; for other regimes this may take many years. What counts in a time perspective is the development stages of regimes (agenda setting, regime formation, implementation and evaluation stages). Thus we need to ask how different development stages of regimes affect their interaction, and thereby their prospects for implementation (Andersen 2002):

- *Synergetic interaction:* Normally a regime in an early stage of development will be influenced by regimes in later stages of development, if their functional scopes are overlapping and they are similar in terms of member states, interests, power and international political weight. This is a typical learning situation, or an ideational interaction, where the diffusion of norms and ideas from one regime to another may result in one regime influencing the material contents of another (Stokke 2000 and 2001b). Such a situation is also fruitful for implementation at the national level, since domestic experiences with the more established regime will be instrumental for implementation of the new regime. Therefore, synergies may emerge.
- *Interaction with barriers:* In situations like the above, but where the regimes are characterized by conflicting interests regarding the overlapping issues, the established regimes will normally be in the position to determine the direction of the newcomer. Such a situation may emerge when the regimes have evolved from different rationales and with different driving forces behind them. Negotiating countries usually strive to get new regimes to be in compliance with already adopted agreements on the same issue. By trying to

8 Helfer (2004) calls the phenomenon *regime shifting*. Whereas his analysis is interesting, the term *regime shifting* is not precise: it is not the regimes that are shifting, but the actors that move their emphasis from one regime to another. Therefore the term *forum shopping* seems preferable.

9 Forum shopping may also shape regime overlap.

rule out conflicts between the two regimes, the parties will seek to harmonize the regime at the earlier stage with the one at the later stage. Depending on interests and power constellations, negotiating the newer regime can thus be difficult and time-consuming. At the national level, the effect will be that the established regime is further implemented, whereas the regime in the earlier stage of development is not ready for implementation. Thereby, structures shaped at the domestic level may hinder later implementation of the early-stage regime.

- *Interaction with strategic opportunities:* The conflict situation described above may also give rise to another type of development if conflicts are not ruled out, in that the regime under formation has the potential to modify the effects of the more established regime. The new regime may introduce rules that limit the scope for implementation of the more established regime. For example, it may provide rules that in effect make exemptions from the established regime for some situations, or provisions that add to those in the more established agreement, thereby limiting its functional scope. There is also the possibility of learning. If a certain period of time elapses between the beginning of the two regimes, and the international discussion has in the meantime reached a certain consensus on overlapping topics of relevance, the newer regime will be in a position to induce a modernization of the more established one. Such a potential to modify an established regime through a new regime represents a strategic opportunity for actors and countries that support the new one. This opportunity may be used in implementation at the domestic levels by interpreting the regime constellation in favor of the new agreement, and shaping domestic policies accordingly.

The output of the single international regime processes and their interaction pertaining to an issue-area is labeled 'regime constellations'. These are the constellations that developing countries have to cope with when implementing international regimes. How then do developing countries respond?

Regime Constellations and Aggregate Effects

The term '*regime constellations*' refers to the entirety of international norms and rules pertaining to an issue-area that result from all the international regimes pertaining to that issue-area, the overlaps of their functional scopes, and the interaction between them. Although the word 'constellation' may otherwise connote something rather static, in this context it is understood and used as something evolving and dynamic. Norms and rules are basically the results of the process of regime formation, and their content is further developed, shaped and molded throughout the process of regime implementation. Also regime interaction is highly dynamic, and regime constellations will change over time, as a result of these dynamics.[10]

---

10 Raustiala and Victor (2004) describe some aspects of what I call 'regime constellations' with their term 'regime complex'. This term is defined as an array of partially overlapping and non-hierarchical institutions governing a particular issue-area (p. 279) characterized by

Aggregate regime effects have been defined as the sum of the national-level effects of the regimes under scrutiny and their interaction (Sprinz et al. 2004, 3). We will modify this definition slightly, because in our setting it appears that aggregate regime effects are the national-level effects of the international regime constellations pertaining to an issue-area.

Aggregate regime effects can be assumed to take varying forms. They may strengthen or weaken the implementation of provisions of the agreements, and they may cause unintended conflicts between the implementation processes involving different regimes. Analytically, we must distinguish between *regime constellations* and *aggregate regime effects.* Careful analysis of the former should enable the prediction of the latter, as we will see.

As the dependent variable is divided in two, we distinguish between aggregate regime effects on policy decisions and on goal achievements. This distinction is important for three reasons:

- International regimes have differing mechanisms to support domestic implementation.
- International regimes have differing enforcement mechanisms to ensure domestic implementation.
- Goal achievements may come about as independent action from, for example, NGOs influenced by international regimes, and not as a direct result of domestic policies (which may also be influenced by international regimes).

Such mechanisms and actions are assumed to be important determinants for the overall aggregate regime effects of international regimes on an issue-area. Therefore it is central to understand them when examining how international agreements deal with the management of PGRFA.

Identifying Mechanisms of Influence

In the process of transforming international agreements into domestic policies, various mechanisms may come into play – from dynamics emerging from power constellations, to the influence of institutional learning or norm diffusion. There are many ways to categorize the theory approaches for explaining the formation and effects of international regimes within this spectrum. For our purposes it is sufficient to distinguish between two main schools as points of departure for selecting the

horizontal, overlapping structure and the presence of diverging rules and norms (p. 305). However, their concept is static and does not analytically grasp the dynamics of regime interaction. In fact, it is largely not informed by the growing body of literature on regime interaction (rather they speak of 'a large hole in the existing body of theory', p. 278). What their article highlights is the need to find a concept to cover the new constellations and dynamics emerging out of the increasing density of international regimes pertaining to specific issue-areas. Such a concept would include single regimes, regime overlap and regime interaction pertaining to an issue-area. Not having found such a concept myself, I will continue to use the term 'regime constellations'.

mechanisms of influence assumed to have the greatest explanatory power for this study:[11]

1. *realist regime approaches*, which focus on power and interests as explanatory factors;
2. *ideational*[12] *regime approaches*, which focus on the diffusion of norms and ideas as explanatory factors.

Both schools agree that power/interests and norms/ideas are important in regime formation and implementation. They diverge as to which they regard as having the most significant effect on regime formation and implementation under certain conditions.

Here I will assume that both sets of approaches have aspects that may help in explaining the effects of international regimes. The WTO system is based on decision procedures where those with the greatest economic power have most influence, so this can be regarded as an expression of the interests of financially powerful states. This in turn means that we can expect the realist regime model to be particularly useful for analyzing the effects of the TRIPS Agreement on the management of PGRFA in our case country, the Philippines.

The CBD, the IU and ITPGRFA were developed within the institutions of the United Nations, where each member state has one vote. To a greater extent the emphasis is on a balance of different interests. Here it is easier to bring up the norms and ideas of financially weak countries in terms of specific issues: they receive more attention, and they have generally more prospects for breakthrough. For that reason, we may expect the ideational regime model to be especially fruitful for analyzing the effects of the CBD as well as the IU and ITPGRFA on the management of PGRFA in the Philippines.

With these two sets of approaches, our attention is directed towards the actors seeking influence on domestic policies pertaining to PGRFA. The effect of their influence is not only a result of their actual deeds. It is also a result of the capacity of the domestic system (including the political will) to respond to such attempts, to balance them against other interests, to develop policies and to implement them. Underdal (1998, 7) and Hanf and Underdal (1998, 163) have underscored the importance of taking account of the capacity factor in regime implementation studies. This is a particularly important issue in developing countries, and must be reflected in a study like this one. Therefore it is essential to include institutional capacity as one of the dimensions in our study.

---

11 Based on Krasner (1982), Hasenclever, Mayer and Rittberger (1996 and 1997) and Rosendal (1999 and 2000).

12 Rosendal (1999, 42) uses the term 'ideational' of an approach to studying international regimes where the main focus is on norms and ideas. Norms and ideas pertaining to the formation and implementation of international regimes can be studied from a positivist angle as well as from a constructivist perspective, so the term is more useful as a common denominator for these approaches than, for example, constructivism alone. It is also preferable to 'knowledge-based theories' (Hasenclever, Mayer and Rittberger 1997), since ideas and norms may be shaped by more than knowledge as such.

The above considerations form the basis for identifying the anticipated mechanisms of influence. Each of the main traditions has been taken as a point of departure for selecting theory perspectives for explaining the influence of international regimes on domestic policies pertaining to PGRFA in the case country:

- The theory of structural power was selected from the realist tradition.
- The advocacy coalition framework was selected from the ideational regime tradition.
- Approaches to institutional capacity were selected from several traditions.

These choices will be further explained in the three following sections, where the anticipated mechanisms of influence will be presented.

The First Mechanism of Influence: Structural Power

Susan Strange (1988) defined *structural power* as the ability to shape and determine the structures of the global political economy within which other states, their political institutions, their economic enterprises and their professional people have to operate. In this sense, she went one step further than the traditional realists, who focused on the power of and between states. Referring to earlier debates on the difference between political and economic power, Strange emphasized (ibid., 24) that the distinction between structural and relational power, that is the power between two actors such as states, would be far more useful for understanding and analyzing power in political economy than the distinction between, for example, economic power and political power. She maintained that it would be impossible to have political power without the power to purchase, to command production, and to mobilize capital. Likewise it would be impossible to have economic power without the sanction of political authority, without the legal and physical security that can be supplied only by political authority.

Like most scholars who have dealt with the issue of power, Strange sought to identify its sources. She asked if it was based on command or coercive force, the possession of great wealth, or on moral authority derived from the proclamation of powerful ideas with wide appeal (Strange 1988, 23). Her answer was that those who exercise authority derive power from all three sources – from force, from wealth and from ideas – but that the extent to which different groups make use of each of these sources will vary according to four fundamental structures of structural power:

- the structure of the security system, defined as a framework of power created by the provision of security by some individuals for others, allowing these individuals to determine and limit the range of choices and options available to the others (ibid., 45);
- the structure of the production system, defined as the sum of all the arrangements that determine what is produced, by whom and for whom, by what method and on what terms, thereby creating wealth (ibid., 64);

- the structure of the finance system, defined as the sum of all the arrangements that govern the availability of credit, and all the factors that determine the terms for exchange of currencies (ibid., 91);
- the structure of the knowledge system, defined as the structure that determines what knowledge is discovered, how it is stored, and who communicates it to whom and on what terms (ibid., 121).

As with everything in the world, all these structures are interrelated, and in certain ways determine each other. An important difference to other approaches that seek to explain power is that Strange's analysis is not limited to the distribution of power between and among states. Strange (1996) highlighted the distribution of power between actors within these four power structures, based on the hypothesis of the declining power of the state and the concomitant rise of globalization.

All other elements of the international political economy – such as transport, trade, aid, energy and welfare – would be secondary structures that are molded by the fundamental structures indicated above (ibid., 28). These primary and secondary structures would in turn be decisive for positions and processes at the multilateral level, that is the negotiation and implementation of international agreements. True realism, she argued (ibid., 25), must recognize the power of non-state authorities over the structures and therefore over some of the outcomes of the system.

In practice, an approach to analyzing structural power regarding one specific issue-area would take the four primary structures of power as its point of departure. The challenge would be to identify the changes within these four structures that have had a substantial impact on the policies of powerful states, on company strategies, on market conditions and on the overall triangular balance of power involving states, companies and markets. On this basis, we could analyze the spin-off effects on the policies of states, on the outcomes of international organizations and on the political economy of related markets. Finally, the distributional consequences could be determined.

In an effort at further developing the theory of structural power, Mytelka (2000, 39–56) has criticized Strange for not defining the knowledge system as a power structure in its own right. Rather, Strange would have it understood in terms of 'dominant' ideas and belief structures that legitimized the authority of key actors in a political system. Whereas Strange would embed transnational corporations within the 'production structure', Mytelka proposes locating them within the 'knowledge structure', because they are key actors in the system of knowledge production that is so decisive for economic and political developments today.

Including in the analysis all institutions that contribute to technological innovation through knowledge production changes the understanding of the knowledge structure significantly. Mytelka points out that one of the most marked current trends within this structure is the privatization of knowledge – notably through the development of international frameworks and domestic legislation on intellectual property rights. As a result, patenting activities have increased dramatically since the 1980s (ibid., 43). Characteristic of this situation is that most patents are held by transnational corporations, not by individual inventors. In the fastest developing sector – biotechnology – five transnational corporations based in the US and in

Europe control more than 95 per cent of the gene-related patents (UNDP 1999, 68, referred to in Mytelka 2000, 43). This represents a sharp contrast to the knowledge structure of only a few decades ago, where public-sector researchers were the main producers of biotechnology-related knowledge. The change began with the advent of genetic engineering, and accelerated as new legislation was introduced to encourage cooperation with the private sector. In the US, the share of public-sector patents in biotechnology sold under exclusive license to the private sector had risen from just 6 per cent in 1981 to more than 40 per cent by 1990 (UNDP 1999, 67, referred to in Mytelka 2000, 43). As Strange (1996, 44) emphasizes, the global shift from state authority to market authority has been largely the result of state policies. Transnational corporations have not 'stolen' power from the state governments – power has been handed to them on a platter.

Mytelka (2000, 51) stresses the importance of understanding how the privatization of knowledge has contributed to the growing dependence of states upon private firms for the generation and application of knowledge. This helps to explain the emergence of a consensual relationship between states and transnational corporations, based on their contribution to the competitiveness of the nation. The close ties that states and transnational corporations maintained throughout the international negotiating processes that led up to international agreements on intellectual property rights have shaped the rules that govern access to and control over knowledge, she maintains. In other words, the transnational corporations developed a structuring potential that increasingly served to shape international rulemaking within the knowledge structure by the close of the 20th century. Mytelka's development of Strange's theory with regard to the knowledge system is highly relevant to this study.

Strange showed the emerging importance of structural power as compared to relational power, but did not specify how the two interact.[13] However, it seems clear that the complex fundamental power structures described by Strange are shaped by the interaction of many factors, where relational power between states is one. For example, relational power between the US and several individual states on each continent was decisive to shaping a new post-9/11 security structure with global reach. Even with selected states within the EU, Washington established individual relations towards this end. On this background, the study of relational power can, in my view, provide a fruitful contribution to the study of structural power. Also the study of structural power may shed light on the relational power between two states.

Structural power is not something that emerges out of nowhere. Someone has to take initiatives. Strange focused on a system perspective, without much precision as to the actors and their roles in the development of structural power. Whether they are labeled *actors*, as realists like to do, or *agents,* as constructivists love to do, it is important to stress that any analysis of the development of structural power will have to deal with them. Transnational corporations are core agents, according to Strange. But they are not the only ones. Also state entities can be instrumental in developing structural power, and there may be other agents as well. Fruitful analysis of the

13 For an interesting discussion of relational versus structural power, see Keohane 2000.

variable of structural power in this study will depend on careful identification of the relevant agents/actors.

Susan Strange has touched upon dynamics in teaching us to identify the changes that have taken place within each of the fundamental power structures. However, her aim is to provide an account of 'checks and balances' with regard to power. On this background, the effects within particular issue-areas can be studied. Analyzing structures or dynamics is a question of choice, and so far most scholars within the realist tradition of regime theory – including Strange – have opted to analyze structures. Grasping dynamics is a difficult challenge, and the theory basis for that within realist theory is limited. It is important to highlight this limitation in connection with the theory propounded by Susan Strange.

Strange's theory of structural power provides a fruitful point of departure for the first mechanism of influence, because it comes closest to reflecting the nature of the issue-area for this analysis from a realist perspective. Transnational corporations are central to the management of PGRFA in most parts of the world, and the internationalization of rules and procedures for this management points to structural power as a very useful framework of understanding.

Furthermore, Mytelka's analysis of the knowledge structure provides a highly relevant operationalization of the overall theory with regard to the emerging systems for worldwide protection of intellectual property rights in biotechnology. In our context, PGRFA is the core element and basis for the production structure. However, seeds themselves are useless without knowledge of their breeding and harvesting conditions and, not least, their uses. Therefore, and for the reasons pointed out by Mytelka, I argue that it is most fruitful to analyze the present topic within the framework of the knowledge structure.

We will examine how agents of structural power may have influenced relevant domestic actors in their efforts towards policy decisions and goal achievements pertaining to PGRFA management, and how international regimes mattered in this context. The analysis will involve three steps:

- First, it is necessary to identify more precisely the parts of the structural power knowledge system relevant for the issue-area. In this context, we ask whether and how structural power can explain the formation and international implementation of the regimes in question.
- Second, it is necessary to identify whether and how this potential structural power affected the actors involved in and with policy decisions and/or goal achievement at the domestic level. Here it would be helpful to identify those actors within the structural power system who were active regarding our case country in addition to the domestic-level actors, and analyze their interaction.
- Third, we will distinguish between the influence of structural power on decision-making and on goal achievement, to assess its overall effect on the management of PGRFA in the Philippines.

Propositions have been developed in order to find out whether and how structural power is relevant as a mechanism of influence. They are based on the assumption that, of the international regimes examined in this study, the TRIPS Agreement seems to be the one that most clearly works through structural power as a mechanism of influence. If so, we should expect to find one or more of the following in the case of the Philippines (propositions):

- P_1: Agents of structural power have sought to exercise influence on the process leading to the adoption of the act of legislation that refers to the TRIPs, that is An Act to Provide Protection to New Varieties of Plants (PVP Act).
- P_2: Agents of structural power refer to the TRIPS and interpret it in line with their own interests when seeking influence over the legislation process.
- P_3: The PVP Act has been formulated largely in accordance with the demands of the agents of structural power, whereas other, diverging, demands are less reflected in its formulations.
- P_4: The PVP Act has been accompanied by measures to ensure a relatively strong institutional framework for implementation, as compared to other relevant Philippine acts of legislation, which have not been influenced by agents of structural power.

The Second Mechanism of Influence: Learning and Norm Diffusion

Whereas hard-core realists will maintain that the analysis of every political decision boils down to the question of power, ideationalists of various colors will direct attention to the influence of knowledge, norms and ideas.[14] Power and economic interests may not always dominate in the world: people can also pursue ideals, seek the truth – and they can learn. Ideationalists have sought to identify situations where such processes can be observed, and to analyze these. They do not reject the importance of power in international relations, but they also see learning and norm diffusion as central factors under certain conditions.

Based on Checkel (1999, 4–6), we may distinguish between two phases of ideationalism. During the first phase, the empirical focus was directed towards later stages of compliance, where most internalization had occurred. Thus, compliance was found to be rule-governed and driven by a certain logic of appropriateness. It followed that social interaction was understood in a rather static sense. As these works were criticized for not addressing the question of why actors comply with social norms, studies of the second phase have been devoted to rectifying that, placing greater emphasis on process and agency. Summing up these efforts, Checkel (1999, 6–9) points to two causal mechanisms through which actors come to comply with normative prescriptions that have been identified by these scholars:

1. *Social protest and coercion* at the domestic and transnational level have been shown to influence state decision-makers. At the national level, NGOs

14 On this, see for example Checkel (1997, 1999 and 2001); Risse, Ropp and Sikkink (1999); Finnemore and Sikkink (1998); Finnemore (1996); Katzenstein (1996); Strang and Meyer (1993); Strang and Chang (1993); Jenkins-Smith and Sabatier (1993); Haas (1992).

and trade unions are examples of organizations that generate pressures for compliance. At the transnational level, non-state actors and policy networks that are united in their support for particular norms serve to mobilize and coerce decision-makers into complying with them. Shaming often is part of this strategy. Compliance in these situations does not, however, imply that the norms have become internalized by the political elites.

2. *Learning* is an important factor influencing political decisions. Learning is here not understood merely as becoming informed in order to make choices within predefined frameworks of preferences (which would correspond to a rational-choice perception). Rather, it is seen as involving complex social processes that may lead to correspondence with normative prescriptions. These processes are explained on the basis of theories from, *inter alia*, cognitive and social psychology, showing that, under certain conditions and when exposed to the prescriptions embodied in norms, individuals may adopt new interests.

The theories of social protest and mobilization describe situations where there is more resistance to learning, whereas theories of learning (also called theories of knowledge and norm diffusion), describe situations where there is more openness to learning, often because of uncertainty regarding the potential effects of regulation. This categorization can be useful to show more concretely what we mean by ideational explanations. On the other hand, in our case, both situations apply, so we cannot limit the ideational mechanism of influence to one of these two categories. As learning may also be the result of social protest and coercion, the label *learning and norm diffusion* for the ideational mechanism of influence refers to both categories.

The theory of Advocacy Coalitions Frameworks was developed to explain how *advocacy coalitions* may bring about policy change under certain conditions. This approach has been chosen for this study for three reasons:

- It is among the 'ideational' approaches that have the clearest actor focus, which is useful for this actor-oriented study.
- The framework comes closest to describing the kind of coalitions found in our case country.
- It is among the 'ideational' approaches assumed to have the greatest explanatory power, as will be further explained below.

The Advocacy Coalition Framework (ACF) was developed by Jenkins-Smith and Sabatier (1993 and 1994; see also Sabatier 1993) to synthesize what they considered the best features of top–down and bottom–up approaches to understanding policy-oriented learning in implementation analysis. The term *advocacy coalitions* refers to coalitions of actors who share a set of policy core beliefs, and advocate policy decisions on the basis of these beliefs. The actors connected to an advocacy coalition may come from governmental entities, private organizations (for example NGOs – Non-Governmental Organizations), trade enterprises or industrial groups.

In addition to the advocacy coalitions there may, depending on the situation, be a further group of actors, which Jenkins-Smith and Sabatier label 'policy-brokers'.

They are the ones who take on the role of mediating between various coalitions in case of conflicting strategies and/or policy goals.

One basic distinction between this and more realist-oriented frameworks is the assumption that the coalitions in question are based on beliefs and not on common interests. Policy core beliefs are the glue that holds coalitions together, as Jenkins-Smith and Sabatier put it (1994, 195). These include basic ontological and normative beliefs, like the perceived nature of human beings or the relative valuation of individual freedom or social equality (ibid., 180). They also include fundamental value priorities, such as the relative importance of economic development versus environmental protection, or the appropriate division of authority between governments and markets (ibid., 181). The authors also define a set of secondary aspects of a coalition's belief system. This covers several beliefs with narrower scope, concerning, for example, the seriousness of the problem or the relative importance of various causal factors in specific localities, policy preferences regarding desirable regulations or budgetary allocations, the design of specific institutions, and evaluations of the performance of various actors.

The explanatory focus of the ACF is policy-oriented learning. In line with Heclo (1974, 306), Jenkins-Smith and Sabatier (1994, 182) understand this form of learning as being relatively enduring alterations of thought or behavioral intentions that result from experience and that are concerned with the attainment or revision of policy objectives. There are three basic assumptions concerning policy-oriented learning:

- Learning is instrumental, in that the coalition members seek to understand the world better, in order to further their policy objectives.
- Coalitions will resist information that may lead to the conclusion that their policy core beliefs are invalid or unattainable.
- The only way that policy core attributes of a coalition are changed is through shocks originating from outside the sub-system of coalitions that are utilized by the relevant minority coalitions in order to induce change.

Therefore policy-oriented learning across belief systems is most likely when the conflict which gives rise to discussion concerns some secondary aspects of one belief system and core elements of another, or when it concerns important secondary aspects of both belief systems.

Jenkins-Smith and Sabatier assume that learning is more likely when information is based on quantitative data and theories, than when based on qualitative material. Furthermore, they assume that problems involving natural systems are more conducive to this form of learning than problems regarding social or political issues. The latter assumption builds on the argument that, when it comes to social or political issues, the critical variables are themselves active strategists, and also because controlled experimentation is more feasible.

In order for policy-oriented learning to take place between belief systems, a forum is necessary. Learning is most likely to ensue if such a forum is prestigious enough to induce professionals from several coalitions to participate, and if it is dominated by professional norms (ibid., 184). A precondition is that the coalitions have the technical resources for participating in such debate.

An evaluation of the ACF based on its application in a range of cases all generally supportive to the framework suggested two additional hypotheses. First, the administrative agencies within a coalition will usually advocate more moderate positions than their allies from outside the administration. Second, even if the accumulation of technical information does not change the views of the opposing coalition, it can have important impacts on policy, at least in the short term, by altering the views of policy-brokers or important government officials.

There are four basic principles for using the ACF (ibid., 178):

- A time perspective of a decade or more is necessary to understand the process of policy change, as the policy core beliefs of advocacy coalitions are very resistant to change, and actor constellations may remain stable over a decade or more.
- It is useful to focus on the interaction among actors from different institutions who seek to influence governmental decisions in a policy area.
- It is important to differentiate between actors within the government – to include an intergovernmental dimension comprising all levels of the government, also sub-national levels.
- Public policies or programs can be conceptualized as sets of value priorities and causal assumptions about how to realize these – in other words, as belief systems.

All these principles are applicable in our study. The approach appears highly relevant for analyzing the situation in the Philippines, given the constellation of advocacy coalitions and how they have worked with regard to biodiversity in the period under study.

Whereas the ACF is relevant in explaining the influence of international regimes on domestic PGRFA management (as well as in explaining regime formation), many aspects still need to be elaborated further, not least because cognitive processes and learning are not easy to identify or analyze empirically.

An important aspect of cognitive processes is that they take place in individuals. One implication of this would be that attention must be paid to the role of individuals – but this is not really taken into account in the ideational approaches to political influence. According to a hard-core rational choice perspective, individuals are destined to choose on the basis of calculations; by contrast, it seems that they are mainly guided by social relations, when it comes to ideationalism. This social determinism leaves little space for explaining why individuals sometimes do break out and shape new ideas and new strategies. Risse (2000) is among those who come closest to addressing this question, when he assumes that individuals or actors might base their behavior on truth-seeking. Where does this truth-seeking stem from? Risse analyzes behavior in the context of social relations, but without examining the individual dimension. Other ideationalists have highlighted the existing gap with regard to the individual dimension, without yet contributing theory approaches of their own.

Moreover, the ACF was not designed on the basis of experience from developing countries, so we must ask whether it is adequate for analyzing the situation in the

Philippines. Developing countries differ greatly as to the level and structure of their civil society engagement, interaction between actors from the state and civil society, and democratic governance. The Philippines is among the developing countries in the world with greatest number of civil society organizations. It is also known for the massive popular protests that led to the fall of President Ferdinand Marcos (among Filipinos called Peoples' Power I) in 1986, and President Joseph Estrada (called Peoples' Power II) in 2001. Perlas (2000) uses the Philippine experience to draw inferences of more general relevance as to how civil society organizations contribute to shaping policies. From the initially exceptional good experience of Philippine civil society organizations with the Philippine Agenda 21 process, Perlas concluded that the belief system or the glue that united the organizations was their cultural foundation – a realm of shared identity. This realm covers the generation of knowledge, meaning, art, ethics, and a sense of the sacred (particularly important in the Philippine context). For Perlas, civil society organizations represent a third force (as compared to the state and the market) necessary to balance the market forces, especially in times when representative democracy seems to be eroding. He goes on to distinguish between different types of organizations in order to identify those that can be labeled as 'true to type' civil society organizations, in contrast to those merely pretending to be such organizations. The legitimacy of civil society organizations comes not from representativity, as this is the legitimacy associated with the parliamentary system. Rather, its legitimacy derives from the strength of its arguments. Perlas' approach provides a useful supplement to capture the special situation in the Philippines with regard to learning and norm diffusion by advocacy coalitions.

It could be argued that good analytical entry points for the present study would be provided by the broad and rapidly increasing literature on civil society and NGOs and their political and cultural influence (for example Cohen and Arato 1992; Rucht 1992; Diani 2001, 209–10; Edwards 2004), with regard to their capacity to mobilize social capital for political change (for example Bourdieu 1986; Coleman 1990; Putnam et al. 1993; Putnam 1995; Uphoff 1998; Rudolph 2000; Edwards et al. 2001), and related to environmental issues (Steinberg 2001). Indeed, this literature offers a highly valuable sounding board for the current work, but it is not directly instrumental for the analysis here.

From the above, we may define this mechanism of influence as the way in which learning and norm diffusion takes place through advocacy coalitions in the Philippines. What we will seek to find out is the extent to which, and how, learning and norm diffusion have contributed to policy decisions and implementation. We start out by identifying the actor networks that can be classified as advocacy coalitions. Thereby the contribution of Perlas will be used to analyze the common ground of the various actors of the advocacy coalition – the glue that unites them. The next step will be to investigate the interaction among actors of the advocacy coalitions and the effect on political decision-making – and on goal achievements – guided by the ACF.

As noted earlier, learning and norm diffusion theories are assumed to have particular explanatory power regarding the effects of the CBD and ITPGRFA, including its predecessor, the International Undertaking on PGRFA. If so, we should expect to find that one or more of the following hold true in the case of the Philippines (propositions):

P_5: Representatives of advocacy coalitions have sought to exercise influence on the process leading to the adoption of the act of legislation that refers to the CBD, that is Executive Order 247.

P_6: Representatives of advocacy coalitions have referred to the CBD and interpreted it in line with their own ideas when seeking influence over the legislation process.

P_7: Executive Order 247 has been formulated largely in accordance with the demands from advocacy coalitions, whereas other, diverging demands are less reflected in its formulations.

P_8: Executive Order 247 has been accompanied by an institutional framework for implementation that is rather weak, as compared to other relevant Philippine acts of legislation that have been influenced by agents of structural power.

The Third Mechanism of Influence: Institutional Capacity

Finally we examine how institutional capacity affects the ability of the political system to shape and implement policies pertaining to PGFRA management. A basic assumption underlying the selection of this mechanism of influence is that institutional capacity is a crucial condition for state actors to formulate adequate and feasible policies and to implement them. In other words, it is a precondition for them to exercise power. Also the way in which other actors engage in policy formulation and implementation will depend, *inter alia*, on the capacity of the state to involve them (Edwards et al. 2001, 18). From the ideational perspective, Walzer (1992, 102–3) has emphasized that a democratic civil society seems to require a democratic state, and a strong civil society seems to require a strong and responsive state. He calls this 'the paradox of the civil society argument'.

Walzer's observation is especially relevant concerning environmental politics in developing countries. NGOs are important actors in pushing for environmental policies and in shaping them[15] – indeed, in many developing countries they represent the only domestic force in this direction[16] – so their prospects for interaction with the state are of central importance.

15 NGOs have historically played important roles in developing international environmental law, as well as in its domestic implementation (Swanson and Johnston 1999, 209; Nagel 2002, 99). Worldwide, the number of organizations labeled NGOs has increased dramatically in recent decades (Princen and Finger 1994). The NGO phenomenon comprises a wide variety of organizations, from local activist groups via national associations to international organizations. NGOs can be categorized along many dimensions, such as scope and range of activities, ideology, size of budgets, staff and offices, professionalism, levels and forms of internal participation, legal status and identity.

16 This does not mean that NGOs are always positive forces in environmental policies – that depends on a range of factors. The point here is that they are often the sole forces working in this area. In their comprehensive study of environmental NGOs in world politics,

The lack of institutional capacity among state actors is a major problem in many developing countries, also in the Philippines, affecting the ability to formulate policies as well as to implement them. State capacity for social control is usually weak in developing countries – a particular concern when it comes to implementing environmental agreements (Hanf and Underdal 1998, 158). This study will be delimited to the question of how institutional capacity affects outputs and outcomes, and not why there is weak institutional capacity. Exploring the latter question would necessitate a thorough discussion of development theories and empirical investigations that, although important, would go beyond the scope of this analysis.

Based on Hanf and Underdal (ibid., 164–5), we may see institutional capacity as a function of properties of the institution itself (including its administrative system) and its relationship to society in general and the other actors involved in, and social groups directly affected by, the policies in question.

Organizational theory could contribute many starting points for defining institutional capacity for our context (see for example Powell and DiMaggio 1991). However, such analyses could easily become too comprehensive for our purposes, so it seems better to select elements from analytical approaches that have been specifically designed for studying the development and implementation of environmental policies.

Based on Hanf and Underdal (1998, 165) and Jänicke (1995), we may distinguish between two main categories of elements constituting institutional capacity. The first category concerns the competence and resources vested in an institution or entity; and the other its capacity to make use of these, which is basically a question of leadership.

Competence and resources The scope and depth of formal competence is a basic determinant for the capacity of an institution or entity to take action (Hanf and Underdal 1998, 165). It refers to the tasks vested with the institution or entity, its responsibilities for taking decisions and for carrying them out. The scope and depth of formal competence refer to the issue-area and types of decisions over which the institution or entity has the competence to make decisions or take action.

Whether an institution or entity is in the position to make use of its competence depends on its personnel and financial resources as well as political will. When assessing staff size as compared to tasks, professional competence is also important, but it is normally difficult to get good data on professional competence.

Financial resources refer to the funds allocated to an institution or entity for it to carry out its tasks. As sources of income are severely limited in many developing countries, Official Development Assistance (ODA) often represents a substantial part of allocations to, for example, environment tasks. ODA may be assumed to contribute to policy formulation in two ways. If the structural power approach provides the most adequate explanatory framework, we would expect to find that ODA is tied to projects directed at achieving specific policies and implementing

---

Princen and Finger (1994, 232) conclude that extensive studies on the political influence of NGOs must include the tensions and conflicts within the NGO community and within the organizations themselves.

them. Donors may also impose conditionality regarding certain policies, making new transfer of funds contingent on progress with regard to such policies. If learning and norm diffusion provide the main explanatory perspective, I assume that ODA is used as a means for developing participative and integrative capacity for policy formulation (see below), and that it is used to support the implementation of the policies thus derived.

Leadership resources Various factors could be accounted as leadership resources. Three aspects appear especially important regarding institutional capacity: the political 'clout' of the leadership (Hanf and Underdal 1998, 165), and their participative and integrative capacities (Jänicke 1995).

By political 'clout' is meant the ability of the leadership to win through with opinions, ideas and suggestions in decision processes, to have action follow intentions, and to get things done. This can be analyzed by simply assessing the performance of the leadership.

Participative capacity refers to the openness of the policy-formulating institution to new inputs from stakeholders, enabling them to participate in debates on policy formulation with inputs. Whereas Jänicke describes this aspect in a normative way as a tool for proponents of environmental policies, it can also be seen as an empirical factor. All kinds of actors may utilize openness for influence, whether they seek or pretend to represent broader masses of the population or the economic interests of a few. I assume that participatory capacity (to the extent that it exists) will be used by actors seeking or pretending to represent broader masses, if learning and norm diffusion are the dominant mechanisms of influence on policy formulation and implementation pertaining to the management of PGRFA. Likewise, I assume that it will be utilized by agents of structural power, if structural power is the dominant mechanism of influence.

Integrative capacity can refer to three forms of co-ordination: (1) Intra-policy co-ordination (Knoepfel 1993), where it is the internal integration of the issue-specific policy field at all levels in the organization. This is identical with vertical unity (see Hanf and Underdal 1998, 165). Typically, this would mean that, for example, all branches of the Philippine ministries of environment and agriculture would work to implement the policy on access and benefit-sharing pertaining to PGRFA, that is that such policy has become mainstreamed into the activities from the top of the ministry/department in question and down to its executive branches. (2) Inter-policy co-ordination means the cross-sector integration of conflicting policies (Carew-Reid et al. 1994). In our case, it would mean that, for example, environmental and agricultural policies currently in conflict concerning the conservation of PGRFA are mainstreamed in this regard. (3) External integration refers to the co-ordination and integration among the actors in an issue-area. ('Network management' is another term describing this element.) In our case, it could mean the way government institutions structure a network and utilize its resources for policy formulation and implementation.

Whereas the theory perspective applied here to determine institutional capacity is considered most appropriate, it is not sufficient for explaining the particular institutional situation in developing countries. Institutional malpractice, or corruption, is widely acknowledged to be a serious problem in many developing countries; further, it is generally known that the Philippines ranks among the countries with the greatest problems of corruption – indeed, this has been openly admitted by the government itself. As detailed information on corruption is normally inaccessible, the empirical evidence in this direction will be limited.

Another aspect not covered by the theory perspective is the question of how acts of legislation fit with the situations they are intended to regulate. de Soto (2000) maintains that an important reason for the lack of economic development in the Third World is that its legislation has been mainly adopted from Western countries, primarily the former colonial powers, and not adapted to the situation and needs of developing countries. This aspect should be taken into account when we analyze the capacity for goal achievement regarding the regulatory acts under study here.

Institutional capacity can be regarded as a mechanism of influence in two ways. First, it is a mechanism in its own capacity, since the extent to which an international agreement is implemented at the national level will depend on the country's institutional capacity to do so. Second, it is indirectly a mechanism, as it is assumed to determine the degree to which, and in what direction, agents of structural power and representatives of advocacy coalitions influence PGRFA management.

In this context we will pay special attention to how the two former mechanisms of influence affect institutional capacity in different phases. We expect to find both mechanisms of influence at work in policy-decision processes, but that they affect goal achievements differently. Agents of structural power are assumed to have means to exercise grater influence on goal achievements than advocacy coalitions have. Also, advocacy coalitions can be fragile networks geared towards particular policy decisions, and motivation may sink or be channeled to other policy goals once a 'victory' is achieved in terms of a policy decision. Finally, work to ensure goal achievement may differ substantially from the normal advocacy work in political-decision processes, so there arises the question of whether advocacy coalitions are equipped for such work.

In the Philippines, the institutional capacity to implement the international agreements pertaining to PGRFA is assumed to be limited, so we will expect the relative influence of agents of structural power and representatives of advocacy coalitions to be high. If so, we should find one or more of the following to hold true (propositions):

P_9: There is little co-ordination between the various branches of the ministries involved in PGRFA management.

P_{10}: Legislative acts pertaining to PGRFA management are largely formulated in line with the demands of the dominant influencing actors.

P_{11}: Advocacy coalitions depend on responsive authorities to win through with their demands regarding policy decisions and their implementation.

P_{12}: Agents of structural power shape and mold institutional structures and processes so as to succeed with their demands regarding policy decisions and their implementation.

Chapter 4

Research Strategy and Methods

Table 4.1 Overview of the research strategy

Steps	Purpose	Strategy
1	**Analysis of explanatory variables:** Determining regime constellations pertaining to PGRFA management	• Identify norms and rules from individual regimes • Identify overlap of functional scopes between regimes • Analyze interaction between regimes • Identify regime constellations • Identify driving forces and mechanisms of influence behind regime constellations as points of departure for explanations
2	**Analysis of dependent variables:** Determining the state of PGRFA management in the case country	• Identify policy coverage and goal achievement with regard to each area of PGRFA management set out in Table 3.1 • Identify compatibility between policies as well as goal achievements pertaining to each area of PGRFA management set out in Table 3.1 • Determine policy conduciveness and goal achievements in terms of indications set out in Table 3.1
3	**Linking dependent and explanatory variables:** Identifying relative aggregate effects of the regime constellations on the state of PGRFA management in the case country	• Identify correspondence between international regimes and national-level policies • Identify causal relations between international regimes and national-level policies within the context of the country's agricultural and environmental politics, and the larger historical and political setting • Determine strength of regime effects • Determine conduciveness of regime effects
4	**Explaining the findings:** Determining mechanisms of influence which explain these aggregate effects	• Select embedded cases and determine units of analysis • Identify driving forces behind policies and goal achievements – and their strategies for influence • Assess pattern matching between strategies for influence and mechanisms of influence proposed in Chapter 3
5	**Concluding the analysis:** Determining the scope for generalization of the findings	• Before starting investigations at the national level, ensure that the case country is representative with regard to central features for the analysis • Ensure that the case-study approach is developed according to the requirements for theory-developing case studies during the analysis of the domestic level (dependent variables). • Determine the conditions under which the results have relevance for other countries

To approach the high level of complexity in this study, a multi-faceted research strategy is required. Table 4.1 provides an overview of the research strategy developed for this book. The very first step, identifying components of PGRFA and determining the criteria for assessment of PGRFA management, was already dealt with in Table 3.1 and provides the basis for this research strategy.

Explanatory Variables: Determining Regime Constellations

In Chapter 3 it was proposed that a careful analysis of regime constellations could enable the prediction of aggregate regime effects. For this, we must have precise methodological tools. Norms and rules are central concepts in the definition of regime constellations, so a core challenge will be to identify such norms and rules systematically on the basis of the empirical evidence. This will provide the foundation for analyzing regime overlap, regime interaction and regime constellations.

Identifying Norms and Rules from Single Regimes

The most important sources of information for determining norms and rules are the texts of the international agreements themselves. However, although these agreements have been adopted and have entered into force, the content of their norms and rules is in a continuous process of further definition and specification. This in turn means that the identification of norms and rules will have to take into account the decisions from all relevant negotiations of the decision-making bodies of these international agreements in a time perspective.

Identifying norms and rules cannot be based on a pure text analysis of provisions and decisions. Interpretation is essential. In the process of interpretation, both the historical and the institutional contexts are central – in addition to analyzing the effects of regime interaction:

- The *historical context* reveals the background to the rules and norms – and thereby the background against which they must be interpreted – and provides an account of the time dimension in regime development, which is especially important in analyzing regime interaction. The historical context is also important for identifying the driving forces behind regime formation, that is the actors who initiate regime formation and who most significantly influence the processes. Analyzing the historical contexts of the three regimes could provide material for at least three books, so a major challenge is to delimit this analysis, keeping in mind the purpose of the exercise. We will concentrate on the key features of the historical background for the most central decisions pertaining to the management of PGRFA and for identifying the driving forces behind these decisions.
- The *institutional context* in which a regime is embedded – or to which it is tied – can be expected to influence the interpretation of the content as well as the strength of enforcement. International regimes are often part of larger institutional frameworks, or they may be single regimes with their own web of ties to various international institutions. Using the classification model on regime types elaborated by Oran Young (see Chapter 3), we will analyze pattern matching between the observations and the classification model, and derive inferences as to implications for regime formation and development.

- *Regime interaction* refers – in this context – to the effects of other international regimes on the norms and rules of the regime in question. It will be analyzed partly as an element in the historical context, and partly in the context of forum shopping and mechanisms of influence, as a basis for our analysis of overlapping norms and rules (see next section).

In seeking to identify norms, we will take as our point of departure each of the overall objectives and principles of the international regimes, and ask how the contexts and processes have affected expectations as to the behavior of the Contracting Parties regarding these objectives and principles. On this basis, we will distinguish between core and secondary norms and rules, as explained in Chapter 3.

In identifying rules, we will take the regulatory rules of the international agreements as points of departure, analyzing their contents and importance in the light of further decisions within the regime during the course of implementation – and informed by the analyses of the contexts and processes mentioned above. On this basis, we will distinguish between core and secondary rules.

Identifying Overlap of Functional Scopes between the Regimes

The next question will be how to identify the overlaps between them, and the effects for the overall regime constellations.

Rosendal (2001) suggests that regimes may be compatible in terms of norms and rules – a situation with synergetic potentials. They may also diverge with regard to norms and rules – a situation with potentials for conflict. If the overlapping functional scopes are compatible in terms of norms, but diverging as to rules, the potential for conflict is considered high. Conversely, if overlapping functional scopes are compatible as to rules, but diverging as to norms, the conflict is not expected to have serious effects. The worst case is if the overlapping functional scopes of two regimes are diverging with regard to both norms and rules – a case of full conflict. If they are compatible in both respects, on the other hand, there are good potentials for synergies. This is illustrated in Table 4.2.

Table 4.2 Overlapping norms and rules of regimes: Scopes for synergy or conflict

Regime A → Regime B ↓	Overlap with norms	Overlap with rules
Overlap with norms	*Least* consequential overlap between the two regimes in terms of synergy or conflict	Quite consequential overlap between the two regimes in terms of synergy or conflict
Overlap with rules	Quite consequential overlap between the two regimes in terms of synergy or conflict	*Highly* consequential overlap between the two regimes in terms of synergy or conflict
Overlap with both norms and rules	*Most* consequential overlap between the two regimes in terms of synergy or conflict	

Rosendal (2001) proposes a further framework for analyzing the contents of diverging norms and rules. This framework may also be used to analyze the contents of compatible norms and rules. In a slightly modified version of her approach, we distinguish between core and secondary norms and rules. The most conflicting case will be if the regimes diverge with regard to both core norms and core regulatory rules, whereas the case with the greatest synergetic potential will be when core norms and core regulatory rules are compatible. (Of course, whether these potentials are actually utilized will depend on interaction.) This, together with the assumed effects of further overlap combinations, is illustrated in Table 4.3. Here the potentials for conflicts or synergies are indicated numerically on a scale from 1 to 5, where 1 is the highest and 5 is the lowest potential for conflict or synergy. These express the assumed strength of the effect, whether conflict or synergy.[1]

Table 4.3 Core and secondary norms and rules: Overlaps and scopes for conflict or synergy

Regime A → Regime B ↓	Core rules	Core norms	Secondary rules	Secondary norms
Core rules	1	1	2	4
Core norms	1	2	3	4
Secondary rules	2	3	3	5
Secondary norms	4	4	5	5

Regime overlap is not static. It evolves and decreases, depending on developments within the regimes in question. For that reason, we will apply a dynamic approach, analyzing regime overlap in the context of the time dimension noted in Chapter 3.

Analyzing the Effects of Regime Interaction and Mechanisms of Influence

Regime interaction is assumed to affect (1) regime constellations at the international level, and (2) their effects in developing countries (see Chapter 3). In this section we will address the former, whereas the latter will be the subject of the next section.

In order to understand the effects of regime interaction on the regime constellations, the time dimension is of great importance. In this context, the driving forces behind the interaction, their mechanisms of influence and practice of forum shopping appear central (see Chapter 3 for further details).

- *The time dimension:* To analyze the time dimension, we will identify the *development stages* of the various regimes, in order to analyze the dynamics of interaction between regimes at various development stages (based on Andersen 2002). There are normally five development stages: (1) agenda-

1 The table is for purposes of illustration, and cannot be applied on a mathematical basis.

setting phase, (2) negotiation phase, (3) adoption of agreement and ratification period, (4) implementation phase, and (5) review/evaluation phase. These frequently overlap, and this too must be taken into account. The development stages of the various regimes will be organized along a timeline to identify the differences. On this basis we will classify whether the regime interaction on main issue-areas was synergetic, was characterized by barriers, or by strategic opportunities, as a background for identifying regime constellations with regard to PGRFA.

- *Driving forces and mechanisms of influence:* Taking the historical analysis as our point of departure, we will identify the main driving forces behind the various regimes and how they affected regime formation in the widest sense. These findings will be compared with the mechanisms of influence noted in Chapter 3, to identify the actual mechanisms of influence at the international level, applying the method of pattern matching detailed below. This identification is also important for our analysis of mechanisms of influence across scales.
- *Forum shopping:* In the context of the time dimension and mechanisms of influence, forum shopping is highly relevant, as it reveals where political 'clout' is vested from time to time – an important factor for regime constellations. Taking the indications of forum shopping from the historical analysis, we will examine how this has evolved in light of the mechanisms of influence.

Identifying Regime Constellations

In identifying the resulting regime constellations, we will carefully identify all instances of regime overlap between the three regimes according to rules and norms, as they concern the management components of PGRFA. This will yield a complex picture of overlapping core and secondary norms and rules pertaining to a range of PGRFA management components. To approach this picture systematically, we must analyze the overlaps between each set of two within the three regimes, three sets of analyses, before bringing together the findings to the final regime constellations (see Figure 4.1).

For each set of analysis, the overlaps will be analyzed systematically in terms of core and secondary norms and rules. All in all, the systematic analysis of regime overlap will be a challenging task due to the many possible combinations, so we will aim to simplify where possible. Table 4.4 illustrates how the instances of regime overlap may appear in one of the three sets of analysis, where some but not all norms and rules overlap in terms of functional scope.

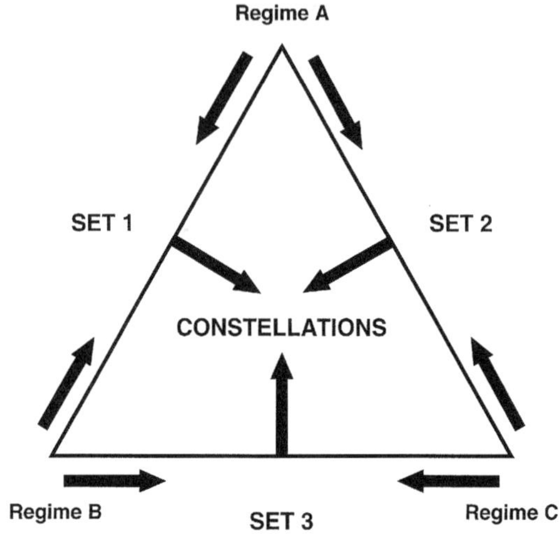

Figure 4.1 Sets of analysis of regime constellations

Table 4.4 Identifying the overlap of functional scopes of two international regimes

Regime I → **<u>SET 1</u>** **Regime II ↓**	**Core norms:** – Core norm A – Core norm B	**Secondary norms:** – Secondary norm A – Secondary norm B	**Core rules:** – Core rule a – Core rule b	**Secondary rules:** – Secondary rule a – Secondary rule b
Core norms: – Core norm X – Core norm Y	Item 1.1 (A+X) Item 1.2 (B+X) Item 1.3 (B+Y)	Item 1.11 (A+Y) Item 1.12 (B+X) Item 1.13 (B+Y)	Item 1.22 (a+X) Item 1.23 (a+Y)	Item 1.31 (a+X)
Secondary norms: – Secondary norm X – Secondary norm Y	Item 1.4 (A+X) Item 1.5 (A+Y)	Item 1.14 (B+X) Item 1.15 (B+Y)	Item 1.24 (a+Y) Item 1.25 (b+X) Item 1.26 (b+Y)	Item 1.32 (a+X) Item 1.33 (a+Y)
Core rules: – Core rule x – Core rule y	Item 1.6 (A+x) Item 1.7 (A+y) Item 1.8 (B+y)	Item 1.16 (A+x) Item 1.17 (A+y) Item 1.18 (B+y)	Item 1.27 (a+x) Item 1.28 (a+y)	Item 1.34 (a+x) Item 1.35 (a+y)
Secondary rules: – Secondary rule x – Secondary rule y	Item 1.9 (A+x) Item 1.10 (A+y)	Item 1.19 (A+x) Item 1.20 (A+y) Item 1.21 (B+x)	Item 1.29 (b+x) Item 1.30 (b+y)	Item 1.36 (b+y)

For each of the identified items we will compare the respective norms and/or rules and analyze their contents and scope for conflicts or synergies in light of the findings from our analysis of regime interaction, taking account of the time dimension, mechanisms of influence and forum shopping. Where there have been changes in the overlaps, their contents or scopes for conflicts or synergies during the period under study, this too will be addressed.

Finally we will take the PGRFA management components (Chapter 3) as points of departure for analyzing the regime constellations pertaining to each of them, that is the constellations of norms and rules with which developing countries have been faced throughout the period under study. With this approach we can grasp not only regime overlaps with regard to each management component in a time perspective, but also as to the overlaps between the various PGRFA management components. This is a critical point, because even within one and the same international regime there may be potentials for conflicts between norms and rules pertaining to different PGRFA management components, as we will see.

At this stage in the analysis, it is important to check whether there are norms or rules that have not been covered by the analysis because they are not overlapping with the functional scopes of other international regimes. In the case of PGRFA management, this can generally be assumed not to be the case, as the international regimes are intimately interlinked. However, this possibility should not be discarded.

Finally, we will take the regime constellations as points of departure for developing assumptions on aggregate regime effects in developing countries. In this process, we will again take regime interaction into consideration, as it is expected to affect regimes not only at the international level, but also across scales to the national level. The mechanisms of influence are likely to be central in this context. We will test the assumptions in the analysis of the dependent variables.

Dependent Variables: Determining the State of PGRFA Management with the Six Cs Model

In order to identify the relative aggregate effects of the international regimes for the management of PGRFA in developing countries, the case-study approach has been selected as a basic strategy. The theory perspectives employed for this purpose, as presented in Chapter 3, necessitated the use of a basically qualitative research strategy in order to establish necessary and sufficient conditions for causal relationships between the variables. In the field of political science, qualitative case studies have emerged as an important means of describing and analyzing political situations and dynamics. The use of qualitative case studies makes it possible to delimit comprehensive and complex fields of analyses to observable study objects. Thus it becomes possible to analyze the entirety of an issue-area – exemplified in one case or a selected number of cases – while also focusing on details that are pivotal to explaining the findings. Furthermore, case studies can be designed so as to determine the conditions under which the findings have general relevance. Further details on the case-study approach, its prospects for generalization of the findings, and the reasons for the choice of the Philippines as a case country are provided later in the chapter.

To measure the state of PGRFA management in the case country, we need to define values for the dependent variables; and in a study based largely on qualitative methods – like this one –measurement is a challenge. However, values are necessary to determine as exactly as possible how the state of policy decisions and goal achievement can be assessed, as basis for our further investigations.

In developing values for the state of domestic policy decisions and goal achievement pertaining to PGRFA management, we will apply the concepts developed in the analytical framework (Chapter 3). We will measure the state of domestic policy decisions as well as of goal achievement as related to each of the six categories determined as core elements of the management of PGRFA, as shown in Table 4.5.

Table 4.5 Distinguishing values of the dependent variables

Core elements in the management of PGRFA	State of policy decisions pertaining to PGRFA management	State of goal achievement pertaining to PGRFA management
In situ management	(value)	(value)
Ex situ conservation	(value)	(value)
Access to PGRFA	(value)	(value)
Utilization of PGRFA	(value)	(value)
Benefit sharing	(value)	(value)
Farmers' rights	(value)	(value)

The crucial question is how to determine the values to be filled in. In the search for a strategy for measuring not only PGRFA management, but also the effects of international agreements on the domestic management of PGRFA, it appeared necessary to develop a model. This model, which I have called the *Six Cs Model on the Measurement of Aggregate Regime Effects*, will be presented and applied in two parts. Part I (the first three Cs) is devoted to measurement on the dependent variable, which is the subject of this section. Part II (the last three Cs) addresses the measurement of the links between the explanatory and dependent variables, which is the subject of the next section.

The first three Cs in the Six Cs Model on the Measurement of Aggregate Regime Effects stand for *coverage*, *compatibility* and *conduciveness.* These are parameters that will be applied in analyzing policy decisions as well as goal achievements in the management of PGRFA. Analyzing coverage is necessary to find out whether the policy decision or goal achievement pertains to the whole issue-area or only to parts of it. Analyzing compatibility is necessary to find out whether the policy decisions or goal achievements pertaining to the issue-area are pulling in the same direction or in different ones, and whether policies that pull in different directions are conflicting. Analyzing conduciveness is necessary to find out whether policy decisions or goal achievements are conducive to the management of the different components of PGRFA, as defined in Table 3.1.

For each of the parameters, a set of values has been developed, highlighted with symbols to illustrate the contents of the values. The symbols will first be applied in the analysis of policy decisions, and thereafter in the analysis of goal achievements. In Table 4.6, the symbols are placed in the middle between the values of policy decisions and goal achievements, for the sake of simplicity:

Table 4.6 Values and symbols of the Six Cs Model, Part I

Coverage		
State of policy decisions: Values	**Symbol**	**State of goal achievements: Values**
Policy decisions cover the respective PGRFA management area highly or fully	**A**	The goals set out in the policies have been achieved to a large extent or fully
Policy decisions cover the respective PGRFA management area partially	**B**	The goals set out in the policies have been partially achieved
There are marginal or no policy decisions pertaining to this issue-area	**0**	The goals set out in the policies have not been achieved – or there were no goals
Compatibility		
State of policy decisions: Values	**Symbol**	**State of goal achievements: Values**
Policy decisions pull in one direction fully or largely, i.e. they are compatible	⇒	Goal achievements pertaining to the issue-area pull in the same direction
Policy decisions pull in opposite directions, i.e. they are incompatible	⇔	Goal achievements pertaining to the issue-area pull in opposite directions
Policy decisions pull in different but non- conflicting directions	=	Goal achievements pull in different but non-conflicting directions
Conduciveness		
State of policy decisions: Values	**Symbol**	**State of goal achievements: Values**
Decisions are largely conducive to the management of the issue-area	+	Achievements are largely conducive to the management of this issue-area
Decisions are largely not conducive to the management of this issue-area	÷	Achievements are largely not conducive to the management of the area
Policy decisions are neutral in terms of conduciveness	≈	Goal achievements are neutral in terms of conduciveness

Coverage of policy decisions may refer to two distinct aspects of policies: either to their scopes, in terms of the range of regulated sub-issues, or to the degrees of freedom in regulation. Which of, and how, these forms of coverage are applied in the analysis will be explained in detail along the way. In measuring the goal achievement, we will take the policy decisions as our point of departure and ask whether the goals set out in the policies have been achieved. In measuring the compatibility of policy decisions, we will seek to identify whether they pull in the same direction. Compatibility of goal achievement will be measured along the same lines. When measuring the conduciveness of the policy decisions, we will again assess all of them, asking whether they are compatible or not; then we see whether the policy decisions, taken together, are conducive to the management of this issue-area. The analysis of goal achievements will follow the same lines.

Table 4.7 Combinations of values of the Six Cs Model (Part I), from 'best' to the 'worst' case

Coverage → **From the 'best case' to the 'worst case': Combinations of Values** **Compatibility and *Conduciveness* ↓**	Full or high coverage of policy decisions/goal achievements	Partial coverage of policy decisions/goal achievements	Low or no coverage of policy decisions/goal achievements
'BEST CASE': Policy decisions/goal achievements pull in **one direction** which is ***conducive*** to PGRFA management	**A ⇒ +**	**B ⇒ +**	**0**
Policy decisions/goal achievements pull in **different non-conflicting directions** which are in sum ***conducive***	**A = +**	**B = +**	**0**
Policy decisions/goal achievements pull in **conflicting directions** but are in sum ***conducive***	**A ⇔ +**	**B ⇔ +**	**0**
Policy decisions/goal achievements pull in **one direction** which is ***neutral*** in terms of conduciveness	**A ⇒ ≈**	**B ⇒ ≈**	**0**
Policy decisions/goal achievements pull in **different non-conflicting directions** which are ***neutral*** in terms of conduciveness	**A = ≈**	**B = ≈**	**0**
Policy decisions/goal achievements pull in **conflicting directions** which are in sum ***neutral*** in terms of conduciveness	**A ⇔ ≈**	**B ⇔ ≈**	**0**
Policy decisions/goal achievements pull in **conflicting directions** which are in sum ***negative*** in terms of conduciveness	**A ⇔ ÷**	**B ⇔ ÷**	**0**
Policy decisions/goal achievements pull in **different non-conflicting directions**, which are in sum ***negative*** in terms of conduciveness	**A = ÷**	**B = ÷**	**0**
'WORST CASE': Policy decisions/goal achievements pull together in **one direction** which is ***negative*** in terms of conduciveness	**A ⇒ ÷**	**B ⇒ ÷**	**0**

Part I of the Six Cs Model enables us to grasp the key variables relevant to determining the state of the management of PGRFA within each management area. Various combinations of values are possible. Some of these combinations indicate that PGRFA – within the respective management area – are well taken care of in the case country; others indicate that they are not. Between these 'best case' and 'worst case' scenarios there is a range of possibilities, as outlined in Table 4.7.

The exact order of these combinations of values from the 'best' to the 'worst' case will depend on the contents of the values. As the Table combines quantitative and qualitative measurements, it is not possible to determine the exact order of the value combinations.

The right-hand column has no values because in these cases there are no policies or goal achievement to assess, or because these are infinitesimal. A situation without policies may be either conducive or less conducive to PGRFA management. (For example, it is better to have no policy than one that is negative to PGRFA management.) As our focus is on the effects of international regimes in our case country, the Philippines, in analyzing the context, we will account for the state of PGRFA management regarding the 'no policy option'.

The values of the first three *Cs* in this model are considered to be neutral to subjective perceptions, which is necessary to account for internal validity. It might be argued that the value of conducive or unfavorable decisions could be influenced by subjective perceptions. However, if we limit the exercise to pattern matching between the texts of the policy decisions and actual data on policy outcome with the indications of conduciveness presented in Table 3.1 – and if we make each assessment explicit – we should be able to arrive at values that are neutral to subjective perceptions. When it comes to the coverage of goal achievement, it could be argued that there are too few nuances: The value of high/full goal achievement is unlikely to be relevant in most cases, and between the full and low/no goal achievement there is a huge range of possibilities. However, distinguishing between values within this spectrum would permit subjective interpretations, and that would be unfavorable to the internal validity of the case studies.

It is not possible simply to fill in the symbols in the first table presented above and then extract an average. The different elements of the management of PGRFA have different qualities, and cannot be measured against each other on a mathematical basis. Also the qualitative assessments leading to the choice of the different values in the various settings rule out using such quantitative methods. Instead, we will need to rely on reasoning in order to make conclusions as to the overall state of policy decisions and goal achievement pertaining to PGRFA management. However, this reasoning will have a solid basis in the measurements undertaken, and will provide a richer picture of the management of PGRFA in the Philippines.

The Six Cs Model of Aggregate Regime Effects does not express the potential variation during the period in question, as including also the time dimension in the model would have made it too complicated and blurred the main intention, which is to show whether and how the case country acts in line with the relevant agreements to which it is bound. We will address the time dimension at three levels: First, it will be analyzed at the outset of the case study, where all legislation and policies are presented (in Chapter 9); second, it will be dealt with in the analyses of policy decisions and goal achievement pertaining to each of the PGRFA management issues; third, we will include sub-values in cases where policies have clearly changed or where there has been a marked change in performance as to goal achievements.

Linking Dependent and Explanatory Variables: Identifying Aggregate Regime Effects with the Six Cs Model

The greatest challenge in developing a qualitative case-study design is to provide for causal inferences – linking the dependent and independent variables. In our case, we need to establish causal links between the relevant international regimes, that is, the regime constellations resulting from the single regimes, their overlap and interaction, and our findings from the case country. The aim is to determine as precisely as possible the aggregate effects in the case country, and in order to do this, we need to define exactly what to measure. Again, it is necessary to define the values, in order to provide for optimally exact analysis and for transparency. The effects of international regimes will be identified according to the defined values as shown in Table 4.8.

Table 4.8 Measurement of aggregate effects of international regimes on PGRFA

Relative effects →	**Relative effects of the IU/ITPGRFA on the state of:**		**Relative effects of the CBD on the state of:**		**Relative effects of TRIPS (27.3.b) on the state of:**		**Aggregate effects of the international regimes**
... pertaining to ↓	**Policy decisions**	**goal achievements**	**policy decisions**	**goal achievements**	**policy decisions**	**goal achievements**	
In situ PGRFA Management	(value)	(value)	(value)	(value)	(value)	(value)	**Inference**
Ex situ PGRFA Conservation	(value)	(value)	(value)	(value)	(value)	(value)	**Inference**
Access to PGRFA	(value)	(value)	(value)	(value)	(value)	(value)	**Inference**
Utilization of PGRFA	(value)	(value)	(value)	(value)	(value)	(value)	**Inference**
Benefit sharing	(value)	(value)	(value)	(value)	(value)	(value)	**Inference**
Farmers' rights	(value)	(value)	(value)	(value)	(value)	(value)	**Inference**

How then should we define the values? There are no shortcuts to one value that will cover all the aspects of regime effects, as these are multifaceted. The degree to which an international regime affects a particular component of PGRFA management does not indicate whether the effect is in accordance with its norms and rules. For example, the provisions of one international regime (regime A) may be developed especially strongly in domestic policy as a response to another regime (regime B). In that case, regime B has a strong effect on the development of this policy as well, but the policy will probably not correspond with the provisions of regime B. Nor can we know whether the regime effects as such are conducive to the management of PGRFA, according to the indications set out in Chapter 3. It is important to distinguish between correspondence and conduciveness, since international regimes or single provisions under these regimes are not necessarily conducive to PGRFA management. There are obviously several dimensions to account for when defining values.

Against this background, Part II of the Six Cs Model has been developed, addressing the following three dimensions (three *Cs*) of regime effects:

- *Causation* refers to the degree to which we can ascribe a policy decision or a goal achievement to an international agreement. It is measured against other factors that explain these outputs and outcomes, to be identified as we proceed in tracing the causes.
- *Correspondence* refers to the degree to which the identified effects are in accordance with norms and/or rules of an international regime.
- *Conduciveness* refers to the degree to which the effects can be described as positive for management of the particular PGRFA area. It is determined on the basis of the indications of conduciveness set out in Chapter 3.[2]

How to determine the values within each of these dimensions is the subject of the rest of the section. For the purpose of overview, the values are presented in Table 4.9 together with symbols that illustrate their contents.

Exploring in depth all effects of international agreements on policies and goal achievements pertaining to all management areas of PGRFA in the Philippines would be a demanding task. To delimit the task, a two-level approach has been selected:

- Level 1: Brief assessment of all effects according to documentation of attribution, but without thorough tracing of the lines of causation.
- Level 2: In-depth analysis of the effects with regard to two policy documents (the EO 247 and the PVP Act) and their implementation in order to trace the lines of causation in detail.

As will be explained below, the two analyses at Level 2 are designed as two embedded case studies. The two policy documents are central for the management of PGRFA in the Philippines and in terms of the implementation of international agreements, so tracing the effects of the regimes on these two regulations and their outcomes should substantially improve our basis for conclusions on regime effects. The Level 2 analysis will be the subject of the next main section, as it is intimately linked with the search for the mechanisms of influence and the method of process tracing. The Level 1 assessment is the subject of this section.

2 It is important to distinguish between the 'conduciveness' of the effects of international regimes (Part II of the Six Cs Model) and the 'conduciveness' of the policies and goal achievements (Part 1). The effects may point in a different direction from the overall conduciveness of the policies and goal achievements.

Table 4.9 Overview of values and symbols of the Six Cs Model, Part II

Causation		
Regarding policy decisions: Values	**Symbol**	**Regarding goal achievements: Values**
Most lines of influence can be traced back to the respective regime	**III**	Most lines of influence can be traced back to the respective regime
Some lines of influence can be traced back to the respective regime	**II**	Some lines of influence can be traced back to the respective regime
Marginal or no lines of influence can be traced back to the respective regime	**0**	Marginal or no lines of influence can be traced back to the respective regime
Correspondence		
Regarding policy decisions: Values	**Symbol**	**Regarding goal achievements: Values**
Full or substantial match between the policy decision and the regime in question	≡	Full or substantial match between the goal achievements and the regime in question
Partial match between the policy decision and the regime in question	–	Partial match between the goal achievements and the regime in question
Low or no match between the policy decision and the regime in question	≠	Low or no match between the goal achievements and the regime in question
Conduciveness		
Regarding policy decisions: Values	**Symbol**	**Regarding goal achievements: Values**
The effect on the policy is largely conducive to PGRFA management	+	The effect on the goal achievement is conducive to PGRFA management
The effect on the policy is largely not conducive to PGRFA management	÷	The effect on the goal achievement is not conducive to PGRFA management
The effect on the policy is neutral in terms of conduciveness	≈	The effect on the goal achievement is neutral in terms of conduciveness

To identify causal relations between international regimes and PGRFA management at the domestic level is basically about tracing lines of influence. By lines of influence, we refer to the stream of influence from an international regime, or a constellation of international regimes, via the actors who transform these impulses into various kinds of action, and to the effect of this action in terms of policy decisions and goal achievements. Distinct lines of influence are, in other words, the streams of influence that can clearly be traced back to their sources, whether these are international regimes (including the actors behind these regimes), or other sources, for example at the domestic level. The details of this method will be further explored in the next main section.

For our more broadly based assessment of causation in management areas, however, we need a shortcut to make the investigations feasible. Instead of tracing the lines of influence, we will seek to identify documentation – for example, the preamble to an act of legislation –that reveals whether a policy can be ascribed to an international regime. We will also seek to identify documentation on the extent to which goal achievements are ascribed to international agreements. It is important to distinguish between policies and goal achievements in this regard. Whether a policy derived from an international agreement is implemented in a developing country will normally depend on a host of other factors. Also there is the possibility that a grassroots initiative may be ascribed to an international agreement, whereas the national policy to which it contributes is not.

To what extent can information on ascription be taken as a reliable source for determining causation? Although a regulation may be 'marketed' as CBD implementation, there can be several factors explaining why it emerged. This means it is important to cross-check documentation with central stakeholders in the form of interviews, and to express uncertainties explicitly. Against this background, our method can be considered reliable for the purposes of this study.

In measuring correspondence between international agreements and domestic policies, we will simply compare the texts of the international agreements and the national policies as expressed in rules and regulations or in other policy documents.

A similar approach will be used for measuring correspondence between international regimes and goal achievements. Here we will have to compare the provisions of the international agreement with the action that has taken place – not with formulations in documents – in order to determine the extent of match.

Correspondence as such does not provide any indication of causation. Whether correspondence is an effect of an international regime will depend on whether there are causal links between the regime and the policy or goal achievement in question. The values suggested in this section can therefore be understood in the context of regime effects only when linked with the values of causation.

The identified effects of international regimes on policy decisions and goal achievements pertaining to PGRFA management can be considered positive, negative or neutral to the management of the issue-area, as compared with the indications of conduciveness set out in Table 3.1. It is analytically important to distinguish the conduciveness of these effects from the conduciveness of the policies and goal achievements (Part 1 of the Six Cs Model), as the effects may point in another direction than the overall conduciveness of the policies and goal achievements.

Measuring the conduciveness of various effects will yield a complex picture. For example, one and the same effect may be conducive regarding policies on benefit sharing but negative as to policies on access to genetic resources. When we differentiate as to management areas and international regimes, this complexity will unfold, in turn giving rise to the question of how to draw conclusions on the aggregate effects of international regimes.

As in all qualitative analyses, it is difficult to determine the precise values of the variables and of their links. The methodology outlined above seeks to capture the richness and complexity of the situation and the developments under study, and is meant to serve as a guideline. The quality of the analysis will depend on the explicit expression of all the important considerations determining its use along the way.

It will not be possible to draw inferences from the sets of values indicated above and to aggregate regime effects for PGRFA management on a mathematical basis. Rather, the determined values will serve as the basis for inferences and conclusions to be spelled out in detail with regard to contents and importance. The values are solely means to reveal the richness and the complexity of the picture in a consistent, systematic and transparent matter, and thus ensure the validity of the analysis.

In Chapter 3 it was assumed that the aggregate effects of international agreements could be predicted by a careful analysis of the relevant regime constellations, that is, the constellation of the relevant regimes in terms of overlapping rules and norms – and the interaction between them. Having analyzed the aggregate regime effects as to PGRFA management, we will compare our findings against the predictions set out in the analysis of regime constellations and assess pattern matching, in order to verify or falsify this assumption in terms of the aggregate effects of international regimes for PGRFA management in the Philippines.

Explaining the Findings: Determining the Mechanisms of Influence

Explaining the findings with regard to the effects of international agreements in the case country is vital to understanding how these agreements work at the national level. Only with a precise analysis of the mechanisms of influence is it possible to identify the entry points for improvements or change. To make such an analysis feasible, two cases have been selected – embedded in the larger case study – for in-depth study.

Selecting Embedded Cases and Units of Analysis

Whereas all issue-areas relevant to PGRFA management were analyzed in terms of the dependent variable, we will need to narrow down our focus when seeking to explain the effects of international regimes. That is the reason for selecting two embedded case studies deemed central to PGRFA management in the Philippines:

Case 1: The process leading up to the Executive Order 247 titled Prescribing Guidelines and Establishing a Regulatory Framework for the Prospecting of Biological and Genetic Resources, Their By-Products and Derivatives, for Scientific and Commercial Purposes and For Other Purposes (18 May 1995), and its implementation until the end of 2004.

Case 2: The process leading up to the Republic Act No. 9168 titled An Act to Provide Protection to New Plant Varieties, Establishing a National Plant Variety Protection Board and for Other Purposes (7 June 2002), and its implementation until the end of 2004.

The case-study approach, including its potentials for generalization, is further elaborated below.

If we are to explain the effects of international agreements on PGRFA management in the case country, we must first define the units of analysis, that is, the entities or process elements that are to be observed as basis for analysis (Frankfort-Nachmias and Nachmias 1992, 53).[3] In investigating how overlapping international regimes affect PGRFA management, a central observation is that action has been taken by various groups of *stakeholders*, such as institutional entities and/or individuals within the authorities, the economy and the civil society.

As Galtung has noted (1971), it is important to distinguish between *parties* and *actors*, although the two terms are often used interchangeably. Parties are categories of people or institutions that may or may not take a stand on a given issue-area. If they do take a stand, and start to seek influence over the processes regarding that issue-area in one way or another, then we may label them actors. Also other entities may emerge as actors, in that they seek to represent the stands of certain parties, or try to help them.

The basic parties in our case can be defined among:

- categories of farmers, such as small-scale farmers using traditional seed varieties and production systems, and large-scale farmers using commercial seed varieties and production systems;
- categories of plant breeders, such as domestic breeders and multinational breeders;
- categories of consumers, such as consumers in the farm communities and urban poor consumers;
- categories of gene bank conservators, whether in domestic gene banks or in international gene banks.

When it comes to actors, the picture is somewhat different, as there are actors seeking to represent several of the parties described above. Some of these actors are identical with the parties, but several are not. The most important actors in our case can be identified among:

- Entities and individuals within the authorities: Parliamentary groups, relevant ministries or entities within these ministries, regional and local authorities or entities, and individuals with specific formal expertise.

3 In defining units of analysis, there are various possibilities, for example categories of individuals, groups, institutions or societies, or process elements, such as perceptions, attitudes or behavior (Frankfort-Nachmias and Nachmias 1992, 52).

- Economic actors or groups or individuals: Individual seed corporations, organizations of seed corporations, farmer federations, bilateral and multilateral donors, various categories of banks, and individuals with a particular involvement.
- Civil society[4] actors: NGO actors and actors from research institutions, including individual members of civil society.

As Galtung (1971) has pointed out, parties have interests, whereas actors have goals. The latter derive from what the actors perceive as the interests of the parties that they seek to help or represent. Analytically, it provides clarity to distinguish between the interests of parties and the goals of actors, because this enables reflections on the legitimacy of the activities undertaken by actors.

Against this background, it seems fruitful to define *actors* as units of analysis for this study, as indicated in the research questions. The ways in which international regimes affect domestic policies are assumed to depend decisively on whether and how actors take initiatives, seek influence and follow up decisions.

Identifying Driving Forces and Strategies for Influence

The core method for identifying driving forces and strategies of influence is called *inductive process tracing*. It is a modified form of what is otherwise known as process tracing, which in turn is a method that helps in tracing processes back to their origins.[5] It was first described as a method by George and McKeown (1985, 35), as 'the decision process by which various initial conditions are translated into outcomes'. Thereby the dependent variable is transformed into several variables expected to have direct effects on the ultimate dependent variable. These sub-variables are in turn translated into further variables that are all considered necessary and sufficient for determining the dependent variable. The process is repeated as long as necessary and feasible, until the involved actors and their influence have been accounted for. Hence, process tracing is a method that allows us to examine multiple observations with a view to assessing causal versus correlating relationships (King et al. 1994, 227).

Process tracing is a way of searching for evidence that is consistent with the overall theory for an analysis, and it is a procedure that may involve interviewing

4 Civil society can be defined as the sphere of social interaction between economy and state, comprising the intimate sphere (especially the family), the sphere of associations (especially voluntary associations), social movements, and forms of public communication (Cohen and Arato 1992, ix). Rudolph (2000, 1766) distinguishes among various categorizations of associations in civil society: political and non-political, deliberative and interest-based, egalitarian and hierarchical, voluntary and ascriptive.

5 *Origin* is in this context a relative term. Process tracing can presumably go on forever, as there will always be another reality behind the situation one traces. Therefore it is theoretically possible to trace such processes back to situations a long time ago, involving psychological, religious, historical, environmental or/and other explanations. For practical reasons, in our context *origin* refers to the more immediate sources of explanations, such as international regimes.

actors and reading their written records (ibid.). The way in which process tracing is described in the literature indicates a deductive method. Interviews are undertaken (where necessary) to provide evidence for already established theoretical assumptions or hypotheses on causation or correlation. For the purpose of our analysis, such a procedure could result in a map that would not fit with the landscape, since our 'landscape' is a highly complex one. Therefore, we will apply process tracing in an inductive manner, asking open questions to a large number of stakeholders from all sides, in order to reveal all the decisive paths of influence. This means we are not at the outset seeking to establish a causal relation between the international regimes and the identified policies and goal achievements. Instead, we are open to all possible explanations that can enable us to determine the *relative* influence of international regimes. In other words, we seek to trace the lines of influence back to their sources, be these international regimes (including the actors behind these regimes), or other sources, for example at the domestic level.

A central task will be to identify the actors who contributed to the changes in policy decisions and goal achievements, and the chain of actors who influenced those actors. Further, we will need to identify the arguments they used, and the contexts in which they understood the issue. Seeking to map this as precisely as possible will enable us to discover the extent to which the actors derived their arguments from the respective regimes, and/or the extent to which they received some sort of support related to policy development and implementation pertaining to the regime in question.

Determining the Mechanisms of Influence

When determining which mechanisms of influence have been at work in our cases, we will use *pattern matching*, a method where an empirically observed pattern is compared with a more predicted one (Skjærseth 1999, 98), such as an hypothesis or a proposition on causation. If the patterns correlate, this means that a necessary requirement for the verification or falsification of the hypothesis or proposition is fulfilled. However, the correlation is not sufficient to explain causation. Only when the findings can be underscored through process tracing is it possible to establish sufficient conditions for verification or falsification. In a case study aimed at generalization, patterns may be related to the dependent as well as the independent variables, or to both. Pattern matching is regarded as a robust strategy to strengthen the internal validity of a case study (ibid., 15). Combined with our method of inductive process tracing, pattern matching will be used to compare the observed patterns of influence with the patterns of the mechanisms of influence, as elaborated in Chapter 3. In order to establish casual explanations in terms of pattern matching, an important requirement is to design precise hypotheses, or in our context propositions. The propositions presented in Chapter 3 will provide the points of departure for our analysis of pattern matching.

Combining the findings from the process tracing with the findings from the pattern matching will enable us to draw inferences as to the actual mechanisms of influence at play among the international regimes in question – and PGRFA management in the case country. This is how we will seek to explain the effects of international regimes.

Determining the Scope for Generalization

As explained above, *case studies* have been selected as a basic strategy for analyzing the effects of international agreements on the management of PGRFA in developing countries. A crucial question is thus what scope one single case country – even with two embedded case studies – can provide for generalization to other countries.

The appropriateness of using case studies has been debated in political science, especially as to the potentials for generalization. Is it really possible to draw general conclusions on the basis of one single case, or some few cases? In retrospect, it seems evident that this debate has been complicated by the lack of distinction between types of generalization. Whereas political science had been relying on quantitative methods for a great part and thus on the logic of statistical generalization, qualitative case studies emerged from the logic of analytical generalization. Case studies aimed at analytical generalization were never meant to represent samples, nor were they aimed at generalizing to populations or universes. The aim was to generalize to theoretical propositions (Yin 1994, 10). Analytical generalization from case studies is regarded as possible, as long as certain methodological requirements are fulfilled. This point will be further elaborated in this chapter.

The distinction between and the comparative advantages of qualitative and quantitative methods have received more attention, and skepticism to qualitative case studies as a research strategy is decreasing. Case studies can very well be used in combination with such other strategies, or elements of these strategies can be used as methods for qualitative case studies, for example in terms of data collection. It is, however, important to be aware of what to use when and how. Combining the two does not change their basic inherent logic and their different scopes for generalization. If, for example, an analysis is based on qualitative methodologies, but information is added on a statistical basis, it is still basically a qualitative study, and conclusions on its general relevance will have to be based on the logic of analytical generalization.

For this study, archival analyses, statistical material, and to some extent historical approaches have provided crucial input in the data collection. Thus, the case studies have been informed by a range of methods in order to diversify the sources of evidence, while remaining qualitative in the analytical approach.

The main purpose is to generalize analytically to other developing countries with similar framework conditions. This involves determining the conditions under which the findings from our case country can have relevance elsewhere.

In addition, the data material allows for some reflections as to empirical generalizations. Data are readily accessible on policies concerning several PGRFA management areas in other developing countries. This allows us to consider whether the findings from the Philippines are representative in this context. Such findings are useful to provide an overall picture of the state of PGRFA management in developing countries, whereas the analytical generalizations can help in explaining the conditions under which such outcomes may be affected by international regimes.

Type of Case Studies

This analysis takes several sets of theories as points of departure for developing propositions[6] with regard to the expected empirical findings, that is, how international agreements affect the management of PGRFA in the case country, the Philippines. The propositions have been developed to enable us to analyze the findings in terms of identifying the conditions under which they may have relevance in other similar situations. Thus the main feature of the analysis is the use of theory to understand empirical material, and it is this direct analytical purpose that guides the work.

Additionally, an overall analytical intention is to contribute to theory development. Thus the theory-developing case study is applied. Such case studies provide the possibility of developing theories with a basis in empirical material that is structured in terms of assumptions or propositions. They can take existing theories or concepts as points of departure for organizing new empirical material, or they can be based entirely on empirical material. The choice depends on whether the aim is to improve existing theory or to develop new propositions. For both choices, the overall objective is not to explain empirical situations or processes but to use them in order to explain theoretical relevance (Andersen 1997, 73). This approach has been chosen because it can best enable us to meet the challenges of the overall aim of the book, which is to explain – on the basis of case studies – how international regimes affect the management of PGRFA in developing countries.

Requirements for Theory-Developing Case Studies

In order for theory-developing case studies to deliver valid and reliable findings, certain requirements must be fulfilled. These are commonly explained as pertaining to construct validity (operational measures, the sources of evidence), internal validity (causal relationship between factors of explanation), external validity (the domain to which the findings of study can be generalized), and reliability (conducting the study so as to allow other scholars to reach the same conclusions) (see for example Yin 1994).

Whereas the first three requirements have been accounted for in the above research design, the latter is discussed in the next main section.

One Case Country as Basis for Generalizations

The selection of one case country has been based primarily on the view that in-depth studies of one country provide better possibilities for understanding the specific causal mechanisms in focus than if the scope were expanded to include a range of countries. This choice involves a trade-off between depth and scope – the single-case country allowing for greater focus on causal mechanisms, whereas the

6 In qualitative analyses it is more appropriate to use propositions than hypotheses. Hypotheses require precise testing, which is possible only on the basis of quantitative methods. In qualitative methods, the aim is to identify the conditions under which particular causal effects may occur. For this purpose, propositions provide a better form.

comparative approach would have provided more room for analyzing the relevance of the propositions.

High quality of research design and multiple replications of the findings are the core strategies for ensuring external validity, but it is not always easy to balance the two within the given limits of a project. Consideration was given to including one or two countries for replication studies, but such an approach would have proven too demanding in terms of data collection and analysis. Here it should be noted that data collection in developing countries is especially time-consuming. As the quality of the case study is crucial to the potentials for generalization, the gains of concentrating on one case country were considered to outweigh the potential gains of a few more possibilities for replication.

The Philippines as Case Country

In order to ensure relevance of the study and feasibility of the work, five categories of criteria were identified for selecting the case country:

1. *Relevance for PGRFA management:* It should be a country relevant with regard to the issue under study, that is, it should possess a relatively high diversity of PGRFA.
2. *Relevance with regard to the international regimes in question:* It should have ratified the CBD, adhered to the FAO International Undertaking on PGRFA, and be a member of the WTO, in order to allow for studying the effects of these agreements and their interplay at the national level. Since the ITPGRFA was not adopted until November 2001 and did not enter into force until 2004, it was not considered necessary that the country should have signed or ratified the Treaty.
3. *Scope for generalization of findings to other countries:* In order to provide for potentials for generalization, it should be a developing country with features similar to many other developing countries in terms of challenges pertaining to PGRFA management, institutional capacity and actor constellations.
4. *Scope for interesting results:* In order to provide for the possibility of learning from experience, it should be a developing country that is considered to have made considerable progress in implementing the CBD, as the only binding international agreement concerning PGRFA management in the period under study.
5. *Working conditions:* It should preferably be a country where the study will be welcomed and informants are positive to sharing their insights, and a country with good access to written information in English.

Relevance for PGRFA management The variety of PGRFA is huge and extremely rich in terms of colors, tastes, smells and uses. The Philippines is one of several centers of diversity of rice in the world, indicated by at least 12 different species of wild rice (DENR and UNEP 1997, 70), in addition to a wide diversity of varieties of *Oryza sativa* (Asian rice) of different colors, sizes, tastes and uses.[7] It is among the richest areas in the world in diversity of fruit trees; also within tuber crops, fiber plants, coconuts and other nut trees there is huge diversity, with several species endemic to the Philippines.[8]

Relevance with regard to international agreements The Philippines was among the first 31 countries to ratify the CBD, which set the Convention into force in 1993. A member of the FAO since 1945, the Philippines adhered to the International Undertaking on PGRFA when it was adopted at the Twenty-second Session of the FAO Conference in 1983. It has been a member of the World Trade Organization since 1 January 1995, and is therefore also bound to the TRIPs as one of the agreements under the WTO. Like most other developing countries, the Philippines is not a member of the UPOV.[9]

Scope for generalization to other developing countries The Philippines has a range of features similar to those found in many other developing countries. Its PGRFA has disappeared at an alarming rate, especially in the 1960s, 1970s and 1980s, due primarily to the introduction of high-yielding genetically homogeneous crop varieties. The greatest diversity of on-farm food crops is found in marginal agricultural areas, often in the highlands, where conditions are not conducive to intensified agriculture and farms are often small. These features are similar in most developing countries, but with variation in scope. The Philippines has national gene banks, but, as in most other developing countries, they lack the financial resources to maintain accessions, provide them with passport data and make them accessible. Institutional capacity to implement international agreements is severely limited, as in so many other developing countries. Actor constellations are similar to those in other developing countries with an active civil society: Organizations from civil society as well as business corporations seek to influence state actors with limited capacity to respond to such influence. The Philippines is probably among the developing countries with the highest number of civil society organizations. This might make it less representative – but in the field of PGRFA there are only a few. Therefore, the situation is not considered to be very unlike that of other developing countries where there are civil society organizations.

There are of course also many differences between the Philippines and other developing countries, not least its unique culture, as well as the particular (and controversial) position of the US as a close ally – a position with deep and complex historical roots. However, what counts in our context are the immediate factors that

7 Due to the 'Green Revolution', the number of varieties grown in the fields has fallen dramatically.

8 The Pili nut (*Canarium ovatum*), for example, is a tasty and popular nut which grows only in some parts of the Philippines, and nowhere else in the world.

9 As of January 2004, 15 developing countries (former East Bloc countries not taken into account) were members of UPOV, and 14 of these were members of UPOV '78, whereas one country was member of UPOV '91. These countries were Bolivia, Brazil, Chile, China, Colombia, Ecuador, Kenya, Mexico, Nicaragua, Panama, Paraguay, South Africa, Trinidad and Tobago, Tunisia (member of UPOV '91), and Uruguay.

can be expected to affect policies on PGRFA management. Here, the Philippines is comparable with other developing countries with such characteristics, although the exact degree of comparability will vary. We can limit the scope somewhat, as it would hardly be reasonable to try to generalize the findings from the Philippines to countries that lack essential state structures, that are repressive concerning civil society organizations, and/or that are among the very poorest of the world's low-income countries.

Scope for interesting results The Philippines was among the most active developing countries under the negotiations leading up to the CBD and was one of the first to adopt legislation on access and benefit sharing, thereby implementing the Convention. Civil society organizations as well as researchers around the world have held up this legislation as a model for other developing countries. There has been no similar development as to the management of PGRFA, although several efforts have been made to promote legislation on farmers' rights in the country (the Philippines has also advocated such regulation at the international level). The Philippines has a range of other interesting policies pertaining to PGRFA management, including the Philippine Agenda 21 – and, more recently, legislation on plant breeders' rights, which is a controversial policy in the country.

Working conditions Working conditions and access to information in the Philippines are very good, as compared to many other developing countries. Most people relevant for interviews speak fluent English, documents are normally written in English, and also meetings and conferences are often conducted in English. People are hospitable and generous with information, and there are normally no problems in getting access to documents once one knows where to find them. On the other hand, getting access to information can be extremely time-consuming – as in most other developing countries.

The Two Embedded Cases

Two embedded cases were selected to enable an in-depth analysis of the mechanisms of influence. *Embedded case studies* refer to case studies with multiple units of analysis that are embedded in larger frameworks. This may be designed from the very beginning, or be the result of a reformulation of a research design for a single case study, where the evidence proved to call for multiple units.

The process of defining one or more cases as a part of a larger context[10] is a crucial methodological step to enable generalization based on the results. Cases are normally selected on the basis of assumptions as to how the independent (explanatory) variables affect the dependent variable(s), and this is also the reason for our choice:

10 See Ragin and Becker 1992; this is also referred to as 'casing' (ibid.; King et al. 1994, 128–38).

Case 1 The processes related to the EO 247, concerns the *aggregate* effects of international regimes pertaining to PGRFA on the domestic regulation of access to genetic resources and the sharing of benefits resulting from their utilization in one developing country with the characteristics given above. It is assumed that the CBD as well as the TRIPS Agreement affected the process decisively.

Case 2 The processes related to the PVP Act, concerns the *aggregate* effects of international regimes pertaining to PGRFA on the domestic regulation of intellectual property rights to plant varieties in one developing country with the characteristics given above. It is assumed that the TRIPS Agreement and the UPOV affected the process decisively, whereas the IU and the ITPGRFA had a limited role. Possibly also the CBD may have affected the process.

The cases can thus be expected to capture the influence of all four regimes in the Philippines. As embedded case studies, the analysis forms part of a broader analysis that allows a rich understanding of the framework conditions under which the processes took place. The external validity of the research design is improved by expanding the number of case studies within the country.

In addition to being embedded, the cases can be described as two single cases.[11] *Single cases* are particularly useful for exploring the relevance of well-formulated theories as to critical aspects, indicating whether the critical case give rise to reformulation of the theory. They may also be appropriate for analyzing events that are extreme or unique, where it is still too early to establish common patterns for similar events. Finally, events that have not been examined before may justify the use of one single case study to explore how such phenomena can be studied (Yin 1994, 40). For example, when studying the implementation of a new type of international environmental agreement, a focus on one single case pertaining to a clear, distinguishable event could be useful to indicate how to approach further cases. Although this book aims at analyzing the effects of international regimes on PGRFA management in the chosen case country, it focuses on two distinguishable processes that led up to two different acts. These are the most evident expressions of the international agreements in the case country. It is necessary to cover both in the case studies, because they are expected to be the most different and are thereby necessary to provide as complete a picture as possible. As such, the two case studies can be analyzed as single cases.

Yin (1994, 46) distinguishes between studies aimed at similar results (literal replication) and those that produce contrasting results (theoretical replication). Literal replication is used to strengthen the inferences on which the propositions are built, whereas theoretical replication is appropriate for testing rival theories, according to Yin. This is aimed at producing contrasting results in order to test rival theories.

11 They are not multiple, since they are only two, and multiple case studies are normally comparative. They are not comparative because they cannot be interpreted in terms of the same variables (see Eckstein 1975, 85).

Three different theories are explored[12] in terms of their explanatory power for the analysis. It is expected that the theory of structural power, in combination with the institutional capacity approach, will have the greatest explanatory relevance for the second case, whereas the theories of learning and norm diffusion, in combination with the institutional capacity approach are expected to be most relevant for the first case. Structural power and learning and norm diffusion – and their different forms of interaction with institutional capacity – are seen as relevant for guiding the explanation of different phenomena, and are as such regarded as rival approaches.

The main difficulty in using embedded case studies is to shift between the levels of analysis. It is easy to lose sight of the larger context in trying to grasp the challenges at the sub-levels of analysis (Yin 1994, 44). All levels selected for analysis must be adequately examined and interlinked.

Determining the Conditions for Generalization

To generalize means to draw an inference from observations of the particular to a broad and encompassing statement or proposition. It is an act of reasoning from the observed to the unobserved, from a specific instance to all instances believed to be like the instance in question.

In this research design, it has been important to establish valid criteria for the selection of case country and cases, in order to be able to define the conditions under which the findings will have general relevance for other developing countries with similar framework conditions. Furthermore, the method must be replicable, in order to account for replication of the method for other possible cases, making it possible to test the findings against other contexts with similar framework conditions.

The development of theory is essential to establish any basis for generalization, whether the intention is to understand the empirical situation within an issue-area, to test theories or to improve or develop new ones. It lies in the very nature of analytical generalizations to provide the conditions under which certain events will occur, in other words to produce theory. Thus, it does not make sense to conduct case studies aimed at generalization without developing a rich theoretical framework for the research.

Sources of Evidence, Data Collection and Use of Data

Sources of evidence, methods of data collection and the uses of data are central to construct validity as well as the reliability of an analysis.

Sources of Evidence and Methods of Data Collection

The main sources of evidence for this analysis are documents and interviews. The documents consulted comprise the texts of the international agreements, as well as

12 The concept of 'testing theories' seems more appropriate for the 'hard' or natural sciences and quantitative social science methods, whereas this book focuses on exploring the theories' explanatory power in qualitative terms.

documents from the negotiations, the meetings of the conferences of the parties, of subsidiary groups and background papers. They also comprise relevant acts of legislation in the case country together with other policy documents, position papers and documents. Furthermore, large numbers of documents and gray literature from the various actors in the case country, pertaining to the processes studied as well as their own way of working, have been assessed. Some of these documents are referenced, whereas parts of the gray literature have provided valuable background information. One problem should be pointed out: After a scandal blew up in the media in the Philippines in 2003, central web-pages documenting the course of events in the second case study were closed down. As I had collected paper copies from these web-pages, and they are central sources of evidence, I have chosen to reference them.

In addition to these documents, secondary literature on the issue-area has been used to establish the state of research as well as to provide context. Statistical material, where available, has been used in analyzing the state of the art regarding PGRFA management in our case country, the Philippines. As developing countries often struggle to provide precise statistical data, the data used have been carefully cross-checked, and any problems will be explicitly stated as we proceed.

Documents from the international agreements have been systematically downloaded and archived, whereas documents and literature from the Philippines have been systematically collected during two field trips, and additional new documents have been provided to me by helpful individuals in the Philippines. It is therefore assumed that most of the relevant documents for the analysis of the international as well as national levels have been collected.

Interviews were conducted with 60 people[13] in the Philippines, in addition to 12 informants at the FAO and in Norway. The Philippine interviewees include representatives from all relevant units in the Department of Agriculture and in the Department of Environment and Natural Resources, in gene banks, relevant research institutions (some of them actors, other observers), all the NGOs directly engaged in PGRFA at the national level, several NGOs engaged in more general biodiversity issues, a few representatives of business actors and of the main donor organizations. Within each category I sought out different people at different institutional levels in order to detect possible nuances in the descriptions and perspectives, and also possibly differing opinions on relevant matters.

For the first interview series in the Philippines, I had prepared some questions, but used a very open approach to the interviews. I was still in the more inductive phase of the work, and therefore followed up the information by tracing it further with more in-depth questions. For the bulk of the interviews (in 2002), however, a semi-structured method was used. I had prepared guidelines for four categories of interviewees: state actors, civil society actors, business actors and observers. In the interview situation, the guideline was used as the point of departure, but was followed up with in-depth questions. If the interviewee had more information on

13 A few of the most central people were interviewed two and sometimes three times, after I had received more information. This is not accounted for in this figure. Some interviews were group interviews, where two or more individuals were interviewed together.

one topic than on another, our talk focused on the former. However, I tried to cover all questions of the guide directly or indirectly (they were answered as part of answers to other questions, or sometimes in documentation received). I also sought to distinguish between ego perceptions and alter perceptions,[14] in order to cross-check the information given. Additionally, it was possible to cross-check some of the information with available documents. In order to ensure reliability, the contents were written out soon after the interview, on the basis of the notes taken, and while the interview was still fresh in memory.

Due to the collected documents, the number of interviews, and the methods of data collection, I regard the information situation as valid in terms of construct validity, and as reliable. There could have been more business actors interviewed. However, I managed to speak with some of the most central actors in the processes under study, as indicated in the position papers from the business actors to the legislation processes, and also by other actors. To increase the construct validity of the case studies, meetings were held with several informants for debriefing, before I left the Philippines. In addition, after returning to Norway I continued to exchange information with some key informants in order to test my conclusions.

There could have been more interviews with representatives from the various international agreements. The main emphasis in this regard was on the Commission on Genetic Resources, at FAO in Rome. However, two research colleagues at the Fridtjof Nansen Institute, and not least our Director, have covered the processes under the CBD and the TRIPS Agreements, as well as UPOV, in considerable detail, and I was fortunate in being able to draw on their experience from these agreements.[15] Such delimitation was pivotal to making the project feasible.

The Uses of the Data

The first interviews in 2000 provided important insights into the situation in the Philippines with regard to PGRFA management, and were decisive to the formulation of research questions, the identification of a theoretical framework and the development of propositions. In this way, the interviewees participated in the

14 This is inspired by the 'EAR instrument' developed by Arts and Verschuren (1999), which is a method aimed at analyzing political influence in complex decision-making processes. It takes three perspectives as points of departure: The ego-perspective (E), the alter-perspective (A) and the researcher's analysis (R). The ego-perspective refers to an actor's view regarding its own role and influence, whereas the alter-perspective refers to this actor's view regarding the roles and influences of other actors. Representatives of all central actors are interviewed along the same lines. The researcher's analysis is a document analysis pertaining to the issue-area to cross-check interview findings. An interesting example of the application of the EAR instrument is provided in Sauvé and Watts 2003.

15 The Director of the Fridtjof Nansen Institute, Peter Johan Schei, was the former head of the Norwegian delegation to the Conference of the Parties to the CBD. Dr. G. Kristin Rosendal has studied the negotiation and implementation of the CBD in depth in her doctoral dissertation (2000) and in her later work. Morten Walløe Tvedt has focused on the TRIPS Agreement, the WIPO and to some extent the UPOV in his doctoral dissertation, which is currently in process.

formulation of the research design for the book. On returning to the Philippines two years later, I met several of the same interviewees again – in addition to many new informants – and experienced their high motivation to contribute with information and insights to the work, as well as warm hospitality and generosity. The information received through the interviews helped me to establish a fuller understanding of the situation. However, there are few references to the interviews in the text. The reason is that I have sought to document all information and inferences rather than referring to the interviews. In other words, the interviews were instrumental in providing me with the necessary insights, for which I sought written documentation in order to confirm the impressions – or test them, when required. I also sought to bring essential findings back to some central informants, in terms of debriefing towards the end of my stay. This also enabled me to test central assumptions and propositions.

Whereas the interviews were crucial for depicting the structures of the highly complex situation in the Philippines, the documents that I collected served to provide the vital data used as sources of evidence in the book, and were analyzed following the basic methods of text analysis. Text analysis was also central to the investigations at the international level, where documents provided the main source of evidence. Also here, interviews were used to gain an overview over the context as well as central questions, and to test various assumptions and propositions.

PART 3
The Constellations of International Regimes Pertaining to Plant Genetic Resources for Food and Agriculture

The purpose with this part of the study is to identify the entirety of international norms and rules on PGRFA management in the period 1992–2004 which resulted from the international regimes pertaining to this issue area, the overlaps in their functional scopes, and the interaction between them. In other words, we will identify the regime constellations pertaining to PGRFA management that developing countries faced during this period.

The histories of each single regime reveal the driving forces behind them, their goals and strategies, influence via other regime processes (regime interaction), and how the specific regime rationales took shape. They also show how norms and rules were molded, resulting in the international agreements, and how the contents of these agreements were further weighted and specified in the course of implementation.

Each of the first three chapters starts by describing the history of the respective regime(s) in question with a particular view to their development related to PGRFA management. In addition, for each regime there will be a sub-chapter on the institutional context, to provide the necessary background for analysis. On this background, each of the first three chapters concludes with a detailed analysis of the core and secondary norms and rules of each regime concerning PGRFA management.

On this foundation, we will analyze regime overlap and interaction in the final chapter of this part, concluding with an overview over the resultant regime constellations with which developing countries are faced. Our first task here will be to summarize the driving forces behind the development of the various regimes pertaining to PGRFA management and their strategies, seeking to identify the mechanisms of influence at the international level. Understanding these mechanisms at the international level is a basis for our analysis of the mechanisms of influence across scales. Furthermore, we will analyze how different actors have used 'forum shopping' as a strategy. Finally, we analyze the regime overlaps and interaction concerning PGRFA management, before deriving conclusions as to the resulting regime constellations, with assumptions as to the likely aggregate effects of these regime constellations in developing countries.

Chapter 5

The International Treaty on Plant Genetic Resources for Food and Agriculture with the International Undertaking on Plant Genetic Resources

The International Treaty on Plant Genetic Resources for Food and Agriculture (ITPGRFA) was adopted at the Thirty-first Session of the FAO Conference in Rome, 3 November 2001. It entered into force 29 June 2004 as the first legally binding agreement dealing exclusively with PGRFA management. Its stated objectives are the conservation and sustainable use of these resources, and the fair and equitable sharing of the benefits arising from their use – in harmony with the CBD – for sustainable agriculture and food security (Art. 1). The most important benefit to be provided from the use of PGRFA, according to the Treaty, is that of access to these vital resources for food and agriculture (Art. 13.1).

The forerunner to the ITPGRFA was the International Undertaking on Plant Genetic Resources (IU), adopted at the Twenty-second Session of the FAO Conference in Rome in 1983 (Conference Resolution 8/83). The objective of the International Undertaking, as stated in its original 1983 version, was to ensure that plant genetic resources for food and agriculture would be explored, preserved, evaluated and made available for plant breeding and scientific purposes. Its two-pronged goal was clear: conservation and access. Unlike the ITPGRFA, the IU was not legally binding. It remained formally in force until the adoption of the ITPGRFA. However, it received little attention after the adoption of the text of the Convention on Biological Diversity in 1992, which – as a legally binding agreement – covered PGRFA, and which set the stage for the renegotiation of the IU that was to result in the ITPGRFA.[1]

In 1996, the Global Plan of Action for the Conservation and Sustainable Utilization of Plant Genetic Resources for Food and Agriculture (GPA) was adopted by the International Technical Conference on Plant Genetic Resources, in Leipzig, Germany, 17–23 June 1996, and was endorsed by 150 countries.[2] In turn, the GPA formed part of the FAO Global System for the Conservation and Utilization of PGRFA established in 1983, covering a range of FAO initiatives – including first the IU and later the ITPGRFA. This whole system may be seen as the FAO regime on the management of PGRFA. For the sake of analytical clarity, we will distinguish

---

1 As proposed in UNEP (22 May 1992): *Nairobi Final Act of the Conference for the Adoption of the Agreed Text of the Convention on Biological Diversity*.

2 Available online at <www.fao.org/ag/AGP/AGPS/GpaEN/gpatoc.htm>.

between the components in the historical analysis, with special focus on the IU and the ITPGRFA, and less on the GPA.

Historical Context: The Winding Path from the 'Common Heritage of Mankind' to the Multilateral System

The history of the IU and the ITPGRFA can be seen as a winding path, from the principle of PGRFA as a 'common heritage of mankind' (IU) via the demise of this principle (Brush 2005 and 2004; Fowler 2004), followed by its limited revival in the Multilateral System on Access and Benefit Sharing (ITPGRFA). The major milestones of this process are presented in Table 5.1, and will be further elaborated.[3]

Table 5.1 Development stages and milestones of the FAO regime on PGRFA

1979 – 1981	**Agenda-setting phase** introduced by publication of a book on PGRFA (Mooney 1979) and following discussions in many countries, leading to a Mexican proposal and a subsequent resolution at the 1981 FAO Conference to draft elements of a legal convention on the feasibility of establishing an international gene bank (Fowler 1994, 180–87).
1981 – 1983	**Negotiation phase** leading up to the adoption of the IU in 1983.
1983 – 1989	**Implementation phase** for the IU: some achievements at the international level in terms of *ex situ* conservation, but little at the national level.
1989 – 1993	**Continued implementation phase**, but gradually diminishing, while a parallel **new agenda-setting phase**[a] takes shape with the adoption of agreed interpretations of the IU in 1989 and 1991 and, after the adoption of the CBD, with the decision of the 1993 FAO Conference to revise the IU in harmony with the CBD.
1994 – 2001	**Negotiation phase**, leading up to the adoption of the ITPGRFA in 2001. Parallel efforts at expanding the Global System on PGRFA, e.g. through the Global Plan of Action on PGRFA, adopted at the International Technical Conference on Plant Genetic Resources in Leipzig in 1996.
2001 – 2004	**Ratification period**, waiting for the necessary number of countries to ratify the Treaty, while preparing for its implementation.
2004 –	**Implementation phase** for the ITPGRFA, following its entry into force on 29 June 2004.

[a] There is no clear review/evaluation phase with regard to the IU. The agreed interpretation in 1991 (Resolution 3/91 of the FAO Conference) states that the conditions of access to PGRFA, which is central in the IU, needs further clarification, which could indicate the need for a review or evaluation. However, the initiative from the negotiating parties to the CBD for revising the IU in harmony with the Convention commenced before such a process could be initiated by the FAO. That is how the agenda-setting phase emerged without any significant review or evaluation of the IU.

3 This presentation is largely based on Andersen 2003.

Agenda Setting: Emerging Awareness on PGRFA Management

FAO engagement in PGRFA began a few years after the organization was established in 1945. As early as 1948 the FAO conducted a conference for 44 member states on biological research in forestry, addressing the use and exchange of genetic material on a global scale (Pistorius 1997, 11). In 1957 the organization began publication of a newsletter on plant genetic resources,[4] and from 1961 onwards, it convened a series of technical meetings and conferences on plant genetic resources. An FAO Panel of Experts on Plant Exploration was established in 1963 to advise the organization and set international guidelines for the collection, conservation and exchange of PGRFA. The conference on plant genetic resources in 1967 formulated several resolutions, which were subsequently adopted by the UN Conference on Environment and Development in Stockholm 1972. In 1973 a new conference interpreted the resolutions of the Stockholm Conference in the context of PGRFA (FAO 1998, 2).

Parallel to efforts within the FAO, in 1974 the International Board for Plant Genetic Resources (IBPGR)[5] was founded under the auspices of the Consultative Group on International Agricultural Research (CGIAR).[6] The IBPGR was located at FAO headquarters in Rome and drew on the staff designated for the FAO program on genetic resource conservation – an institutional structure that was characterized as an 'historical anomaly' by IBPGR itself (Fowler 1994, 184). Collecting missions were accelerated, and gene banks were constructed and expanded at national, regional and international levels (FAO 1998, 2). The erosion of PGRFA had been increasing at an unprecedented rate due to the 'green revolution', and the efforts of the IBPGR were vital to saving plant varieties in danger of extinction.

This meant that the IBPGR had considerable power over the direction of conservation activities and was assumed to have the authority to designate certain gene banks for holding particular collections. Only 15 per cent of the samples

---

4 It has been published continuously since then. Originally it was called *FAO Plant Introduction Newsletter*. Since 1978 it has been issued jointly by the FAO and the International Plant Genetic Resources Institute (IPGRI) under the title *Plant Genetic Resources Newsletter*.

5 The IBPGR was transformed into the International Plant Genetic Resources Institute (IPGRI), which is part of the CGIAR (see next footnote), in 1974. In 2006, IPGRI and the International Network for the Improvement of Banana and Plantain merged, and changed the name to Bioversity International. In this book, we will continue using 'IPGRI', as this was the name in most of the period studied and it is still the most widely known.

6 The CGIAR was founded in 1971 on the initiative of the Ford and Rockefeller Foundations to unite privately-funded international agricultural research centers (IARCs) into one network. As an informal association of public and private donors that support the IARCs, it is a donor-led group that has provided a forum for discussion of research priorities and co-ordination of funding (FAO 1998, 248). As parts of the network, the IARCs have their own governing bodies. The United Nations Environment Program, the FAO, the United Nations Development Program and the World Bank co-sponsor the system, and the CGIAR is headquartered at the premises of the World Bank in Washington DC. Efforts are being made to increase the representation of developing countries (ibid., 249).

collected were designated for storage in developing countries, whereas 85 per cent were stored in industrialized countries and in the gene banks of the International Agricultural Research Centers (IARCs) of the CGIAR (Fowler 1994, 184). There were some obvious reasons for this choice. First, gene banks in the South were few. Second, they were often unable to ensure the safety of seed accessions at the level of gene banks in the North, due to shortage of electricity, lack of institutional capacity as well as financial constraints. Most genetic resources samples provided from the IARCs went to countries in the North, and the IARCs were, together with the IBPGR, heavily criticized by NGOs for serving private interests in the North (see, among others, Fowler and Mooney 1990, 182–93). The IBPGR and the IARCs did invaluable work in saving fast-eroding plant varieties from extinction, but they also deprived developing countries of the possibilities to control their own genetic resources.

The disparities between the North and the South regarding PGRFA were addressed in the book *Seeds of the Earth: Private or Public Resources?* by Pat Roy Mooney[7] in 1979. The book provoked much concern and discussion and is held to have given a decisive push to a multilateral process towards international regulation of the management of PGRFA (see Fowler 1994, 180). At the FAO Conference in 1981, the Mexican delegate proposed that the elements of a legal convention on the feasibility of establishing an international gene bank should be drafted, to be reported back to the FAO Conference in 1983 (ibid., 187).

From the very first discussions in the FAO, there was major disagreement as to what the FAO should be striving for. Most developing countries were concerned about the loss of diversity in PGRFA, combined with the North/South disparity of designated storage facilities and related issues of access and control. They supported the Mexican proposal to draft a legal convention (Fowler 1994, 186). Most industrialized countries, fronted by the United States, Great Britain and Australia together with the seed industry, were skeptical to an international regime of this type, since they feared the politicization of the issue and the possible loss of control over gene bank-conserved PGRFA to the FAO (ibid., 188).

Thus the agenda-setting phase started out with the question of how to improve the utilization of PGRFA, but discussions were soon dominated by the problem of genetic erosion and how to ensure efficient collection and conservation of these vital resources. In the wake of conservation efforts, new challenges arose, since developing countries lost control over genetic resources – mostly to the North. When intellectual property regimes on plants emerged in the latter, the developing countries became increasingly concerned about how access to PGRFA could be ensured under such conditions. This set the stage for the negotiations that led to the International Undertaking on Plant Genetic Resources, with FAO-member developing countries as driving forces.

7 Pat Roy Mooney and Cary Fowler founded the Rural Advancement Foundation International (RAFI) in 1984, together with Hope Shand. This was probably the most influential NGO during the negotiations in the FAO on crop genetic resources in the 1980s and the early 1990s. RAFI has since changed its name to the Action Group on Erosion, Technology and Concentration (ETC Group). Mooney and Shand still front the organization.

The International Undertaking on Plant Genetic Resources: Negotiations and Adoption

During the negotiations the major conflict line went between those in favor of intellectual property rights over improved varieties of plants, and those in favor of unrestricted access to all plant varieties (Fowler 1994, 187–91). The United States and representatives from the seed industry were the leading proponents of the former stance, and developing countries of the latter. This is worth noting, since most developing countries later changed their positions on access in order to provide for benefit sharing from the use of genetic resources, which required stricter regulation of access as well.

When the IU was adopted in 1983, it attracted wide support. Soon 113 countries had signed. However, some of the countries most central for the management of PGRFA did not adhere to the Undertaking – among them Australia, Canada, Switzerland and the US.

The adoption of the IU can be seen as a partial victory for the developing countries. It was a victory because it was achieved despite the opposition of major industrialized countries led by the US, which also chaired the FAO at the time of its adoption. It was partial, because the new agreement ended up as a legally non-binding Undertaking, and was not signed by industrialized countries important to the international management of PGRFA. This severely limited its prospects for realization.

The IU: Objectives, Tasks and Institutional Structure

The objective of the 1983 International Undertaking (IU) was to ensure that plant genetic resources for food and agriculture would be explored, preserved, evaluated and made available for plant breeding and scientific purposes. Its dual aim was clear: conservation and access. The IU was based on 'the universally accepted principle that plant genetic resources are a heritage of mankind and consequently should be available without restriction' (Art. 1). Its main tasks were:

- exploration of varieties of crops, their wild relatives and non-domesticated food plants in IU signatory countries;
- preservation, evaluation and documentation of plant genetic resources in the areas of their natural habitat in the major centers of genetic diversity as well as outside, in gene banks and living collections of plants;
- ensuring open access to the plant genetic material preserved.

These tasks were to be carried out by the states signatories to the International Undertaking, and also by multilateral actors such as the FAO and the institutions of the CGIAR. The implementation of the IU indicated close international co-operation in the form of international gene banks and information-sharing networks, steps to enhance the performance and numbers of gene banks, and efforts to encourage financial support and capacity-building.

Along with the IU, the Commission on Plant Genetic Resources (CPGR) was established by the Twenty-second Session of the FAO Conference in Rome in 1983 (Conference Resolution 9/83). The CPGR was an intergovernmental body charged with ensuring implementation of the IU and especially monitoring the operation of international arrangements on PGRFA management. In 1995, the Commission was renamed the *Commission on Genetic Resources for Food and Agriculture* (CGRFA) and its mandate expanded to cover all components of biological diversity of relevance to food and agriculture, including animal genetic resources, forestry and fisheries. Its advisory role for the FAO was strengthened, as was its co-ordinating function with regard to other multilateral institutions.[8] As of January 2007, a total of 168 countries and the European Community were members of the CGRFA, which holds regular sessions every second year and extraordinary sessions as required.

Two intergovernmental working groups have been established to assist the Commission in its work: one on plant genetic resources for food and agriculture and one on animal genetic resources. Work on forestry and fishery has been postponed due to capacity constraints. The Intergovernmental Technical Working Group on Plant Genetic Resources was established in 1997 and consists of representatives from 27 countries, elected by the Commission. The group meets upon the decision of the Commission. The CGRFA and its subsidiary bodies are supported by a secretariat of two permanent professional and two to three temporary professional staff, with additional support staff.

According to its statutes,[9] the CGRFA is to review all matters relating to FAO policy, programs and activities related to genetic resources for food and agriculture, and to recommend measures for developing a comprehensive global system on these resources. Furthermore, the CGRFA serves as an intergovernmental forum for negotiations, like those leading to the ITPGRFA, and a body overseeing co-ordination with other relevant international agreements as well as with other international, governmental and non-governmental bodies dealing with the conservation and sustainable use of genetic resources.

The CGRFA is among the core components of the Global System for the Conservation and Utilization of PGRFA (FAO 1998, 256).[10]

---

8 FAO Conference Resolution 3/95, 1995.

9 FAO Council Resolution 1/110, 1995.

10 Further components of the Global System developed since 1983, analyzed in Andersen (2003), are as follows: the International Fund for Plant Genetic Resources (which never materialized, although some minimal contributions were received); the International Network of *Ex Situ* Collections under the Auspices of the FAO; the World Information and Early Warning System on Plant Genetic Resources for Food and Agriculture; the International Code of Conduct for Plant Germplasm Collecting and Transfer; report on the State of the World's Plant Genetic Resources for Food and Agriculture; the Global Plan of Action for the Conservation and Sustainable Use of PGRFA.

Implementation of the IU

The implementation of the IU at the international level is not directly relevant for our purposes.[11] However, one outstanding achievement is important for our understanding of actor constellations and driving forces: this is the *International Network of Ex Situ Collections under the Auspices of the FAO*. In 1994, the 11 centers of the CGIAR with gene banks,[12] plus one additional center, concluded a series of agreements to place their germplasm collections under FAO auspices. These agreements constitute the cornerstones of the International Network: some 600,000 accessions of plant varieties (FAO 1998, 83), including information about them, were made available for plant breeding and direct use under formally agreed terms – under the auspices of the FAO. This had been a central demand from developing countries. The main point was not availability, because much of the material had been available under similar terms for years, but control: management was now to be under the auspices of a multilateral organization.

In 1998, the accessions made up between 20 and 50 per cent of all genetic material conserved in gene banks (FAO 1998, 280).[13] The CGIAR collections contain the highest genetic diversity of PGRFA among the gene banks in the world, due to its emphasis on land-races and wild relatives (Fowler and Hodgkin 2004; Fowler, Smale and Gaiji 2001).[14] They are also known for their well-maintained and documented collections. To ensure continued availability, the Centers agreed that any recipient of the genetic resources should not claim ownership to them or seek intellectual property rights over them or related information – thereby also meeting the demand from developing countries on this issue. However, this obligation does not extend to new varieties of plants resulting from the use of genetic resources from the gene banks.

---

11 See Andersen (2003) for a detailed analysis.

12 Among the best-known gene banks of the International Agricultural Research Centers of the CGIAR are: Centro Internacional de Agricultura Tropical (CIAT) for beans, cassava, tropical forage crops, and rice in Latin America; Centro Internacional de Mejoramiento de Maiz y Trigo (CIMMYT) for maize and wheat in developing countries; Centro Internacional de la Papa (CIP) for potato, sweet potato, and Andean root and tuber crops; International Crops Research Institute for the Semi-Arid Tropics (ICRISAT) for sorghum, pearl and finger millet, chickpea, pigeonpea and groundnut; International Institute of Tropical Agriculture (IITA) for cassava, maize, cowpea, soybean, yam, banana and plantain; and the International Rice Research Institute (IRRI) for rice in developing countries.

13 Such estimates depend on the data available, as well as methods of calculation. In this case, the FAO argued that the extent of redundancy within the CGIAR base collections is low. Therefore they included an estimated 20 per cent duplicates as compared to an estimated total of 1 to 2 million unique accessions in the world. Three years later, Fowler, Smale and Gaiji (2001), using a different method, concluded that the CGIAR Centers maintain approximately 12 per cent of the accessions held in *ex situ* conditions worldwide.

14 Whereas land-races and wild relatives comprise only approximately 16 per cent of national *ex situ* collections of PGRFA, they make up 73 per cent of the CGIAR collections (FAO 1998, 94).

At the national level there have been few achievements of any real substance. The most comprehensive review of the national-level situation was finalized in 1998 (FAO 1998). It concluded that about 40 per cent of all countries (industrialized as well as developing countries) had relevant national programs and co-ordination mechanisms (ibid., 223). However, these programs were focused mainly on *ex situ* conservation, with poor institutional linkages to utilization efforts. Only 27 of 154 countries reported the inclusion of *in situ* conservation in their national programs, and 26 countries reported that utilization was an integral part of the program (ibid., 202).

Agreed Interpretations of the IU: New Agenda-Setting Parallel to Implementation

A great problem with the IU was that major industrialized countries of crucial importance for the international management of PGRFA had not signed. In the controversies on control over genetic resources in the 1980s, there were deep conflict lines between the parties. The United States, as well as many firms in the seed and pharmaceutical industries, feared reduced access to genetic resources, whereas developing countries and many NGOs were highly skeptical to intellectual property rights to such resources (Fowler 1994, 196).

Even at the first Session of the Commission on Plant Genetic Resources in 1985, proposals for modifying the text of the Undertaking in order to attract more signatories[15] were considered.[16] The Commission recommended that the Secretariat prepare a paper for consideration by the Commission at its next session, analyzing countries' reservations to the IU and delineating possible courses of action, including suggestions for possible interpretations of the text to increase acceptance (paragraph 13). At the following FAO Conference, countries were urged to spell out their reservations to this non-binding international agreement.[17] Various members indicated that their national legislation, including plant breeders' rights, determined the degree to which they could adhere to the Undertaking. At its Second Session in 1987, the Commission on Plant Genetic Resources agreed to adopt practical measures to ensure wider adherence to the IU,[18] and established a Contact Group composed of 17 members to work out proposals.[19] Also a working group had been established to initiate negotiations on agreed interpretations of the controversial parts of the IU, including the interpretation, clarification and recognition of plant breeders' and farmers' rights. This was noted 'with satisfaction' at the Session of the FAO Council in June 1987.[20] However, there remained deep controversies on these issues between the major OECD countries on the one side, and the group of developing countries and their NGO supporters on the other.

15 At that time there were still only 74 signatories.

16 FAO Report CPGR/85/REP (1985), paras 12 and 13.

17 FAO Report C 1985/REP (1985), para. 291.

18 FAO Report CL 91/14 (1987), para. 12.

19 Ibid, para. 34.

20 FAO Report CL 91/REP (1987), para. 104.

These controversies were also fuelled by another process, the Uruguay Round of the General Agreement on Tariffs and Trade (GATT), which ultimately led to the WTO, where intellectual property rights had been brought into the negotiations by the US. There was heated controversy regarding, *inter alia*, intellectual property rights to PGRFA, and many developing countries protested against such negotiations (see Chapter 7).

With the deepening controversies, William Brown, then chair of the US National Board for Plant Genetic Resources (prior to that he had been president of Pioneer HiBred), initiated contact with the Keystone Center in Colorado, seeking to establish a dialogue on plant genetic resources between international stakeholders (Fowler 1994, 197). The Keystone Approach was to invite stakeholders as individuals, in order to reduce conflict levels and seek dialogue, to keep the discussions off the record, and to produce a report on the basis of consensus only. These discussions took place from 1988 to 1991, chaired by Prof. M.S. Swaminathan, who also led an Interim Steering Committee that gave direction to the dialogues. Facilitators were the staff of the Keystone Center. At its three sessions, the Center gathered 92 stakeholders altogether, many of them central to the negotiations, from 30 countries, and this was important in framing the international discussions on such issues as farmers' rights, common heritage of mankind, international funding and to some extent intellectual property rights (Keystone Center 1991). Some of the proposals from the Keystone dialogues found their way into FAO Conference Resolutions on the International Undertaking in 1989 and in 1991, initiated to meet the reservations of core industrial countries to the IU.

The first two resolutions proposed in 1989[21] recognized the rights of both donors of technologies and donors of germplasm to be compensated for their contributions through the simultaneous recognition of plant breeders' and farmers' rights. The Conference recognized that both resolutions were intended to lay the foundations for an equitable and lasting global system for sharing the costs and benefits of the world's plant genetic resources for present and future generations. The resolutions were adopted by consensus on 29 November 1989 as *Resolution 4/89 'Agreed Interpretation of the International Undertaking'* and *Resolution 5/89 'Farmers' rights'*, to be annexed to the International Undertaking.

These two resolutions were the results of tense negotiations. There had been fierce resistance to plant breeders' rights among developing countries, and the agreed interpretations providing for the acceptance of such rights could be adopted only with the simultaneous recognition of farmers' rights. In this way, the opponents of plant breeders' rights gained recognition of farmers' rights in exchange for something that already existed, that is plant breeders' rights.

The resolution on farmers' rights (FAO Resolution 5/89) stands as a milestone, as it seeks to outline the contents and implications of this concept. The international community pledged itself to ensuring that farmers would receive full benefits from the improved use of genetic resources and to supporting the continuation of their contributions to the conservation and sustainable use of plant genetic resources by providing sufficient funds and assistance.

21 FAO Report C 1989/REP (1989), paras 105–9.

In 1991 a new annex to the IU was adopted (FAO Conference Resolution 3/91).[22] This time, the FAO Conference stated that the concept of genetic resources as the heritage of mankind, as applied in the Undertaking, was subject to the sovereignty of the States.[23] Against this background, FAO members responded that the conditions for access to plant genetic resources required further clarification.[24] The original purpose of the IU – to ensure unrestricted access to genetic resources – was no longer clear, nor was the principle of these resources as being the common heritage of mankind. New factors had been introduced, which complicated the follow-up and limited its prospects of implementation.

In the same annex to the IU, the FAO Conference also decided to set up an international fund to implement farmers' rights. This was a direct result of the Keystone dialogues, where the need for such a fund had been noted and generally accepted by all. Participants from industry ensured that such a fund would be a way to recognize farmers' rights, and not to compensate farmers – and Third World participants were so surprised by this consensus that they did not start a debate on this precondition (Fowler 1994, 201). However, despite the fine words, only a few contributions were received and it never materialized as envisioned.

All in all, the three resolutions probably marked the beginning of the end for the IU. The driving forces behind its initiation had made such substantial compromises in order to get central OECD countries onboard that its very substance had become blurred.

New Negotiation Phase: Harmonizing the IU with the CBD

The Commission on Plant Genetic Resources did not get the possibility to clarify the uncertainties concerning the objective of availability of PGRFA under the IU, before the CBD was adopted next year. This was to be a decisive event also for the development of the IU regime. Whereas the IU was the first international agreement (albeit non-binding)[25] to address the conservation and sustainable use of PGRFA, the CBD became the first legally-binding international agreement ever to address the sustainable management of biological diversity worldwide. As we will see in Chapter 6, the CBD was developed as a stand-alone convention and as a framework convention. An important aspect of the framework convention approach was that details with regard to issue-specific areas were to be negotiated in separate protocols.

22 FAO Report C 1991/REP (1991).

23 This principle was first voiced at the 1972 United Nations Conference on the Human Environment in Stockholm in the form that states have sovereign rights to exploit their natural resources in accordance with their own environmental priorities (Stockholm Declaration, Principle 21). It had become an important principle during negotiations leading to the CBD, as a reaction to the breakthrough for intellectual property rights on genetic resources in the Uruguay Round of the GATT negotiations leading to the TRIPS Agreement. See next chapters.

24 FAO Report C 1991/REP (1991), paragraph d.

25 According to the United Nations Treaty Guide, the term 'Agreement' can be used for legally binding as well as non-binding agreements (see: <http://untreaty.un.org/English/guide.asp>).

The negotiators assumed that the management of PGRFA could be negotiated in the form of such a protocol.

At the Conference for the Adoption of the Agreed Text of the Convention on Biological Diversity in May 1992, the Nairobi Final Act was adopted.[26] In this context, a resolution on the interrelationship between the CBD and the promotion of sustainable agriculture was adopted on 22 May (Resolution 3). This resolution deals particularly with the importance of plant genetic resources for food and agriculture, and urges that ways and means be explored to develop complementarity and co-operation between the CBD and the Global System for the Conservation and Sustainable Use of Plant Genetic Resources for Food and Sustainable Agriculture, established under the FAO with the IU as a central component. Finally, the resolution recognizes the need to seek solutions to two outstanding matters concerning plant genetic resources:

1. access to gene-bank collections acquired prior to the entry into force of the CBD
2. the question of farmers' rights.

At its next session,[27] in November 1993, the FAO Conference accordingly requested the FAO Director-General to provide a forum for negotiations for harmonizing the International Undertaking with the Convention (FAO Resolution 7/93). The negotiations should include considerations of the issue of access to PGRFA on mutually agreed terms, including *ex situ* collections not addressed by the Convention; and on the issue of realization of farmers' rights. The CPGR followed up with a mandate and a proposed process.[28] This formed the point of departure for the lengthy negotiations that finally resulted in the adoption of the International Treaty on Plant Genetic Resources for Food and Agriculture in 2001.

The task of revising the IU to harmonize with the CBD was a challenging one. The CBD addresses the conservation and sustainable use of biodiversity from an environmental perspective and with emphasis on the fair and equitable sharing of benefits. By contrast, the IU focused on sustainable agriculture and food security, for which access to genetic resources is a vital condition, and more important than any other benefit sharing. The CBD promoted a bilateral approach to access and benefit sharing, but this was considered impractical with regard to the use of PGRFA (Fowler 2001; Andersen 2001). Breeding often requires access to a wealth of plant varieties from different countries, and bilateral agreements with all these countries would impose heavy burdens on the breeders, and hamper plant breeding for food and agriculture – especially since many breeders regarded the first generation of laws on access and benefit sharing as overly restrictive and bureaucratic (ten Kate and Laird 1999, 17–33 and 293–312), a bilateral approach to PGRFA appeared unattractive.

26 UNEP (22 May 1992), *Nairobi Final Act of the Conference for the Adoption of the Agreed Text of the Convention on Biological Diversity*.

27 The Twenty-seventh Session of the FAO Conference.

28 FAO CPGR-Ex1/94/3 (1994), *Revision of the International Undertaking*.

Behind these rather objective arguments lay the interests of the negotiating countries. Although all negotiators agreed on the need for accessibility, there was much suspicion among developing countries towards developed countries, since the latter were the ones who were expected to reap the greatest financial benefits from access and could impose intellectual property rights on their products. Therefore, the majority of developing countries fought for the prohibition of intellectual property rights on PGRFA managed under the Treaty and emphasized benefit-sharing arrangements, whereas most OECD countries advocated open access to PGRFA, and minimal, if any, restrictions on intellectual property rights.

The Global Plan of Action

Probably no-one could have imagined that it would take a decade from when the renegotiations were first addressed until the new regime on PGRFA could be adopted – and almost three more years until the agreement entered into force. Interest in the IU decreased once the negotiations leading to the ITPGRFA were initiated. However, the urge to take action with regard to PGRFA management was pressing. This was the background for the *Fourth International Technical Conference on Plant Genetic Resources,* held in Leipzig, Germany (the 'Leipzig Conference'), where delegates from 150 countries met. In a declaration from the meeting, the representatives noted that major gaps existed in national and international capacities to conserve, characterize, evaluate, and sustainably use plant genetic resources.[29] They also stated that access to and the sharing of both genetic resources and technologies were essential for meeting world food security and needs of the growing world population. On this basis, the representatives adopted the *Global Plan of Action for the Conservation and Sustainable Utilization of Plant Genetic Resources for Food and Agriculture*.

The Global Plan of Action was prepared with the active participation of 154 countries. Each country prepared comprehensive reports on the state of plant genetic resources for food and agriculture in its territories. These reports were then compiled and analyzed in a comprehensive and detailed report covering biological, technical and institutional concerns. This *State of the World's Plant Genetic Resources for Food and Agriculture* (FAO 1998)[30] not only represented the main edifice of the Global Plan of Action and the most important reference work on plant genetic resources for food and agriculture since Jack Harlan's *Crops and Man* in 1975. It also showed a process that galvanized policy-makers and practitioners for the management of these vital resources throughout the world. As such, the report stands as a major achievement in itself.

The Global Plan of Action provides a framework for the identification of priority areas by the countries, and support for capacity enhancement towards these ends. Priority activities are to be identified within *in situ* conservation and development,

29 FAO (1996b), *The Leipzig Declaration.*

30 The report was compiled and produced by a team at the FAO co-ordinated by Cary Fowler and David Cooper.

ex situ conservation, the utilization of plant genetic resources, and institutional development and capacity enhancement.

According to a survey carried out in 2000,[31] *in situ* conservation was still not receiving the necessary attention at the national level. There were various small initiatives, mainly taken by NGOs, but nothing of any appreciable scale. On the other hand, several countries had come far in compiling inventories of plant genetic resources for food and agriculture. Highest priority was given to *ex situ* conservation and increasing the number of accessions. However, only half the countries had properly trained technical staff, a circumstance that clearly could threaten the quality of those missions. Only a few countries reported that they were regenerating threatened accessions, a major threat of genetic erosion. As to the utilization of plant genetic resources for food and agriculture, generally little progress could be noted.

The Global Plan of Action is detailed and focuses on the actions necessary to achieve sustainable management of the world's remaining plant genetic resources for food and agriculture. However, it has not yet thrived, and its potential as a catalyst for action is still largely underutilized.

Continuing Negotiations Leading to the ITPGRFA

During the course of long and troubled negotiations, a multilateral system on access and benefit sharing was formed, intended to address the concerns of the opposed parties. A list was established of crops and forage plants to be incorporated in the multilateral system – to the extent that they were under the control and management of the Contracting Parties and in the public domain – the so-called *Annex 1 list*. In this way the countries, in exercising their national sovereignty, delegated responsibility for facilitating access and benefit sharing to a superior body, composed of all the Contracting Parties (see below). However, it was not easy to decide which plants should be included in this list. Although the countries agreed that the list should cover crops that were important for food security and for which there was interdependence among countries,[32] these criteria were not precise enough to guide the selection of crops without controversies (Fowler 2004). Latin American countries argued that pears were important for food security, but that tomatoes and groundnuts were not. African countries refused to include forage grasses, which resulted in the Latin American countries withholding 'their' forage grasses, although they had originally been inclined to include them in the list (ibid.). China refused to include soybeans, and several other plants were kept outside the system. In addition, it was a great challenge to develop a multilateral system that would be strictly in harmony with the CBD and be agreed to by governments in exercise of their sovereign rights. The multilateral system was to be the cornerstone of the ITPGRFA – and it was the last stone to be laid, finally enabling the adoption of the new treaty.

The issue of farmers' rights was another hot topic during the negotiations leading to the ITPGRFA. Most developing countries, but also some industrialized countries

31 Presented in FAO/CGRFA (2002b), *Country Progress Report* CGRFA–9/02/6.

32 It was already well known that most countries are dependent on other countries for genetic resources pertaining to their food crops, as further elaborated in Chapter 6.

like Norway, advocated comprehensive and internationally binding recognition of farmers' rights, whereas countries like the US and Australia did not support this stand. In 1999 the heated debates resulted in a compromise, which was the final text of the ITPGRFA on farmers' rights.[33]

The Adoption of the ITPGRFA and its Entry into Force

When the *International Treaty on Plant Genetic Resources for Food and Agriculture* was finally adopted in November 2001, many observers had almost given up. Negotiations had been extremely difficult, core provisions in the text were in brackets right up to the last bell and it seemed impossible to unite all fronts on joint solutions. As full consensus proved impossible, the matter had to be put to the vote. At the Conference, 116 countries voted in favor of the Treaty and 2 countries abstained (Japan and the United States). The International Treaty could finally be adopted by the FAO Conference at its Thirty-first session, 3 November 2001, as the first legally binding agreement dealing exclusively with PGRFA – and as the first international treaty of the new millennium.[34] Since then the US has revised its policy and signed the Treaty.[35] The ITPGRFA entered into force on 29 June 2004.[36]

The objectives of the International Treaty are the conservation and sustainable use of plant genetic resources for food and agriculture, and fair and equitable sharing of the benefits arising from their use – in harmony with the CBD – for sustainable agriculture and food security (Art. 1). As present and future agriculture and food security depends decisively on access to PGRFA, the Treaty is designed with a particular view to providing for the conservation, sustainable use and access to these resources, with access seen as a core benefit.

The contents of the ITPGRFA can be grouped into four main components or tasks, to be discussed in detail below:

- ensuring the conservation and sustainable use of PGRFA nationally and internationally (Part II);
- promoting the recognition of farmers' rights nationally and internationally (Part III);
- facilitating access to the PGRFA of specified plants on a multilateral basis (Part IV, Articles 10–12);

---

33 A thorough analysis of the recognition of farmers' rights in the ITPGRFA is found in Batta Bjørnstad (2004). Further analyses of the ITPGRFA provisions on farmers' rights are provided by the Farmers' Rights Project, of which Regine Andersen is the project leader. See <www.farmersrights.org>.

34 The Cartagena Protocol on Biosafety to the Convention on Biological Diversity (Annex to CBD-COP decision EM-I/3), not dealt with in this book, was adopted in 2000, but as a protocol to the CBD it is a part of an already established regime.

35 The US has also signed the CBD, but has not ratified.

36 An interesting analysis of the contents and prospects of the ITPGRFA is found in Fowler (2004). Explanations on the background and contents of the ITPGRFA are presented in Moore and Tymowski (2005).

- providing for the sharing of benefits from the use of these plants on a multilateral basis (Part IV, Articles 10–11 and 13).

Institutional design and financial matters A Governing Body composed of all Contracting Parties is to oversee and promote the full implementation of the Treaty (Art. 19). This function includes a wide range of tasks, such as providing policy guidance, adopting plans and programs, adopting and reviewing the funding strategy, maintaining communication with the Conference of the Parties to the CBD and other relevant institutions and generally to monitor progress in Treaty implementation. Each Contracting party has one vote in the Governing Body, which is to meet biannually. The Secretary of the Governing Body is to be appointed by the FAO Director-General, with the approval of the Governing Body, and will be assisted by staff as required (Art. 20).

Implementation of the ITPGRFA is to be financed through a funding strategy, to be developed and adopted by the Governing Body (Art. 18). However, no funding obligations on the Contracting Parties are specified. As funding proved to be the bottleneck for IU implementation as well as negotiations on the ITPGRFA and its follow-up in the interim period, the formulation of Article 18 could result in a continuation of this problem.

The ITPGRFA does not offer any detailed provisions on procedures for compliance, and the negotiating parties did not deal with the issue of non-compliance (Art. 21). Instead these issues were postponed to the First, then to the Second, and finally to the Third Session of the Governing Body. Disputes between Contracting Parties on implementation of the Treaty are to be solved by negotiation, if necessary by involving a third party (Art. 22). If negotiation is not successful, the Treaty provides for arbitration with its own set of procedures (Annex II to the ITPGRFA). The ITPGRFA does not provide for any sanctions against offending parties in case a dispute cannot be solved. This is a major difference between the ITPGRFA and the TRIPS Agreement, as the latter has a strong enforcement mechanism.

The Institutional Context: Embedded in the FAO

The IU, all related resolutions, and the ITPGRFA were adopted by the FAO Conference, as provided for in the FAO Constitution, Article XIV: 'The Conference may, by a two-thirds majority of the votes cast and in conformity with rules adopted by the Conference, approve and submit to Member Nations conventions and agreements concerning questions relating to food and agriculture.' Thereby the two agreements are included in the overarching institutional structure of the FAO with its broader context of principles and practices, and are clear examples of embedded regimes.

The FAO

The FAO is an intergovernmental organization and a specialized agency of the United Nations, linked to the UN Economic and Social Council. It was established in 1945 in Quebec City, Canada, with headquarters in Washington DC until 1951, when it moved to Rome. As of March 2007, 189 countries and the European Union were members of the FAO. The FAO Conference is its highest decision-making body, and it meets biannually. All members are represented with one vote each. As set out in the Preamble to the FAO Constitution, all member states agree to promote the following objectives:

- to raise the levels of nutrition and standards of living of the peoples under their jurisdiction;
- to ensure improvements in the efficiency of production and distribution of all food and agricultural products;
- to improve the conditions of rural populations;
- thus contributing towards an expanding world economy and towards ensuring humanity's freedom from hunger.

The IU and the ITPGRFA in the FAO

As both the IU and the ITPGRFA were/are regimes embedded in the FAO, we must expect their objectives to be congruent with the objectives of the mother organization. The conservation and sustainable use as well as access to PGRFA are means toward achieving the overall objectives of the FAO. Thus, the rationale for the two regimes has fundamentally to do with food production and food security in the world. Both the IU and the ITPGRFA should be seen as basically agricultural agreements.

The ITPGRFA was negotiated under FAO auspices, but during the course of the negotiations several options were open as to how it should be interlinked with other institutions. During the second half of the 1990s, it was proposed in the Conference of the Parties to the CBD that the new international agreement on PGRFA could be adopted as a protocol to the CBD (see for example decision III/11, paragraph 18 of the Conference of the Parties to the CBD). It was only towards the very end of the negotiations that a decision was made to adopt the ITPGRFA as an international treaty *within* the framework of the FAO, under Article XIV of the FAO Constitution (FAO Resolution 3/2001; ITPGRFA, Preamble, last paragraph; ITPGRFA, Art. 1). The FAO Director-General is the Depositary of the Treaty (Art. 34). Provision is also made for close co-operation with the Commission on Genetic Resources for Food and Agriculture under the FAO (Para. 17.3), which co-ordinates the Global System for Conservation and Utilization of Plant Genetic Resources, including an update of the *State of the Worlds' PGRFA* (see FAO 1998). These provisions are indications of ties with the FAO, showing how the ITPGRFA is embedded in the mother organization.

Links between the ITPGRFA and the CBD

Whereas the ITPGRFA is housed in the FAO, it is also linked with the CBD, but in another way. Article 1 sets out that its objectives are to be attained by closely linking it to the FAO and the CBD. It is formulated to be in harmony with the CBD and there are institutional links for information exchange and co-ordination between the two

(Paras 19.1[l] and [m], and 20.5). For the Conference of the Parties to the CBD, the ITPGRFA is the long-awaited core instrument for implementation of its provisions as they relate to PGRFA. As such, the ITPGRFA is in some ways nested within the CBD, but as a separate regime. This does not change the basic rationale behind the ITPGRFA as an agricultural treaty, but it indicates the links with the CBD. As we will see in later sections, these various rationales are important when it comes to identifying norms and rules as well as for institutional overlap and interaction – and their aggregate effects.

Resulting Norms and Rules for the IU

Although the IU and the ITPGRFA are compatible in terms of norms and rules, there are important differences between them, indicating changes in the expectations concerning PGRFA management. We will therefore analyze the norms and rules of the two regimes separately.

As the IU is not legally binding, none of its norms can be defined as expectations with regard to what the signatories 'must' do. However, some of the norms can be understood as expectations concerning what the signatories 'ought to' do, and these are labeled 'core norms'. Similarly, other norms can be understood as expectations regarding what the signatories 'should' or 'could do' – that is, 'secondary norms'.

Core Norm: Exploration, Conservation and Evaluation of PGRFA

All countries signing the IU committed themselves to explore and preserve known land-races or cultivars in danger of extinction, as well as wild relatives of cultivated plants and non-domesticated plants for food and raw materials. These commitments were confirmed and strengthened through the Global Plan of Action for the Conservation and Sustainable Utilization of Plant Genetic Resources for Food and Agriculture. Signatories further agreed to conserve and maintain PGRFA in farmers' fields as well as in gene banks and other *ex situ* facilities, including documentation of their qualities, growing conditions and uses.

In practice, complying with the norm of the maintenance of genetic diversity in farmers' fields is a demanding task that involves all levels of agricultural bureaucracy in a country, as well as state agricultural research centers and farmers' extension services. It also involves a substantial number of farmers in different areas of the country. In many developing countries, the norm of *in situ* management of PGRFA conflicts with mainstream agricultural policies, where the objective since the 'green revolution' has largely been to reduce the use of farmers' varieties to a minimum, and replace them with high-yielding genetically homogeneous varieties. Thus, implementing the new norm would necessitate complementary policies and incentive structures to support farmers who maintain genetic diversity in their fields. Such a shift in thinking across all levels of stakeholders would in itself be demanding. One might question whether the negotiators were aware of the implications of their decisions.

Also gene bank conservation is challenging. It requires good storage facilities that are protected against earthquakes and other natural disasters, with reliable electricity supply to ensure stable temperatures at –18° C for base collections. Also required are data systems for passport data and adequate know-how with regard to all these functions. Collecting samples of crop varieties is also a skill and a science where adequate data collection on the use of the variety in question, its growing conditions and other information is of crucial importance.

Whereas conservation of PGRFA is a core norm under the IU, it cannot be expected to enjoy top priority in the agricultural policies of the South. Developing countries generally have to aim for the highest possible production rates in order to meet the food demands of a growing population and to provide for exports to earn foreign currency to repay foreign debt. Implementing PGRFA conservation therefore requires support from countries in the North.

Core Norm: Availability of PGRFA

The core intention of the IU and the efforts described to ensure conservation of PGRFA is to make these vital resources available for food and agricultural production. This norm was originally based on the principles of PGRFA as the heritage of mankind, indicating that access to these resources should be unrestricted and that their use as a resource base of heritable characteristics was a right held by all (Fowler 1994, 227). Governments adhering to the IU were to allow access to samples of plant varieties and to permit their export to other countries adhering to the IU, when the resources were requested for the purposes of scientific research, plant breeding or genetic resource conservation. Such samples should be made available free of charge, on the basis of mutual exchange or on mutually agreed terms. This understanding was revised in 1989, when minimum restrictions were accepted on the free exchange of PGRFA to conform to countries' national and international obligations, along with the possibility of imposing charges. In 1991 the 'common heritage of mankind' principle was interpreted in the context of the accepted principle in international law that states have the sovereign right to exploit the natural resources within their territories according to their own political priorities. In other words, countries would remain committed to providing access to genetic resources, as long as this was in line with their own political priorities and their other national and international obligations. In practice, the situation was highly uncertain after the biannual FAO Conference in 1991, as was also confirmed in FAO Resolution 3/91 (first section, point d): 'Conditions of access to plant genetic resources need further clarification'.

Secondary Norm: Acceptance of Plant Breeders' Rights

The acceptance of plant breeders' rights as compatible with the IU was set out as an agreed interpretation of the Undertaking in 1989, and refers to UPOV as it was revised in 1978 (see sub-chapter 7.1).[37] The 1978 Act did not restrict farmers' use of

37 This was two years before the revision in 1991, which considerably strengthened intellectual property protection for plant breeders (see section on UPOV below).

their harvested crops from protected varieties, nor were there any restrictions on the use of such varieties for breeding purposes. Therefore, plant breeders' rights would not conflict with the provisions of accessibility. However, this understanding was controversial, since developing countries felt it unfair that plant breeders should be remunerated for contributing the last few pieces in the chain of breeding work that farmers had been carrying out for hundreds or even thousands of years. That is why this could be accepted only along with the next secondary norm.

Secondary Norm: Recognition of Farmers' Rights

Farmers' rights as they pertain to PGRFA are defined as 'rights arising from the past, present and future contributions of farmers in conserving, improving, and making available plant genetic resources, particularly those in the centers of origin/diversity' (FAO Conference Resolution 5/89). These rights were vested in the International Community, in trust for present and future generations of farmers. However, what exactly farmers' rights should be rights to, was not clarified in the texts: only that farmers were to benefit fully from their contribution to maintaining the heritage of PGRFA, and were to be supported in their continuation of these efforts. This would require sufficient funds for PGRFA conservation, *inter alia* through the fund to be established following the 1991 resolution, through assistance to their protection and conservation of PGRFA, and through their full participation in the benefits derived from the improved use of PGRFA. This was the prevailing norm on farmers' rights from 1989, throughout the period under study in this book and until the entry into force of the ITPGRFA.[38]

Regulatory Rules of the IU

As the IU was not legally binding, none of its rules were mandatory, and thus there are no 'core rules'. As optional rules they must be categorized as secondary rules, according to the definition of core and secondary rules given below.

The governments adhering to the IU agreed to *organize or arrange for missions* to explore plant varieties (Art. 3). They further (Art. 4) agreed to adopt appropriate *legislative and other measures to preserve PGRFA* in areas of their natural habitat as well as in gene banks and living collections, if necessary through international co-operation. Thereby plant genetic resources should be evaluated and fully documented. Finally, the governments agreed to *allow access to samples of PGRFA* under their control, and to *permit their export*, for the purpose of scientific research, plant breeding or genetic resource conservation, without restrictions until 1989 (Art. 5), and thereafter with minor restrictions and charges, subject to national policies. These were the regulatory rules of the IU directed at domestic implementation measures. Further implementation measures related to international co-operation, and are thus not relevant for analyzing domestic implementation of the IU.

38 For a more thorough presentation of farmers' rights as recognised under the IU, see Andersen (2005a).

Resulting Norms and Rules for the ITPGRFA

When it comes to the ITPGRFA, the distinctions between core and secondary norms are not obvious from the objectives. The following distinctions are based on an analysis of the Treaty text, in light of the negotiation history, as presented in the previous sections.

Core Norm: Facilitated Access to PGRFA

Access is not explicitly addressed in the objectives of the ITPGRFA as set out in Article 1. However, the core of the International Treaty is the Multilateral System of Access and Benefit Sharing (Part IV). The Contracting Parties are obliged to facilitate access to all plant genetic resources covered by a list of 35 food and 29 forage crops (Annex 1 crops) which are in the public domain and under their management and control.[39] Access shall be provided upon a Standard Material Transfer Agreement that was adopted at the First Session of the Governing Body. As such, providing access to PGRFA under the Multilateral System is an expectation with regard to what the Contracting Parties 'must' do, that is, it is a core norm. The first paragraph on benefit sharing in the Multilateral System (Art. 13.1) states that facilitated access to PGRFA is in itself a major benefit, indicating that access is the most important aspect also of benefit sharing. A further argument supporting the stand that facilitated access to PGRFA under the Multilateral System belongs to the core norms of the Treaty, is the fact that the objectives formulated in Article 1 are aimed at the overall objectives of sustainable agriculture and food security.

The ITPGRFA does not, however, provide for facilitated access to all PGRFA:

- *Annex 1 crops and forage plants:* These include such important food crops as rice, wheat, maize, rye, potatoes, beans, cassava and bananas. Other important crops, including soybeans, tomatoes, oil palm, cotton, sugarcane, cocoa, groundnuts, many vegetables and important tropical forage plants, are not included. The Parties to the Treaty are obliged to facilitate access to all PGRFA under their management and control and within their public domain, which are listed in Annex 1, for all legal and natural persons under the jurisdiction of any Contracting Party (Para. 12.2).[40] In addition, the Parties have agreed to encourage other holders of PGRFA in their country (private actors) to include in the Multilateral System those resources under their control that are listed in Annex 1 (Para. 11.3). Also the International Agricultural Research Centers (IARCs) of the Consultative Group on International Agricultural Research (CGIAR), which have signed agreements in accordance with the ITPGRFA,

39 Parties may also voluntarily make other crop genetic resources available on the same terms and conditions as under the Multilateral System.

40 Whereas the countries have to provide access to the listed resources, this does not mean that they cannot provide access to other genetic resources. Each country is free to provide access to any, and as many, additional PGRFA under similar terms and conditions as under the Multilateral System.

are obliged to provide facilitated access for all Contracting Parties to materials listed in Annex 1 in accordance with the Multilateral System (Para. 15.1. [a]).[41]

- *Non-Annex 1 material collected before the entry into force of the ITPGRFA and after the entry into force of the CBD:* Contracting Parties to the ITPGRFA are not obliged to provide access to such material under the Treaty. Access is regulated internationally by the CBD, that is, on a bilateral basis. Whether access will be provided, and on what conditions, depends on domestic policies and procedures in the respective country. Material stored by IARCs shall, however, be made available in accordance with the provisions of a Material Transfer Agreement (MTA) pursuant to agreements between the IARCs and the FAO, as regulated in the Treaty (Para. 15.1. [b]).
- *Non-Annex 1 material after the entry into force of the ITPGRFA:* Access to this material is internationally regulated by the CBD, which means that access can be facilitated on a bilateral basis. As for materials kept by the IARCs, these can be made available through agreements between the IARCs and 'the country of origin or the country that has acquired the material in accordance with the CBD' (Para. 15.3).[42]
- *Non-Annex 1 material collected before the entry into force of the CBD:* There is currently no international law that regulates access to such material from countries directly. It is thus up to the national government to decide whether to make this material available, and on what terms. If access is denied or severely restricted, the effects for the further development and use of these resources and their contribution to food and agriculture may be severe. However, material stored by IARCs is to be made available in accordance with the provisions of a Material Transfer Agreement (MTA) pursuant to agreements between the IARCs and the FAO (Para. 15.1. [b]).[43]

As this overview indicates, the multilateral system has solved the difficulties of facilitated access to PGRFA for some important crops, but not for all, and the future for the international management of the diversity of these resources is highly uncertain. As for the multilateral system itself, its efficiency and effectiveness depends on the more detailed procedures of implementation, which were still under development as this book was finalized.

41 The Contracting Parties are also obliged to provide facilitated access for the International Agricultural Research Centers (IARCs) of the Consultative Group on International Agricultural Research (CGIAR) which have signed agreements with the Governing Body in accordance with the Treaty (Art. 15.2) for all crops listed under the Multilateral System. There is no such obligation with regard to material not listed in Annex 1. The IARCs have no mandate to ensure *ex situ* conservation of these crops under the ITPGRFA.

42 For a background and discussion of the problems related to 'country of origin', see Chapters 2 and 6.

43 These agreements were signed on 16 October 2006, and a first report on experiences with the agreements was presented for the Governing Body at its Second Session (FAO IT/GB-2/07/Inf.11 2007).

Core Norm: Conservation and Sustainable Use of PGRFA

The conservation and sustainable use of PGRFA are the primary two objectives of the ITPGRFA, set out in Article 1. However, the wording of the related provisions differs slightly, indicating that expectations as to the sustainable use of PGRFA are generally stronger than those related to conservation.

The ITPGRFA stipulates that Contracting Parties shall develop and maintain appropriate policies and legal measures that promote the sustainable use of PGRFA (Para. 6.1), a formulation that indicates an obligation for all Contracting Parties, and thus a core norm. Compliance with this norm refers to *inter alia* diverse farming systems, research that enhances and conserves biological diversity, plant breeding with the participation of farmers in developing countries, broadening of the genetic bases of crops, increase of the range of genetic diversity available to farmers, expanded use of local and locally adopted crops and underutilized species, wider use of diversity of varieties and species in on-farm management, conservation and sustainable use and the adjustment of breeding strategies and regulations on variety release and seed distribution (Para. 6.2).[44]

The provisions on conservation (Art. 5) focus mainly – but not solely – on *ex situ* conservation of PGRFA and are less mandatory. Here the Contracting Parties shall, 'subject to national legislation, and in co-operation with other Contracting Parties where appropriate', promote an integrated approach to the exploration, conservation and sustainable use of PGRFA (Para. 5.1). Suggested measures are the improvement of *ex situ* conservation of plant varieties – including crop wild species, and support to farmers for on-farm management and conservation of PGRFA. Seen together with the objectives of the Treaty (Art. 1) and the provisions on sustainable use above, conservation as such appears to be a core norm and not a secondary one. Also, the mandatory provision that Contracting Parties shall provide access to Annex 1 material for IARCs associated with the ITPGRFA (Para. 15.2) indicates the emphasis given to *ex situ* conservation of the plant varieties included in the Multilateral System. And finally, the sustainable use of PGRFA necessitates the conservation of these resources, and the distinctions between conservation and sustainable use are gradual – which provide additional arguments for considering these two issues together as a core norm.

Secondary Norm: Fair and Equitable Benefit Sharing

The fair and equitable sharing of benefits arising from the use of PGRFA is included in the objectives of the ITPGRFA (Art. 1), which might indicate that it is a core norm. Also, several of the provisions on benefit sharing in Article 13 are mandatory.

---

44 Sustainable use of PGRFA is a way of maintaining PGRFA *in situ* on-farm, and, as such, it has to do with conservation. However, strictly speaking, *in situ conservation* of PGRFA is not possible, since these resources and their traits are in constant change as living organisms. The term *in situ* on-farm *management* would be more appropriate. The provisions elaborated above address such management, and thus what is often termed *in situ* on-farm conservation.

However, there are two points in particular which indicate that benefit sharing should be seen as a secondary norm:

- In Article 13, the Contracting Parties recognize that facilitated access to PGRFA in itself constitutes a major benefit, and that all other benefits accrue from that. This is the first provision on benefit sharing, and it indicates that the issue is subordinate to that of access to PGRFA. Thus, benefit sharing is to be seen as a secondary norm, according to our definition below.
- The issue, as addressed in this more general setting, is rather new. Benefit sharing was not a topic in the IU, and was first adopted as a principle with regard to farmers' rights in FAO Conference Resolution 5/89, which endorses the concept of farmers' rights, and where one aim is to ensure 'full benefits to farmers'. In the ITPGRFA, however, the principle is not only accounted for in Article 9 on farmers' rights, but is also included in the overall objectives for the Treaty and in provisions on the Multilateral System. This is new and reflects the mandate of developing the ITPGRFA in harmony with the CBD, where benefit sharing is a central norm. As such, this history indicates that benefit sharing was not the primary objective of the FAO regime on PGRFA but was added in the process of harmonizing it with the CBD, so it must be regarded as a secondary norm.

This does not mean that there was little motivation for provisions on benefit sharing among those negotiating the ITPGRFA. Particularly among the developing countries this was an important aspect – but it was problematic, due to the bilateral approach to benefit sharing under the CBD. Such an approach was seen as not appropriate to PGRFA management (Fowler, Smale and Gaiji 2001; Fowler 2001; Andersen 2001), as will be further explained in the next main section on the CBD, and thus a more appropriate approach had to be found. When the negotiators finally agreed on the Multilateral System, everyone agreed that benefit sharing was an important dimension of the new Treaty.

The sharing of benefits arising from the use of PGRFA from the Multilateral System comprises four main elements, all related to the Annex 1 species (Para. 13.2):

- *information exchange:* the sharing of inventories and other PGRFA-related information, of technologies pertaining to the growing and utilization of PGRFA and other relevant research findings;
- *transfer of technology* for the conservation, characterization, evaluation and use of PGRFA;
- *capacity-building* in developing countries and countries with economies in transition, through education and training in, and facilities for, the conservation and sustainable use of PGRFA, and by carrying out research together with institutions in these countries;
- *sharing of monetary and other benefits from commercialization*, including payment and partnerships in research and technology development.

The three first items of benefit sharing are of a more general nature, accruing from the utilization of PGRFA at large and not linked to the utilization of any particular resources. By contrast, the final item is linked directly to the commercialization of specified PGRFA.

The benefits are primarily to flow to farmers, especially in developing countries and countries with economies in transition, who conserve and sustainably utilize PGRFA (Para. 13.3). It is emphasized that implementation of the provisions on benefit sharing – together with a funding strategy (Art. 18) – constitute vital conditions for realizing the Global Plan of Action on the Conservation and Sustainable Use of PGRFA, which is mentioned as an important supporting component for the Treaty in Article 14 (Para. 13.5).

Secondary Norm: Limitations for Intellectual Property Rights

Access to, and transfer of, technology are to be provided on terms consistent with the adequate and effective protection of intellectual property rights, as set out in the ITPGRFA (Para. 13.2 [b. iii]). There is also another provision, intended to limit the scope for protecting PGRFA with intellectual property rights. Recipients of PGRFA from the Multilateral System are not to claim any intellectual property or other rights that limit the facilitated access to these resources, or their genetic parts or components, in the form they are received from the Multilateral System (Para. 12.3. [d]). However, this provision leaves broad scope for interpretation. It is not specified exactly what constitutes the 'public domain', or what the formulation 'in the form received from the Multilateral System' means. Great uncertainty surrounds the formulation 'in the form received', as there is no definition of the extent of genetic changes necessary for genetic resources to be considered as no longer 'in the form received'.[45] Is a minor, scarcely recognizable change to a genetic part of a plant received from the Multilateral System enough to qualify for intellectual property rights over it – or how much change must there be? This question was the subject of heated debates under the negotiation of the Treaty, but was not settled by the negotiators. The Standard Material Transfer Agreement (SMTA)[46] for the Multilateral System adopted at the First Session of the Governing Body of the ITPGRFA in 2006 did not solve this uncertainty.

As there are highly diverging expectations among the Contracting Parties with regard to the provisions on intellectual property rights, this norm will be labeled 'secondary' according to our definition in section 3.5.2.

---

45 Also the isolation of a gene may be considered a change that qualifies as 'another form than the one received'.

46 In this context, a Material Transfer Agreement is an agreement between the provider of a genetic resource, for example a gene bank, and the receiver, stipulating the conditions upon which the material is to be transferred. The Standard MTA, or SMTA, is a standardized agreement where all conditions following from the provisions of the ITPGRFA have been taken into account, and which can be used equally by relevant actors in all countries that are Parties to the ITPGRFA and by international gene banks which have entered into agreements related to the ITPGRFA.

Secondary Norm: Recognition of Farmers' Rights

In the Preamble to the ITPGRFA, the Contracting Parties affirm that the past, present and future contributions of farmers in all regions of the world – particularly those in centers of origin and diversity – in conserving, improving and making available these resources, constitute the basis of farmers' rights. They also affirm that the rights recognized in the Treaty to save, use, exchange, and sell farm-saved seed and other propagating material, to participate in relevant decision-making, and in the fair and equitable sharing of benefits arising from the use of PGRFA are fundamental to the realization of farmers' rights.

Article 9 of the ITPGRFA explicitly states that responsibility for the implementation of farmers' rights, as they relate to the management of plant genetic resources for food and agriculture, rests with the governments. Certain measures to protect and promote farmers' rights are suggested. These encompass the protection of relevant traditional knowledge, equitable benefit sharing, participation in decision making, and the right to save, use, exchange, and sell farm-saved seeds and propagating material. These suggestions are not legally binding, and governments are free to choose the measures they deem appropriate, according to their needs and priorities. Thus, we will consider the recognition of farmers' rights to be a secondary norm.

Two other paragraphs of the ITPGRFA contain provisions related to the realization of farmers' rights – and both of these are legally binding.[47] The first (Para. 13.3) states that farmers who contribute to maintaining plant genetic resources for food and agriculture are entitled to receive benefits arising from the Multilateral System of Access and Benefit Sharing established under the Treaty. The latter (Para. 18.5) ensures that funding priority will be given to the implementation of agreed plans and programs for farmers in developing countries who conserve and sustainably utilize plant genetic resources for food and agriculture. To the extent that there are any funds to distribute, these two Articles may provide incentives for the promotion of farmers' rights, since the funds will be directed to those countries where farmers are active in conserving and utilizing PGRFA. In such a case, the two legally binding Articles may function as a vehicle for implementing the optional Article 9.

The provisions on farmers' rights reflect heated negotiations. The compromise mentions all the main topics raised in the negotiations – but as optional measures and not as rights. The preamble underscores the importance of farmers' rights, but cannot alter the fact that the regulatory rules of Article 9 are optional. This makes it impossible to assess compliance with Article 9, as any policy can be defended as 'appropriate', according to the needs and priorities in a given country. Therefore, a traditional approach to the monitoring of compliance is not applicable, and this situation has created substantial uncertainty as to how the Governing Body of the International Treaty can follow up on implementation of this Article (Andersen 2005a, 2005b).

---

47 Also other provisions are relevant for the implementation of farmers' rights, such as Art. 5 on conservation, Art. 6 on sustainable use, Art. 7 on national commitments and international cooperation and Art. 8 on technical assistance.

A starting point for follow-up of the provisions on farmers' rights can be found in the Preamble to the Treaty, which highlights the necessity of promoting such rights at the national and international levels. From these formulations, it is reasonable to conclude that the role of the Governing Body of the International Treaty is to promote the realization of farmers' rights at the national as well as international levels, and that Article 9 provides some guidelines for this work. Further provisions of the ITPGRFA supporting this stand are Paragraph 19.3, which states that the Governing Body shall promote the full implementation of the Treaty, and Article 21 on the role of the Governing Body in ensuring compliance with the provisions[48] of the Treaty.[49]

Regulatory Rules of the ITPGRFA

The distinction between core and secondary regulatory rules follow more or less the same patterns as between core and secondary norms.

Rules pertaining to facilitation of access to Annex 1 PGRFA The Contracting Parties are obliged to take the necessary legal steps or other appropriate measures to provide access for other Contracting Parties to Annex 1 PGRFA under their management and control and in the public domain (Art. 12). They shall invite and encourage natural and legal persons within their jurisdiction who hold PGRFA to include genetic material listed in Annex 1 in the Multilateral System. Access shall be provided in an expeditious manner, free of charge or with minimal charges, with all available passport data, and pursuant to a standard material transfer agreement (MTA) to be adopted by the Governing Body (Art. 12). All these rules are mandatory, and are thus to be regarded as core rules.

An important 'supporting component' is the co-operation with the Consultative Group on International Agricultural Research (CGIAR), its International Agricultural Research Centers (IARCs) and other similar institutions. In Article 15 the Parties call upon the IARCs to sign agreements with the Governing Body of the ITPGRFA (see below) to ensure such co-operation. Such agreements should ensure that the IARCs make available PGRFA covered by the listed plants for the multilateral system, in accordance with the provisions of the ITPGRFA. Furthermore, they shall make PGRFA other than those listed for the multilateral system and which have been collected before the entry into force of the ITPGRFA, available in accordance

48 There is an important difference between the use of the term 'provisions' and the term 'obligations' in this context: Whereas 'provisions' refer to all provisions in the Treaty, 'obligations' would refer only to legally binding 'shall'-provisions. Thus, the Governing Body has the responsibility to ensure compliance with Art. 9.

49 At its Second Session in 2007, the Governing Body adopted a resolution on farmers' rights in which Contracting Parties are encouraged to submit views and experiences on the implementation of farmers' rights, and the Secretariat was requested to collect these as a basis for an agenda item at the Third Session – to promote the realization of farmers' rights at the national level (see report from the Second Session at <www.planttreaty.org>). This resolution represents a clear step by the Governing Body towards taking on international responsibility for promoting the realization of farmers' rights at the national level.

with the provisions of the MTA currently in use. This MTA is to be amended by the Governing Body (see below) no later than at its second session, and must be seen in relation to the Standard MTA mentioned above. In return, the Contracting Parties shall ensure facilitated access for the IARCs to PGRFA listed in Annex 1 of the Treaty, which are under management and control of the Contracting Parties and in the public domain. Finally, after the entry into force of the ITPGRFA, the IARCs shall make available for access material obtained that is not part of the multilateral system, on terms to be mutually agreed between the IARCs and the country of origin, or the providing country that has acquired the material in accordance with the CBD.

Rules pertaining to the conservation and sustainable use of PGRFA The provisions on conservation address the exploration, collection, characterization, evaluation, documentation and subsequently the conservation of PGRFA, as well as monitoring their maintenance. This is to be achieved through the promotion of such activities, subject to national legislation and as appropriate (Art. 5). Whereas these regulatory rules are rather secondary, in that they are not totally mandatory, the regulatory rules concerning sustainable use are more binding. Here appropriate policy and legal measures shall be developed (Art. 6) – core rules, according to our definition. Measures are suggested, for example for developing and maintaining diverse farming systems, strengthening research, participatory plant breeding, broadening the genetic bases of crops, the expanded use of underutilized species and adjusting breeding strategies. Also the Contracting Parties shall, as appropriate, integrate activities for the conservation and sustainable use of PGRFA in their development policies and programs (Art. 7). The Global Plan of Action is referred to in the Treaty as a supporting component (Art. 14), and the Parties 'should' provide for its effective implementation. Also the provisions to encourage the further development of international plant genetic resources networks (Art. 16) and strengthening the Global Information System on PGRFA (Art. 17) stand as supporting components to the provisions on the conservation and sustainable use of PGRFA.

Rules pertaining to the fair and equitable benefit sharing (Art. 13) The Contracting Parties shall make available information, facilitate access to relevant technologies, and develop or strengthen capacity-building schemes in developing countries (Art. 13). Furthermore, they shall take measures to achieve commercial benefit sharing. Here the Standard Material Transfer Agreement (SMTA) is the key instrument (Para. 13.2. [d]). It requires that a recipient of material from the Multilateral System who incorporates this material in a product that is commercialized shall pay an equitable share of the benefits from this commercialization to a financial mechanism under the ITPGRFA. If, however, the recipient makes the product available without restrictions to others for further research or breeding, payment is not mandatory but shall nevertheless be encouraged. The Governing Body is to determine the levels of payments. The rules on benefit sharing are largely mandatory, and are thus to be regarded as core rules. Whereas the first three mechanisms may provide for substantial benefit sharing if/when implemented, there is more uncertainty with regard to the last mechanism. Most multinational seed corporations are self-sufficient in propagating material for the major commercial crops (ten Kate and Laird 1999), and are unlikely to need much access to the multilateral system. Thus, the income in terms of benefits to be shared from the commercial use of PGRFA will probably be limited, at least in the first decades.

Rules pertaining to limited possibilities to claim intellectual property rights The Contracting Parties are to take the necessary *legal or other appropriate measures* to provide for access to PGRFA of Annex 1 crops in accordance with the condition that recipients shall not claim any intellectual property or other rights that limit facilitated access to the PGRFA, or their genetic parts or components, in the form received from the Multilateral System (Paras 12.2 and 12.3 [d]) (see above). Whereas this provision is mandatory, it is not clear what it actually implies, as discussed under secondary norms above. Therefore this rule will be considered a secondary rule here.

Rules pertaining to the recognition of farmers' rights The ITPGRFA does not oblige the Parties to any regulatory rules, since responsibility for realizing farmers' rights is referred to national governments. Each Contracting Party should, in accordance with its needs and priorities, as appropriate, and subject to its national legislation, *take measures* to protect farmers' rights, including the protection of traditional knowledge, benefit sharing and participation in decision-making (Para. 9.2). Since these measures are optional, as seen from the formulations introducing them, they must be understood as secondary rules. Whether the Contracting Parties will take steps to realize farmers' rights is thus dependent on the political will within the governments, on pressures from interest groups and/or civil society organizations, and on how the issue is followed up in the Governing Body of the ITPGRFA.

Overview of Norms and Rules of the IU and the ITPGRFA

The norms and rules of the IU and the ITPGRFA are summarized in Table 5.2

Table 5.2 Overview of norms and rules of the IU and the ITPGRFA

	International Undertaking	**International Treaty**
Core norms	• Exploration & conservation of PGRFA • Availability of PGRFA	• Conservation of PGRFA • Sustainable use of PGRFA • Access to PGRFA
Secondary norms	• Recognition of plant breeders' rights • Recognition of farmers' rights	• Fair and equitable benefit sharing • Prohibition of intellectual property rights (IPR) on genetic material from the Multilateral System in the form received • Recognition of farmers' rights
Core rules to be implemented by governments	No mandatory rules, since the IU is not legally binding	Policy and legal measures to ensure the sustainable use of PGRFA, and to provide access to Annex 1 material under the management and control of the Contracting Parties and in the public domain Measures to ensure benefit sharing from the use of Annex 1 material, in the form of information exchange, technology transfer, capacity building, and monetary and other benefits from commercialization
Secondary rules to be implemented by governments	Voluntary legislative and other measures to explore and preserve PGRFA *in* and *ex situ*, to evaluate and document them and to provide access to them	Policy and legal measures to explore, conserve, evaluate and document PGRFA Legal or other appropriate measures to ensure that recipients of PGRFA from the Multilateral System do not claim intellectual property rights on material in the form received Voluntary measures to protect and promote farmers' rights

Chapter 6

The Convention on Biological Diversity

The Convention on Biological Diversity (CBD) was the first international treaty to address the conservation, sustainable use and equitable sharing of benefits derived from the utilization of biological diversity in general. It was opened for signature at the UN Conference on Environment and Development (UNCED) in Rio in 1992, and entered into force on 29 December 1993. In its general approach to biological diversity, its provisions are aimed at all types of biodiversity. In this chapter we will examine the CBD as it relates to PGRFA management, especially as regards developing countries.

Historical Context: Agrobiodiversity and the Long Wait

To understand the norms and rules of the CBD as they relate to PGRFA management, we need to understand the main features of the formation process as well as the decisions after its entry into force, which set out the interpretation and guidelines for implementation. The CBD is designed with a view to its further formation process. Thus, implementation and regime formation go hand in hand. In the following, these development stages will be substantiated (see Table 6.1).

Confluence of International Dialogues on Biodiversity and the CBD Agenda-Setting

It is not easy to determine when the agenda-setting phase actually began. Swanson (1999, 307) maintains that the CBD is the result of 'the confluence of a number of international dialogues that have existed for several decades'. He divides these dialogues into five partly interrelated categories: (1) the natural parks and protected areas movement; (2) the movement working for mechanisms of environmental funding (for example debt for nature swaps); (3) the movement working for the sustainable use of natural resources; (4) the PGRFA conservation, accessibility and farmers' rights movement; and (5) the movement against intellectual property rights in bio-prospecting. Swanson concludes that the CBD is derived from all these movements and agreements and represents a 'snapshot' of the state of these various negotiations at the time of the Rio Conference (ibid., 309).[1]

---

1 Opinion differs about Swanson's analysis as to how important some of these movements were during the negotiations and whether the process can be described as a 'confluence'. Leader of the Norwegian delegation, Peter J. Schei (see interview list) remarks that there were also disagreements between them.

Table 6.1 Development stages of the CBD in terms of PGRFA management

1981 – 1988	**Agenda-setting phase,** introduced by a resolution at the World Conservation Union (IUCN) General Assembly in Christchurch, New Zealand, requesting that the IUCN Secretariat analyze the conditions for an international arrangement for the conservation, accessibility and use of biological resources.
1988 – 1992	**Negotiation phase,** starting with the decision by the Governing Council of the United Nations Environment Program (UNEP) to establish an Ad Hoc Group of Experts on Biological Diversity to investigate the desirability and possible form of an umbrella convention to rationalize current activities in the field. PGRFA were addressed. Formal negotiations started in 1991; the CBD was adopted on 22 May 1992.
1992 – 1993	**Ratification period:** The CBD was opened for signature at the United Nations Conference on Environment and Development (UNCED) in Rio de Janeiro on 5 June 1992; and 153 countries and the EU signed it during UNCED. By October 1993, 30 countries had ratified: this was the number necessary for its entry into force three months later.
1993 –	**Implementation phase (with reviews/evaluation included)** introduced by the entry into force of the CBD on 29 December 1993. During the implementation phase, more details were/are worked out in terms of decisions of the Conference of the Parties. Details with regard to PGRFA management were developed through a thematic work program on agricultural biological diversity, adopted by the Conference of the Parties in 2000. Other details pertaining to PGRFA management have been worked out later.

In this sense, we might trace the beginnings of the CBD back to 1902, when the first international agreement concerning biological diversity was adopted, the Convention on the Protection of Birds Useful to Agriculture (Raustiala and Victor 1996, 18).[2] Certainly the UN Conference on the Human Environment held in Stockholm in 1972 also marked a decisive step in the framing of ideas on how to address the biodiversity issue internationally, and it resulted in *inter alia* the establishment of UNEP, the institution that later gave birth to the CBD.[3] Likewise the launching of the World Conservation Strategy in 1980 by the IUCN, UNESCO, FAO, and the Worldwide Fund for Nature (WWF) was an important landmark. However, I choose to see the beginning of the agenda-setting phase as the first time that the idea of an international instrument on biological diversity was launched – which was in 1981.

2 Superseded in 1950 by the Paris Convention for the Protection of Birds (Raustiala and Victor 1996, 44).

3 As an outcome of the Stockholm Conference, UNEP was established in 1972. Its Governing Council identified the conservation of nature, wildlife and genetic resources as a priority area in 1973 (McGraw 2002, 11).

The IUCN adopted a resolution in its General Assembly in 1981, requesting the IUCN secretariat to analyze 'the technical, legal, economic and financial matters relating to the conservation, accessibility and use of these [biological] resources with a view to providing the basis for an international arrangement and rules to implement it' (De Klemm 1982, 120).

The next year, a world treaty to protect wild genetic resources inside and outside protected areas for the future was proposed at the IUCN Third World Congress on National Parks and Protected Areas, held in Bali (McGraw 2002, 10). In this context the Congress also asked the IUCN to investigate the possibilities for international instruments to regulate the commercial exploitation of wild genetic resources. Thus, it was the regulation of conservation and exploitation of wild genetic resources that formed the point of departure for agenda setting towards the CBD. Agricultural biodiversity was not initially an issue.

In the following years the IUCN worked on the idea. In 1985, its General Assembly adopted a resolution (Resolution 16/24) with principles to guide drafters of a global agreement on biodiversity. Its Environmental Law Center started working on a text, and a draft convention was circulated to various governments and NGOs for comments (McGraw 2002, 10). As a result, UNEP and several states became interested in the idea of a global biodiversity convention.[4]

The 1980s were probably the decade when environmental concerns were most in focus in international dialogues. The UN General Assembly adopted the World Charter for Nature in 1982, and the World Commission on Environment and Development produced its report *Our Common Future* in 1987.[5] This was a decade of mass mobilization for the environment in many industrialized countries. Environmental strategies and guidelines were developed by donor agencies and they sought implementation around the world. This political environment was probably conducive to the inception of more formal negotiations towards a global convention on biodiversity.

In 1987 UNEP published its *Global Environment Perspective to the Year 2000*, addressing the challenges involved in the conservation and sustainable use of biological diversity. In the same year, the USA sponsored a resolution within the UNEP Governing Council for an all-encompassing convention on the conservation of species (McGraw 2002, 11). Subsequently, the Governing Council in 1988 decided to establish an Ad Hoc Group of Experts on Biological Diversity mandated to investigate the desirability and possible form of an umbrella convention to rationalize current activities in the field. This marked the beginning of the negotiations and the conservation approach to non-domesticated biodiversity. The idea of an 'umbrella' institutional design was still prevalent.

---

4 Information given by Françoise Burhenne-Guilmin (IUCN Environmental Law Centre) to Désirèe McGraw during an interview in Montreal, 23 October 1996 (McGraw 2002, 10).

5 In the report, the Commission proposed a species protection convention, with reference to the draft distributed by the IUCN, which should clearly state that genetic resources belong to the common heritage of mankind. A convention should be followed by appropriate funding arrangements, to which the whole world should contribute, for the purpose of compensation (World Commission on Environment and Development 1987, Norwegian edition, p.123).

From Expert Groups To Speedy Formal Negotiations towards UNCED

The Ad Hoc Group of Experts met in Geneva three times between 1988 and 1990. The group recommended that a new internationally binding instrument dealing specifically with biodiversity should be established, since there was a range of practical, political and legal obstacles to an umbrella convention that could cover existing international efforts for biodiversity conservation. The group also touched some of the issues that were to be more thoroughly addressed later on, such as the links to development, transfer of technology, access to genetic resources and the specific situation of indigenous peoples (see Rosendal 2000, 95; McGraw 2002, 12). Norwegian representatives were the first to address the issue of agricultural biodiversity, proposing that it be included in an international agreement.[6] During this period, developing countries maintained that they could not agree to genetic resources being considered as the 'common heritage of mankind', and reaffirmed their national sovereignty over them. This was not least a response to the emerging intellectual property rights systems in many countries as well as to the negotiations on intellectual property rights in the GATT Uruguay Round that led to the TRIPS agreement (see Chapter 7). The United States and other powerful industrialized countries were in leading positions in this forum, and the developing countries sought to counterbalance the output of those negotiations with provisions under the CBD.

It became evident that a new convention would not get a 'classical' conservationist design, but would have to cover a wider range of interrelated issues. This was also confirmed in the August 1990 decision of the Governing Council on an international instrument on biodiversity '... within a broad socio-economic context, taking particular account of the need to share costs and benefits between developed and developing countries and ways and means to support innovation by local people...'[7]

On the basis of the above, a second expert group was formed, the Ad Hoc Group of Legal and Technical Experts on Biological Diversity. It met three times until mid-1991, when it was renamed the Intergovernmental Negotiating Committee for a Convention on Biological Diversity. This marked the start of the formal negotiations, which took place in the next four meetings of the Committee. The number of delegates at meetings expanded rapidly to representatives of some 80 countries (Rosendal 2000, 93). The final meeting culminated in the adoption of the Agreed Text of the Convention on Biological Diversity on 22 May 1992 at UNEP

6 Norwegian delegates, among them Peter Johan Schei, Bente Herstad, Jan Petter Borring and Ivar Baste of the Norwegian Ministry of the Environment made repeated suggestions and presented draft text on the inclusion of agricultural biodiversity as the process moved from IUCN via UNEP and to the negotiation stage that resulted in the finalization of the CBD (source: Mail from Jan Petter Borring, 10 April 2007). The IUCN was hesitant to expand the agenda of a new agreement. After a letter to the UNEP Director from the Norwegian delegation, UNEP decided to include the issue in the draft agreement (source: Interview 13 August 2004 with Peter Johan Schei, head of the Norwegian delegation during CBD negotiations and at the first six meetings of the COP).

7 UNEP Report UNEP/GCSS.II3 (1990), Annex I, p. 42.

Headquarters in Nairobi. It was opened for signature at the Rio Conference on 5 June, and entered into force on 29 December 1993.

The formal negotiations took less than a year to reach UNCED in 1992.[8] During this period, dynamics evolved rapidly. For example the US, the country that had taken the first formal initiative to the negotiations, was to be the worst laggard, since the Convention was directed towards broader and more controversial issues than nature conservation as such (Rosendal 2000, 141). This was also the time when the all-encompassing agenda of the CBD, covering all types of biological diversity, including PGRFA, was confirmed. Particularly the Nordic countries advocated this stand (ibid., 95).

In addition to the Agreed Text on the CBD, the Nairobi Conference adopted the Nairobi Final Act on the adoption of the Agreed Text, with four related resolutions as well as declarations from several states. Resolution 3 of the Nairobi Final Act addresses the interrelationship between the CBD and the promotion of sustainable agriculture. It recalls *inter alia* the international consensus achieved in other international forums on the urgent need for action for the security and sustainable use of PGRFA. Furthermore, it refers to the recommendations from the Preparatory Committee of the UNCED that policies and programs of priority for the conservation of PGRFA, as integrated in agricultural policies, should be adopted by the year 2000. Such policies and programs of action should cover the promotion of crop diversification, the utilization of poorly known but potentially useful crops, capacity building for the utilization of PGRFA in specialized institutions as well as in farmers' communities, the regeneration of *ex situ* collections and the establishment of *ex situ* networks. Further, the resolution noted the need to strengthen the FAO Global System for the Conservation and Sustainable Use of PGRFA and supported the Fourth International Technical Conference on the Conservation and Sustainable Use of PGRFA, then at the planning stage. It confirmed the great importance of the provisions of the CBD for the conservation and utilization of PGRFA, and urged that complementarity be sought between the Global System of the FAO and the CBD. It recognized the need to seek solutions to outstanding matters concerning PGRFA that could not be solved in Nairobi – on access to *ex situ* collections not acquired in accordance with the CBD,[9] and the question of farmers' rights, and referred them to the FAO.

Resolution 3 of the Nairobi Final Act is important because it clearly states that PGRFA is covered by the CBD, while leaving to the FAO the responsibility for outstanding matters as well as further details on how to implement the CBD with regard to PGRFA. At the time, probably no one imagined that those negotiations were not to be finalized until almost ten years later. With hindsight, we can say that the resolution marked the beginning of a difficult path for the Conference of the Parties to the CBD. On one hand, they should include the management of PGRFA in their work. On the other hand, they were to await the results of the negotiations in the

8 The CBD was seen as UNCED's major instrument for realizing its Agenda 21 ambitions on biological diversity.

9 This refers to genetic resources acquired before the entry into force of the CBD, or not acquired from countries of origin of the genetic resources.

FAO, and avoid intervening in those processes. Moreover, there was still resistance among some of the Parties to the CBD as to including PGRFA in the work of the COP. As we will see in the next sub-section, during the first 10 years not much was achieved with regard to the inclusion of PGRFA. However, in light of history and the institutional situation during most of this period, the results that were achieved can also be regarded as notable.

Ratification and Implementation

After 153 countries and the EU had signed the CBD during UNCED in 1992, the ratification period took a little more than a year. By October 1993, the required 30 countries had ratified, enabling the CBD to enter into force in December of the same year. The CBD is aimed at addressing all facets of biological diversity (except for man), including PGRFA, within its three objectives: the conservation of biological diversity, the sustainable use of its components, and the fair and equitable sharing of its benefits.

Institutional arrangements for implementing the CBD include the establishment of a Conference of the Parties (COP) to review domestic implementation (Art. 23), a Secretariat (Art. 24) and the Subsidiary Body on Scientific, Technical and Technological Advice (SBSTTA). The latter is a multidisciplinary body open to participation by all Parties, consisting of government representatives competent in the relevant field of expertise, and acting under the authority of the COP. In addition the CBD includes provisions for the adoption of protocols (Art. 28) and for the relationship between the Convention and its protocols (Art. 32).

While the Parties are to implement the various provisions of the Convention, the COP negotiates decisions providing further details concerning its implementation, as the next sub-chapter will explain in greater detail. In order to identify the norms and rules of the CBD pertaining to PGRFA management, and as a basis for analyzing regime interaction and their aggregate effects, close analysis of the international implementation of the CBD related to PGRFA is required. We will therefore examine the COP meetings to see how the Parties have addressed this issue and what they have decided. Two questions will guide our analysis here:

1. What has the COP done to *explicitly address* the issue of PGRFA?
2. How has the COP sought to *mainstream* the issue of PGRFA into its decisions on implementing the various provisions of the CBD?

Explicitly Addressing PGRFA Management

In examining the COP measures explicitly directed at PGRFA, we will look at the first to the seventh meetings of the COP in chronological order.[10]

In line with the Nairobi Final Act, a medium-term work program was adopted at COP I in 1994, suggesting that the second COP meeting in 1995 consider the

---

10 The investigations are based on the Secretariat of the CBD (2001) *Handbook* and on the official reports from the COP meetings thereafter.

relationship between the FAO Global System for Genetic Resources and the CBD (Decision I/9, Annex 5.9). The aims were to be informed about progress in the revisions of the IU and its consistency with the CBD, preparations for the 1996 International Technical Conference on the Conservation and Utilization of PGRFA to be held in Leipzig, and about developments regarding *ex situ* collections of plant genetic resources. Furthermore, it was suggested that the conservation and sustainable use of agricultural biological diversity should be considered at the next COP, within the context of the Convention's three objectives and its provisions (Decision I/9, Annex 6.3).[11]

At COP II in 1995, the Parties recognized that agricultural biodiversity has a special nature, with distinctive features and problems needing distinctive solutions (Decision II/15). They furthermore declared their support for the ongoing processes in the FAO Commission on Genetic Resources for Food and Agriculture with regard to harmonizing the IU with the CBD – which they considered should be finalized as soon as possible – and with regard to the Leipzig Conference on PGRFA. Thereby they requested that the outcome of the Conference be presented at the next COP, and that the *State of the World's PGRFA* and the *Global Plan of Action for the Conservation and Sustainable Use of PGRFA* be made available at that meeting. In a statement to the Leipzig Conference, the COP declared the importance of the objectives of the CBD and urged all forums on related issues to work together to achieve the objectives. They also recognized the substantial contribution that could be made by the FAO, with its experience and skills concerning PGRFA.

Noting the slow progress in the renegotiation of the IU, the COP III, after the Leipzig Conference in 1996, proposed a multi-year program of activities on agricultural biological diversity. The aim was to promote the positive effects and mitigate the negative impacts of agricultural practices on biological diversity in agro-ecosystems and their interface with other ecosystems, as well as to promote the objectives of the CBD with regard to PGRFA (Decision III/11). The program should identify and assess relevant ongoing activities and existing instruments at the international and national levels, identify priority issues for further development of the program, carry out case studies on the issues identified, share experiences, and transfer knowledge and technologies.

The Executive Secretary of the COP was requested to invite the FAO to provide inputs to the first item, and the importance of avoiding any duplication of work was stressed. The multi-year program also encouraged Parties to develop national strategies, programs and plans to ensure CBD implementation in the area of PGRFA. This included proposals to monitor and evaluate agricultural practices with regard to the state of PGRFA, as well as various measures to create positive incentive structures and to encourage the development of technologies and farming practices conducive to the sustainable use of PGRFA. In this context, the contribution of the Global Plan of Action on the Conservation and Sustainable Use of PGRFA was welcomed, and a call was made for the effective and speedy completion of the revisions of the IU, including a strengthening of the FAO Global System on PGRFA.

---

11 The latter was also decided for coastal and marine biological diversity (Decision I/9, Annex, point 5.3).

The main components of the multi-year program seem very similar to the two main components of the Leipzig Conference that same year: In the *State of the World* report the required assessments had already been made, and the measures to be undertaken by national governments had already been adopted in greater detail as the *Global Plan of Action*. Without close co-operation with the FAO, duplication of work would seem inevitable.

The SBSTTA was to review the progress of the program at its third meeting in 1997 and report the results back to the COP. It prepared a report stating that significant progress had been made,[12] addressing all items envisaged in the multi-year program.[13] By then it had also adopted a Memorandum of Co-operation with the FAO, and the FAO had reported extensively to the SBSTTA on its activities.[14]

The Executive Secretary had invited all Parties to report on their activities and instruments pertaining to PGRFA management, and to identify issues and priorities to be addressed at the national level. Only eight countries responded, however – Belarus, Canada, Greece, Latvia, Norway, Thailand, the US and Venezuela. Only one year earlier, 154 countries had provided comprehensive reports to the FAO *State of the World Report on Plant Genetic Resources for Food and Agriculture* (FAO 1998)[15] on more or less the same questions. The findings from the eight country reports were compiled, without the inclusion of any information from the FAO report, in a separate report to the COP.[16] In other words, duplication of work was not avoided, despite the co-operation with the FAO. Particularly the results with regard to national-level assessments were marginal.

At COP IV in 1998, the importance of speeding up implementation of the Multi-year Program was highlighted (Decision IV/6). The SBSTTA reported back the results of the assessment,[17] and recommended a comprehensive program of work based on the multi-year program. The COP endorsed this program, consisting of four mutually reinforcing program elements (Decision V/5, annex B):[18]

- comprehensive assessment of status and trends of the world's PGRFA, its underlying causes as well as local knowledge of its management: to be finalized by 2010, building upon data from other sources, such as the FAO;
- adaptive management measures in terms of management practices, technologies and policies to promote the positive and mitigate the negative effects of agriculture on PGRFA: to be identified, studied and lessons to learn promoted;

---

12 Secretariat of the Convention on Biological Diversity (2001), 476–8 (UNEP SBSTTA Recommendation III/4).

13 Secretariat of the Convention on Biological Diversity (2001), 439–45 (UNEP SBSTTA/III/6).

14 UNEP/CBD Report SBSTTA/3/Inf. 21 (1997).

15 The data were compiled in 1995 and 1996, presented at the Leipzig Conference, and published in 1998.

16 UNEP/CBD Compilation SBSTTA/3/Inf. 9 (1997).

17 UNEP (1999), UNEP/CBD/SBSTTA/5/Inf. 10.

18 Based on citation in Secretariat of the CBD (2001, 240–1).

- capacity building for farmers, indigenous and local communities and their organizations for sustainable management of agricultural biodiversity: to be achieved by establishing local-level forums and regional networks covering at least 1000 communities by the year 2010;
- mainstreaming in terms of integrating national plans or strategies on PGRFA management in sectoral and cross-sectoral plans and programs.

The COP recognized the contribution of farmers, indigenous and local communities to the conservation and sustainable use of agricultural biodiversity and the importance of agricultural biodiversity to their livelihoods, and emphasized the importance of their participation in implementing the program of work (Decision V/5, Paragraph 5). In an appendix to the same decision, the scope of agrobiodiversity was specified.

The program of work was further clarified and adopted at COP VI in 2002.[19] Additional details, with a timetable for implementation and a reporting schedule, were adopted (Decision VI/5). Thereby responsible entities were determined. The FAO was made lead agency in several activities, to ensure avoidance of duplication of work. As part of the decision, Parties and Governments were for the first time invited to provide reports on the implementation of the program of work on agrobiodiversity, and the Executive Secretary was requested to prepare a draft format for such reports for consideration at COP VII. Further, COP VI congratulated the FAO on achieving the ITPGRFA (Decision VI/6). The Conference appealed to the Parties to sign the Treaty and decided to establish links of co-operation between the governing bodies of the two agreements and between their secretariats.

COP VII in 2004 decided to review progress with regard to the program of work on biological diversity (now referred to as the 'strategic plan') at COP IX in 2008 (Decision VII/21). It invited Parties and other Governments to mainstream the management of agricultural biodiversity into their policies, and invited civil society organizations and other NGOs to assist the Parties in their capacity building towards this end (Decision VII/3). International organizations and funding institutions were invited to support the development and use of agrobiodiversity, and Parties and other Governments were urged to ratify the ITPGRFA.

We can note how the language changed considerably in the course of those ten years. First, the COP wanted to be informed about the progress on the IU negotiations, and then it supported the process. The year after, it called for speedy and effective completion of the process. In fact it seems that it could not really wait for finalization, since it adopted the multi-year program on agricultural biodiversity, although this did not produce any significant results in terms of domestic implementation and had elements of duplication. Later it urged the completion of the IU negotiations, and emphasized the importance of speeding up the work on agricultural biodiversity. Then it adopted the program of work on PGRFA, which also proved to be slow. When the ITPGRFA was finally adopted, the COP offered its congratulations, and the program of work could be finalized and co-ordinated with the FAO.

19 The other working programs concern marine and coastal biodiversity, forest biodiversity, the biodiversity of inland waters, and dry and sub-humid lands.

Mainstreaming PGRFA into CBD Implementation

In this section, we will look at how PGRFA management has been mainstreamed into decisions pertaining to the various articles of the CBD. The negotiating parties and later Conference of the Parties have struggled with the mainstreaming of PGRFA ever since it was first introduced in the negotiation text. First, there was fierce resistance to including it in the Convention, expressed by some of the most powerful negotiators, among them the US. Second, PGRFA is a component of biological diversity that has different living conditions, and therefore requires different treatment in certain regards. This made it difficult to mainstream along with the other components of biological diversity. Third, the negotiations on the revisions of the International Undertaking were not concluded until November 2001, and the COP of the CBD, which had initially invoked the negotiations and urged for their completion, awaited their results – not wishing to intervene in the FAO processes.[20] The investigation of the follow-up of the relevant CBD Articles in the meetings of the COP reveals this dilemma in more detail.[21] We will focus on those Articles that are directly relevant for the management of PGRFA in developing countries: Articles 6 to 10, and 14 to 16.

Developing and mainstreaming policies pertaining to conservation and sustainable use of biodiversity (Art. 6) At its second meeting, in 1995, the COP referred to guidelines developed by UNEP, the World Resources Institute and IUCN for national biodiversity planning (Miller and Lanou 1995), which were largely designed for non-domesticated resources. They further requested that the Parties integrate elements of all the thematic work programs into their national strategies and sector plans (Decision II/7), including those on agricultural biological diversity. As such, this was an approach to mainstreaming. However, it was followed up by little guidance. There was no reference to the FAO International Undertaking, which would have provided more guidance on PGRFA management.[22]

Identifying and monitoring components of biological diversity with a view to the potential need for measures to support their further existence (Art. 7) The management of PGRFA has been addressed by the COP in an integrative manner along with other types of biological diversity (for example Decision III/11, paragraphs 9, 15 (a) (g) (m) (n), and 16), including reference to the FAO Global Plan of Action. Much emphasis has been put on efforts to enhance taxonomy, launching a Global Taxonomy Initiative (COP IV, Decision IV/1 D, annex). Also the management of PGRFA is integrated in these efforts, by addressing the lack of taxonomic information on pollinators (Decision V/5, paragraph 15 b).

20 Interview 13 August 2004 with Peter Johan Schei, head of the Norwegian delegation during CBD negotiations and at the first of six meetings of the COP.

21 The investigation is based on Secretariat of the CBD (2001) *Handbook*.

22 The Global Plan of Action on the Conservation and Sustainable Use had not yet been adopted.

Establishing and regulating protected areas (Art. 8 a–c) Articles 8 to 10 on conservation and sustainable use should be seen together. In Article 9 on *ex situ* conservation, the negotiators implicitly addressed agricultural biodiversity, whereas Article 8 on measures to maintain *in situ* conservation of genetic resources was meant to be devoted to non-domesticated biodiversity.[23] This division of labor was due to the consideration of the negotiating parties that PGRFA was not grown *in situ*, since the plants had been taken out of their once-natural habitats.[24]

Subsequently, the COP made no reference to agricultural biodiversity in its discussions and decisions on protected areas related to the first three paragraphs of Article 8a–c. All other thematic areas of biological diversity were addressed. It might be argued that agriculture is not relevant for protected areas. However, PGRFA is also often maintained by people living in areas to be declared as protected areas, and guidance could be useful as to how to integrate such aspects in protected-area management. In practice there is frequently conflict between the need to protect non-domesticated biological diversity and for people to maintain their livelihoods in these areas, including maintenance of PGRFA.

Protecting ecosystems, natural habitats, viable population of species, and buffer zones (Articles 8d and 8e) The COP includes all thematic work areas in its discussions and decisions, except for agricultural biodiversity. Again, one can understand why agricultural biodiversity was not included in all these considerations, but it could have been relevant to include something on viable populations of threatened crop varieties.

Rehabilitating degraded ecosystems and recovery of threatened species (Art. 8f) The COP makes some reference to agricultural biodiversity, to the separate programs of work on agricultural biodiversity referred to above. In other words, PGRFA is sorted out and reference is made to other contexts – but it is not mainstreamed.

Maintaining traditional knowledge (Art. 8j)[25] The follow-up of this Article – on measures to maintain relevant knowledge, innovation and practices of indigenous and local communities, promoting their wider application and encouraging the equitable sharing of benefits arising out of such application – is highly relevant with regard to the IU and ITPGRFA, due to its parallels with farmers' rights. The issue of Article 8j is complex in itself and touches on most of the provisions in the CBD, the thematic work programs, and related topics like intellectual property rights. An Ad Hoc Open Ended Working Group on Article 8j and Related Provisions was established at COP IV in 1998 (Decision IV/9); it presented a highly detailed work program, adopted at COP V in 2000 (Decision V/16). As there is no particular reference to any category of biological diversity, all of them are equally addressed in the work program. However, from the COP it is generally perceived that Article 8j and forest biodiversity are

---

23 According to Peter Johan Schei (interview 13 August 2004).

24 Ibid.

25 Paragraphs *g* to *i* are not considered directly relevant for PGRFA management as delimited for this book.

closely related (Secretariat of the CBD 2001, 125). Therefore, the COP requested the Executive Secretary to provide advice and information on the relationship between indigenous and local communities and forests in the UN processes on forests (see Decision II/9, paragraph 2 a). The COP has nevertheless encouraged Parties to develop national strategies, programs and plans to empower their indigenous and local communities and build their capacity for *in situ* management of agricultural biological diversity (Decision III/11, paragraph 15 f).

Until 2002, no reference was made to recognition of farmers' rights, as adopted at the FAO Conference in 1989 (FAO Resolution 5/89, see section in IU/ITPGRFA above). The Executive Secretary was requested to examine, together with the FAO, the implications of the ITPGRFA on the issues under Article 8j (COP VI WG II/CRP.9/Rev.1). These results were presented at COP VII (COP/VII/Inf. 18) and will be analyzed in Chapter 8, since they pertain to the overlaps between the two agreements. On the other hand, when it comes to the request for consultation and co-operation with other international forums on participatory mechanisms and on the effects of intellectual property rights, many institutions are listed – but not the FAO system. The working group reported on the integration of its tasks in the thematic work programs. Thereby it emphasized the need for further action with regard to agricultural biological diversity, in particular in promoting farming practices and information exchange to assist farmers and indigenous and local communities to transform unsustainable agricultural practices to sustainable ones, and to increase productivity. These formulations fail to acknowledge the important role that farmers, with their traditional knowledge, have in maintaining agricultural biodiversity.

In their outline of the composite report on the status and trends regarding implementation of Article 8j, the status of traditional knowledge of PGRFA is included as one element. All in all, we may conclude that the Working Group has increasingly tried to integrate the management of PGRFA and farmers' rights into its considerations – but has not managed to do this.

Developing or maintaining legislation to protect threatened species and on mitigation of adverse effects on biological diversity (Articles 8k and 8l)[26] These issues have been addressed by the COP with regard to forests (Decisions III/9, paragraph 9 a; IV/7, annex, paragraph 31; V/4, paragraphs 11, 12 and 14), and marine and coastal biodiversity (Decision IV/5, annex, C, operational objective 1.3, paragraph c). Agricultural biodiversity has not been dealt with in the framework of implementing these provisions, but relevant aspects have been addressed in the programs on agriculture, in terms of the misuse of agro-chemicals (Decision III/11, paragraph 15 l). The program also has a general formulation on promoting the positive effects and mitigating the negative impacts of agricultural systems and practices on biological diversity in agro-ecosystems, with particular attention to be paid to the needs of farmers and indigenous and local communities (Decisions V/5, Paragraph 26, Annex, A, paragraph 2 a; B, Activity 2.3).

---

26 Article 8m is not directly relevant with regard to the obligations of developing countries.

Such formulations do little to change the general impression that Article 8 has been formulated to address the *in situ* conservation of non-domesticated biological diversity, and that its provisions and follow-up have not adequately addressed the *in situ* conservation of PGRFA as an integral part of the Article. The implicit division of labor between Article 8 and Article 9 makes it difficult to see conservation efforts related to domesticated and non-domestic biological diversity in context.

Undertaking measures to complement in situ conservation with appropriate ex situ conservation (Art. 9) Until 2000, the COP did not explicitly address implementation of this Article. However, the issue was dealt with in the context of other implementing measures, such as those on taxonomy (elements relevant to *ex situ* collections) and on access and benefit sharing (information on *ex situ* collection acquired prior to the entry into force of the CBD). *Ex situ* conservation has also been addressed in the context of capacity-building measures. Following a request from institutions housing *ex situ* collections, COP V decided to address the proposal for a Global Strategy for Plant Conservation at COP VI. Thereby the important role of initiatives like the Global Plan of Action for the Conservation and Sustainable Utilization of PGRFA were recognized (Decision V/10).

At COP VI the Global Strategy for Plant Conservation was adopted (Decision VI/9). It was not intended as a program of work, like the thematic work programs under the CBD, but as a framework for activities already under way, with reference to various institutions, among them the ITPGRFA and the Global Plan of Action for PGRFA. This is why the Strategy has outcome-oriented targets, instead of providing for activities and outputs. The Executive Secretary of the CBD is to work together with the Contracting Parties, and other international institutions and stakeholders to ensure their participation and the implementation of the strategy. The Strategy applies to plant genetic diversity, plant species and communities and their associated habitats and ecosystems (Paragraph 9). PGRFA are addressed directly in the targets:

- By the year 2010, at least thirty per cent of productive lands – agricultural, grazing and timber production lands – are to be managed in a way consistent with the conservation of plant diversity. This target includes the conservation of crops, pasture, tree species and genetic diversity in the production system, of other plant species in the production landscape that are threatened, unique or of particular value, and the use of management practices to avoid significant adverse impacts on plant diversity (Target 6).
- By the year 2010, seventy per cent of the genetic diversity of crops and other major socio-economically valuable plants are to be conserved, along with the associated indigenous and local knowledge (Target 9).
- By the year 2010, thirty per cent of the plant-based products shall be derived from sustainably managed sources: here the term 'plant-based products' refers to food, timber, paper, fiber, ornamental, medicinal and other plants for direct use (Target 12).

The adoption of the Global Strategy represents a major step forward in international efforts for the conservation of plant genetic diversity. Nevertheless, several questions remain with regard to the details of the strategy. For example, the size of productive lands can be measured in very different ways, particularly when including forests used for timber extraction. To measure achievements related to the target of seventy per cent of genetic diversity of crops, a baseline year will have to be established and baseline data collected, in turn requiring clarification related to quantification of different types and sizes of products. On the other hand, the targets do provide clear indications that PGRFA have finally been mainstreamed in the strategy. COP VII in 2004 decided that targets 6, 9 and 12 should be integrated in the thematic program on agricultural biological diversity (Decision VII/10). It further invited the FAO-CGRFA to consider how the Global Plan of Action for the Conservation and Sustainable Use of PGRFA can contribute to the strategy.

Integrating considerations on the sustainable use of components of biological diversity into national decision-making and undertake efforts to support such use (Art. 10) This issue has also been addressed directly and indirectly in other Articles, but was particularly dealt with at COP V in 2000, including the issue of tourism and its impact on biological diversity. Sustainable use was defined as a cross-sectoral issue, since it relates to all use or intervention in biological diversity, and is one to be addressed through the ecosystem approach (Decision V/24). The latter is a strategy adopted at COP V (Decision V/6), to be applied in implementing the CBD. It describes the integrated management of land, water and living resources so as to promote conservation and sustainable use in an equitable way, with twelve complementary and interlinked principles related to the management of ecosystems and five operational guidelines. The last of these operational guidelines indicates that the management of agrobiodiversity may not initially have been seen as an integral part of the approach (Decision 6, Annex C, Section 5):

> As the primary framework of action to be taken under the Convention, the ecosystem approach should be fully taken into account in developing and reviewing national biodiversity strategies and action plans. There is *also* a need to integrate the ecosystem approach into agriculture, fisheries, forestry and other production systems that have an effect on biodiversity. (emphasis added)

This shows that agriculture is not seen as an integral part with regard to the national biodiversity strategies and action plans, even though part of its production is actually the reproduction of agricultural biodiversity. Agricultural biodiversity is not addressed in the mainstream activities, but is seen as an additional issue. Shortly before this decision, however, the COP adopted a decision on the work program on agrobiodiversity. This decision called for the mainstreaming of agricultural biodiversity in agricultural strategies and action plans, as well as in wider strategies and plans for biological diversity (Decision V/5, Annex B, Paragraph 4.1). We may conclude that it did not prove easy to harmonize the decisions of the COP.

Introducing procedures for impact assessment of proposed projects which potentially may affect biological diversity negatively, and measures to minimize such adverse impacts (Art. 14)[27] The COP dealt with these provisions at its fourth meeting, and invited all Parties and governments to exchange information and share experiences in this regard, for example in the form of reports to be synthesized by the Executive Secretary (Decision IV/10/C). At COP V the Parties were invited to implement Article 14 and provide further details in this regard. At both meetings, collaboration was proposed with international and multilateral organizations. Several organizations were named, and all of them, except for the International Association for Impact Assessment, were working on non-domesticated biological diversity. No mention was made of the FAO or CGIAR institutions, even though they are probably among the most central institutions in competence on the adverse impacts on biological diversity in agriculture. This does not mean that the Article does not apply to agricultural biodiversity, but that there is a lack of awareness of its implications for the management of agricultural biodiversity. However, the COP has invited the Parties to integrate environmental impact assessments into the work programs on thematic areas, like agricultural biodiversity (Decision V/18, Paragraph 1a).

Creating conditions to facilitate access to genetic resources,[28] *on mutually agreed terms and subject to prior informed consent and developing legislative or other policy measures on fair and equitable benefit sharing (Art. 15)* This is one of the most debated articles of the Convention, and is highly relevant with regard to PGRFA. How it has affected PGFRA management will also be discussed in several contexts later in this and the next chapter, as well as in Part IV on domestic response. At this point, we will merely highlight what the COP has decided of relevance for PGRFA management. First of all, the COP has made frequent reference to the International Undertaking and to the question of ex situ collections. Thereby it notes the linkages between Article 15 and the further development and implementation of the work of the FAO on the Global System on PGRFA (Decision III/15). It was especially important to bring solutions to the question of access to ex situ collections acquired prior to the entry into force of the CBD with the renegotiation of the IU (Decision III/15, Paragraph 7 and Decision V/26, Paragraph 8). At COP VI the adoption of the ITPGRFA was welcomed and the links between the multilateral system and Article 15 were addressed. In addition, the Global System for PGRFA and other FAO activities are addressed in the thematic work program. This will be presented in the next section, since it is related to the second question above.

The central challenge for the COP with regard to Article 15 was to get implementation underway while at the same time working out the details on how

27 The three articles before Article 14 were not dealt with in the COPs with regard to PGRFA, and will thus not be examined here. They specify that the Contracting Parties are to develop *incentive measures* for the conservation and sustainable use of biological diversity (Art. 11), and to establish and promote relevant *research and training activities* (Art. 12) as well as *public education and awareness* (Art. 13).

28 That is, genetic resources for which they are countries of origin or which they have acquired in accordance with the CBD (see discussion under norms and rules, next section).

to put into practice the provisions. The first decisions encouraged governments to include access and benefit sharing in their national action plans on biological diversity and to develop relevant policies and legislation, including the identification of national authorities to provide information on the granting of access to genetic resources (Decisions III/9, Paragraph 2c and III/15, Paragraph 5 and 6). In 2000, a working group on access and benefit sharing was established to develop guidelines on implementing Article 15 (Decision V/26, Paragraph 11).

The sixth meeting of the COP adopted the Bonn Guidelines on Access to Genetic Resources and Fair and Equitable Sharing of Benefit Arising out of their Utilization (Decision VI/24). These guidelines are voluntary and are meant to assist in domestic efforts to establish policies and legislation on access and benefit sharing. The guidelines specify the roles, functions and tasks necessary to enable authorities to provide access to genetic resources with prior informed consent on mutually agreed terms, and about the conditions for such access and the sharing of benefits. They suggest elements for Material Transfer Agreements (MTA), as well as a list of monetary and non-monetary benefits. They also provide draft elements for an action plan for capacity building in this regard. Finally, the COP addressed the role of intellectual property rights (to be further elaborated on in Chapter 8).

It is explicitly stated that the guidelines are meant to be without prejudice to the access and benefit-sharing provisions of the ITPGRFA. The Bonn Guidelines were seen as the first step in an evolutionary process that would lead to a more binding regime on access and benefit sharing. COP VII decided that an international regime on access and benefit sharing should be elaborated and negotiated by the ad hoc open-ended working groups under the CBD on access and benefit sharing, and Article 8j (Decision VII/19/D). Thereby it should integrate the concerns arising from Article 8j as well as Article 15. Detailed Terms of Reference were adopted. Further discussions on this link follow in the section on norms and rules, and in Chapter 8, on the interaction between the IU/ITPGRFA and the CBD.

Provide access to and transfer of technology (Art. 16) This is the last article with direct relevance for the domestic management of PGRFA in developing countries. It specifies that access and transfer of technology is to be facilitated or provided to developing countries under fair and most favorable terms. Intellectual property rights are addressed in three paragraphs. First, Article 16 states that access and transfer of technology subject to patents and other intellectual property rights shall recognize and be consistent with the effective and adequate protection of these rights. Second, it obliges each Contracting Party to take legislative, administrative and policy measures to ensure that developing countries that supply genetic resources are provided access to and transfer of the technologies in which these resources are used, and to ensure they are protected by patents and other intellectual property rights, if appropriate. Third, the Contracting Parties shall co-operate at the national and international levels to ensure that intellectual property rights are supportive of, and do not run counter to, the objectives of the CBD.

The COP addresses the issue of access to and transfer of technology apart from the issue of intellectual property rights in its follow-up of Article 16. With regard to the access to and transfer of technology, however, the COP provides little guidance to the Parties on domestic action for implementation of Article 16. There is some relevant guidance in decisions addressing other issues, for example on access to genetic resources (Decisions V/26 A) and on *ex situ* collections (Decisions V/26 C). Otherwise, the COP endorsed that the issue should be dealt with under the thematic work programs (Decision III/16). The multi-year program on agricultural diversity comprises the sharing of experience and the transfer of knowledge and technologies (Decision III/11, Paragraphs 1f, 8 and 19), and requests that the CBD clearing-house mechanisms be used to promote this development by facilitating contact among relevant stakeholders. At COP VI, the Parties requested the Executive Secretary to commission a review to assess the current and potential role of the clearing-house mechanism, including its role in facilitating the transfer of technology and know-how (Decision VI/18). It was recommended that the tool-kit of the mechanism be further developed.

At COP VII a comprehensive and detailed program of work on technology transfer and scientific co-operation was adopted (Decision VII/21). It seems that, prior to COP VII, the provisions on access to and transfer of technology had received relatively little attention, compared to other central provisions of the CBD. One reason might be that far more attention has been paid to the provisions on intellectual property rights under the same article.

Intellectual property rights have caused heated debates since the early days of the CBD negotiations. The issue has been discussed at all COP meetings, and several decisions have been made. At the first four meetings of the COP, decisions centered around information gathering on the effects of intellectual property rights on the management of biological diversity, and on co-ordination with the activities of the TRIPS Council (see Secretariat of the CBD 2001). With regard to agrobiodiversity, the work program adopted at COP V in 2000 (Decision V/5) addressed intellectual property rights. It noted that the Parties should identify appropriate policies to support beneficial practices with regard to access and benefit sharing related to intellectual property rights on PGRFA, but no further details were supplied on what such policies might entail.

At COP VI some more details were worked out as to the role of intellectual property rights in implementing access and benefit-sharing arrangements (Decision VI/24 C). The Parties were invited to encourage the disclosure of the country of origin of genetic resources in applications for intellectual property rights. It was hoped that this would contribute to greater compliance with the provisions on prior informed consent as a requirement for access to genetic resources. Also the disclosure of the origin of relevant traditional knowledge was encouraged.

Greater co-operation was also envisaged with the World Intellectual Property Organization (WIPO). The Executive Secretary was requested to invite WIPO to prepare a technical study for the COP on methods consistent with the treaties administered by WIPO (see Chapter 7) to disclose information necessary for determining and evaluating prior informed consent. The COP also proposed that WIPO assist in gathering more information on intellectual property rights over genetic resources and related knowledge. The provisions on intellectual property rights are further discussed below.

Summary of Central Features of CBD Implementation as to PGRFA

The CBD applied to all biological diversity, but during its first 10 years was designed largely to apply to non-domesticated diversity. PGRFA-related concerns were not adequately mainstreamed in the general decisions of the COP, and a real program on agricultural biodiversity was not in place until 2002. One important reason for this situation was the ongoing negotiations under the FAO.

A central question is how this enforced wait impacted on domestic implementation of the CBD in developing countries. How did domestic measures for the management of biological diversity in general, if they were designed with a view to non-domesticated biodiversity, affect the management of PGRFA? This will be analyzed in the case studies.

Whereas the first 10 years of the CBD may seem lost for the CBD-induced domestic management of agricultural biodiversity, they were useful in several other ways: The Conference of the Parties pushed for completion of the negotiations under the FAO. They discussed the issue and thereby improved their understanding of the role and special characteristics of agricultural biodiversity. Information was compiled and several studies prepared. All this contributed to heightened awareness on the importance of agricultural biodiversity, and on its linkages with the greater context of biological diversity. Seen as a preparatory phase, the first 10-year period was probably fruitful.

The sixth meeting of the COP in 2002 marked the 10-year anniversary of the CBD, and a range of guidelines and concrete decisions were made, representing a great step forwards in implementation of the CBD. A few months prior to that meeting, the ITPGRFA was adopted at the FAO. This meant that the COP to the CBD could start working together with the FAO on matters related to the management of PGRFA, as had been envisaged ever since the Nairobi resolutions in 1992. In several decisions, PGRFA-related concerns were integrated, and this development was presented at the seventh COP meeting in 2004. The efforts regarding agricultural biodiversity will be reviewed at the ninth COP meeting in 2008.

Institutional Context: Nesting in Biodiversity

The negotiations leading up to the CBD were initiated and organized by the United Nations Environment Program.[29] However, the Convention is not embedded in UNEP in the way the IU/ITPGRFA was/is embedded in the FAO. The Depositary of the CBD is the UN Secretary-General, and not UNEP (Art. 41). The Secretariat was to be provided by the Executive Director of UNEP on an interim basis, but only until the first meeting of the Conference of the Parties (Art. 40). At this first meeting a secretariat was established (Art. 24), located in Montreal and not in Nairobi, where UNEP is headquartered. In other words, UNEP helped in giving birth to the CBD, but since the first meeting of the Conference of the Parties, the CBD has been a rather independent institution.

---

29 UNEP Decisions 14/26 1987; 15/34 1989; 16/42 1991.

The CBD occupies a challenging position: It is not only a stand-alone agreement with its own measures of implementation to be followed up, but also a framework convention for those protocols and other agreements that are regarded as implementing measures (for example the ITPGRFA) (Andersen 2007, 161–3). In addition, it is to maintain close co-operation with a great number of other relevant international institutions. With its broad mandate, covering all biological diversity on Earth except for human genetic resources, and with its complex institutional structure, its agenda is indeed comprehensive. The way PGRFA has been dealt with under the Convention must be understood in this context.

An important feature of the institutional context is the financial structure for implementing the CBD. For this purpose, the Global Environment Facility (GEF) was taken as point of departure. GEF was initially set up in 1991 as a three-year pilot program, jointly implemented by United Nations Development Program (UNDP), UNEP and the World Bank (Stokke and Thommessen 2003, 265). As an interim financial arrangement (Art. 39 of the CBD), GEF was designated to provide the institutional structure for the CBD financial mechanism, together with that of the United Nations Framework Convention on Climate Change, on the condition that it was fully restructured. It currently provides funding for projects on biological diversity, climate change, international waters, and ozone layer depletion. Since 2002, its mandate includes the funding of projects on persistent organic pollutants. The Conference of the Parties to the CBD decided that the restructured GEF would continue to serve as the institutional structure to operate the financial mechanism on an interim basis (Decisions I/2, paragraph 2, and decision II/6, paragraph 1). A memorandum of understanding with the GEF Council was adopted at the third Conference of the Parties (Decision III/8, annex). This is how the CBD is also linked to the GEF and the implementing bodies of the GEF – UNDP and the World Bank – an additional component in the complex web of the CBD network.

By 2000, the GEF had supported only four small projects on PGRFA;[30] despite a slight increase after that, it never played a major role in this field. The CBD Conference of Parties has repeatedly urged GEF to accept more PGRFA-related projects.

Resulting Norms and Rules of the CBD

Although the objectives of the ITPGRFA are similar to those of the CBD, this does not mean that the norms as such are the same. The two agreements are based on different rationales and interests, which indicates that the weighting of their norms may differ. Also the rules are quite different, due to the all-encompassing agenda of the CBD.

30 FAO/CGRFA Report CGRFA-9/02/6 (2002).

Core Norm: Conservation and Sustainable Use

The conservation of biological diversity and the sustainable use of its components constitute two of the three core objectives of the CBD. Although the regulations on conservation and sustainable use (Articles 6–14) are expressed in a less mandatory form – with formulations like 'Each Contracting Party shall as far as possible and as appropriate' or 'in accordance with its particular conditions and capabilities' – the sheer number of provisions and the negotiation history indicate that these objectives are central. As we have seen, the need to save biological diversity from eradication through conservation constituted the point of departure for the negotiations towards the CBD and the sustainable use of the components of biological diversity followed in their wake. These two objectives are therefore clearly core norms, as defined in Chapter 3.

Although the two objectives are dealt with as one norm here, they could also have been analyzed as two separate norms, since they can be complementary as well as conflicting. They are complementary in terms of being two strategies aimed at saving biological diversity. They may conflict in situations where conservationists seek to save nature by excluding people from any use of its resources even if such use is deemed sustainable. When it comes to PGRFA, however, conservation and sustainable use are two sides of the same issue. It is not possible to conserve plant varieties under natural conditions *in situ* and/or on farm, since their genetic traits are under continuous change due to environmental conditions and breeding efforts. Thus, it is possible to maintain and manage the *diversity of PGRFA* only through sustainable use. Conservation of PGRFA is possible in gene banks, but in the long run also gene-bank accessions change their traits due to regeneration under differing environmental conditions. These two approaches to maintaining PGRFA diversity are necessary and complementary, not conflicting, and are therefore considered as one norm in our context.

The CBD itself does not specify what conservation and sustainable use mean with regard to PGRFA. We have seen from the decisions of the COP that the obligations remained vague until 2002, when the Global Strategy on Plant Conservation established targets and instruments for the conservation and sustainable use of PGRFA in cooperation with FAO.

Core Norm: Fair and Equitable Benefit Sharing

The fair and equitable sharing of benefits arising from the use of PGRFA is one of the three core objectives of the CBD, which indicates its importance as a core norm. Also the negotiation history detailed above indicates that this is indeed a core norm of the PGRFA. During the negotiation of the CBD, a basic motivation for developing countries were the prospects of fair and equitable benefit sharing. After all, most genetic resources originate from developing countries, whereas corporations in rich countries have used these resources for their own profits, without any form of remuneration to the countries from which they originated. Bio-prospecting without arrangements for benefit sharing was considered 'bio-piracy' and a form of new imperialism. If developing countries were to conserve biodiversity, as demanded by industrialized countries, this would have to compensated through benefit sharing, and thus benefit sharing was seen as a major incentive for conservation efforts. Whereas benefit sharing can be indirect –not linked to the transfer of genetic resources but to co-operation in broader terms – the focus of the discussions under the CBD was on direct benefit sharing, that is, linked to the transfer of genetic resources. This will be our focus in this section, but we return to the issue of indirect benefit sharing in the concluding chapter.

As the norm on benefit sharing concerns the use of all biological diversity, the crucial question for our analysis is what this implies with regard to PGRFA. Three questions are particularly relevant: (1) What kinds of benefits are there to be shared from the use of PGRFA? (2) Who will be the beneficiaries of benefit-sharing arrangements on the use of PGRFA? (3) What will they possibly gain?

The Bonn Guidelines provide a list with examples of benefits, distinguishing between monetary and non-monetary benefits. The former include access fees, up-front payments for bio-prospecting, milestone payments during the process of bio-prospecting, and payment of royalties, as well as license fees in cases of commercialization, and special fees to be paid to trust funds supporting conservation and sustainable use of biodiversity. They may further include salaries and preferential terms, research funding, joint ventures, and joint ownership of relevant intellectual property rights. Non-monetary benefits may include the sharing of research, research collaboration, access to *ex situ* facilities and databases, participation in product development, technology transfer, capacity building, access to scientific information, and contributions to local economies. The scope for benefits could seem promising, and has provided a central rationale for such arrangements.

However, there are substantial differences as to the potential benefits from domesticated and non-domesticated genetic resources. An estimate comparing the commercial seed industry with the pharmaceutical sector shows huge differences with regard to the monetary potentials for benefit sharing (FAO 1998, 290). The benefits to be returned to source countries from the commercial exploitation of plant genetic resources for food and agriculture through plant varieties would hardly cover the transaction costs involved. By contrast, the benefits to be returned to a source country of a plant-derived pharmaceutical product might reach several million dollars under a bilateral agreement. One reason for this is that a plant variety is normally derived from a huge number of parent varieties, with an even greater number of potential countries of origin, whereas pharmaceutical products are often derived from a single plant variety that stems from one country (Fowler and Hodgkin 2004). In addition, commercial breeders often have in their own gene banks the PGRFA they need, collected prior to the CBD and outside the public domain, for which they do not need to pay. Also, plant varieties are under continuous development, and the time period within which one variety may be sold tends to be shorter than that of a successful medicine. FAO (1998, 290) estimates indicate that few monetary benefits can be expected from the commercial utilization of PGRFA.

Another comprehensive study on PGRFA underlines that the demand by many commercial plant breeders for major crop genetic resources for food and agriculture is limited, since their own gene banks and their national gene banks have been sufficiently stocked to last for some decades to come (ten Kate and Laird 1999, 137–

42). This means that developing countries could expect few bilateral arrangements on benefit sharing related to PGRFA during the next few decades, but the picture might change in the long run.

More important for developing countries is the fact that they are not net providers but net receivers of PGRFA. Palacios (1998) and Fowler, Smale and Gaiji (2001) document that individual developing countries in general receive more PGRFA than they provide.[31] For example, countries in Southern Africa are between 65 per cent and 100 per cent dependent on main food crops that originated outside the region. Most countries were found to be more than 90 per cent dependent. Ethiopia, generally considered the richest in PGRFA in Africa, was estimated to be between 28 and 56 per cent dependent on PGRFA from other regions. In general, developing countries anywhere on the globe are dependent on PGRFA originally from other regions in the South. What these figures suggest is that all countries would benefit from the smoothest possible access to PGRFA, and that benefit-sharing arrangements without benefits are more of a burden than no arrangements at all.

Another example refers to a selection of 15 developing countries surveyed over a period of almost 20 years with regard to their exchange of PGRFA with the international gene banks in the CGIAR system. On average, they received four times as many samples from the gene banks than the number they provided to the gene banks during this period.[32] Fowler, Smale and Gaiji refer to the historic transfer of genetic resources, which could be labeled 'bio-piracy' according our current understanding of the historic relations between South and North and the value of genetic resources. However, they maintain, these historical transfers cannot be compared with the recent and current transfers. The former were in most cases aimed at crop introduction, whereas recent and current transfers are aimed at crop improvement. Therefore, they conclude, it can be argued that continued facilitated access to PGRFA in today's world is a 'win–win' situation for all, and one especially important for developing countries as it provides them with improved cultivars. This argument goes to the core of the heated debate on access and benefit sharing: What is there to win, and what is there to lose?

Through the historic transfers of PGRFA for crop introduction, and through the gene banking in past decades, countries of the South have lost control over the PGRFA originating in their territories. The ever-stronger systems of intellectual property rights on resources originating from developing countries add to this picture. The access and benefit-sharing regulations under the CBD are not sufficient to provide fair and equitable shares of the monetary and other benefits accumulated by multinational seed corporations from their use of PGRFA.

---

31 This is also the case for industrialized countries, even if they receive less due to their own well-filled gene banks.

32 Gene banks are able to provide more samples back to countries than they receive, since they multiply the seeds. Thus, one sample from one country may benefit many countries, which is how each country can gain more than it provides.

On the other hand, developing countries lose twice over when they restrict access in order to provide for benefit sharing,[33] if they get few, if any, benefits in return and at the same time obstruct each other (and even researchers in their own countries) in obtaining access to these vital resources. Continued access to PGRFA is more vital to present and future food security – and to the maintenance of these resources – than any monetary benefit sharing can ever be. Thus, benefit-sharing regimes for domesticated PGRFA would have to be designed with a particular view to ensuring continued access to these resources. As the CBD comprises all biological diversity alike – also when regarding benefit sharing – the consequence is most likely that countries will provide for benefit-sharing arrangements with regard to biodiversity on a general basis, without taking into account the specific conditions for PGRFA.

Secondary Norm: Access to Genetic Resources

Access to genetic resources is not among the three objectives of the CBD, but is mentioned as one of several benefits arising from the use of genetic resources (Art. 1). This is one indication that access may be regarded as subordinate to benefit sharing under the CBD, and thus a secondary norm. This indication is, however, not sufficient, as a similar constellation was found in the ITPGRFA, where access was deemed a core norm. A central question thus concerns the expectations the Convention actually sets out with regard to facilitation of access.

The CBD reaffirms the sovereign rights of states over their biological resources, which is an established principle in international law (Art. 15). On this basis, it assigns to national governments the authority to determine access to genetic resources. Each Contracting Party is to create conditions for the facilitated access to genetic resources under its sovereignty to other Contracting Parties, in terms of legislation and other institutional arrangements. No restrictions are to be imposed that run counter to the objectives of the Convention. This wording could indicate that the facilitation of access is a core norm, in the sense of expectations as to what the Contracting Parties 'must' or 'should' do, as defined in Chapter 3. However, the CBD also stipulates that 'Access, *where granted*, shall be on mutually agreed terms' (Para 15.4, my emphasis), which indicates that there is no obligation to facilitate access – another criterion for secondary norms, as defined in Chapter 3. As set out in Article 15, facilitation of access is to be subject to prior informed consent of the Contracting Party providing the resource, and the Parties shall provide for benefit-sharing arrangements upon mutually agreed terms. In practice this means that access will not be granted unless the Contracting Party providing the resource is satisfied with the conditions for the transfer negotiated with the recipient of the resource. This is a clear indication that access is subordinate to benefit sharing. All in all, these indications seem necessary and sufficient for viewing access as a secondary norm under the CBD.

---

33 The provisions on access to genetic resources in order to provide for benefit sharing, as set out in the first generation of such acts, have proved in many cases to be bureaucratic, overly restrictive and time-consuming (ten Kate and Laird 1999, 17–33 and 293–312).

According to legal expert Gudrun Henne (1998, 141), sovereignty over genetic resources can mean two things: (1) sovereignty over genetic resources that originated in the country, or (2) the sovereignty over genetic resources that are found in the country. The CBD is a clear case of the former instance (Henne 1998; Fowler 2001; Andersen 2001):

- Only Contracting Parties that are countries of origin of genetic resources are entitled to exercise sovereignty over these resources for the purpose of the CBD (Art. 15, Para. 1).
- Only Contracting Parties that are country of origin or have acquired genetic resources in accordance with the Convention can be providers of these resources (Art. 15, Para 3).
- 'Country of origin of genetic resources' is defined in the Convention (Art. 2) as '*the country which possesses those genetic resources in* in situ *conditions*'.
- 'Genetic resources' is understood as '*genetic material of actual or potential value*'.
- 'Genetic material', in turn, refers to '*any material of plant, animal, microbial or other origin containing functional units of heredity*'.
- '*In situ* conditions' are defined as '*conditions where genetic resources exist within ecosystems and natural habitats, and, in the case of domesticated or cultivated species, in the surroundings where they have developed their distinctive properties*'.
- The term 'distinctive properties' is not defined in the CBD.
- A distinction is made between 'country of origin of genetic resources' and 'country providing genetic resources', with the latter defined as '*the country supplying genetic resources collected from* in situ *sources, including populations of both wild and domesticated species, or taken from* ex situ *sources, which may or may not have originated in that country*'.

Thus, from the formulations in Convention text, it is by no means possible to understand 'country of origin of genetic resources' as synonymous with 'source country for such resources' or with 'provider country'. It is the 'country of origin', and not the source or providing country, that is to decide on access to genetic resources and that is empowered to negotiate with regard to mutually agreed terms for benefit sharing related to access (Art. 15). It is also the 'country of origin' that is preferred as the site for *ex situ* conservation[34] (Preamble and Art. 9). The Bonn Guidelines on Access and Benefit Sharing confirm and strengthen this understanding: 'Providers should only supply genetic resources and/or traditional knowledge when they are entitled to do so' (Decision VI/24, II/C/16/c/i).

Henne (1998, 144) argues that the principle of country of origin of genetic resources as the country with sovereignty over these resources introduces a new form of access regulation to natural resources in international law: Countries possessing

34 '*Ex situ* conservation' refers to '*the conservation of components of biological diversity outside their natural habitats*', where 'habitat' means '*the place or type of site where an organism or population naturally occurs*' (CBD, Art. 2).

genetic resources within their territory delegate this right to those that are countries of origin. This may give rise to problematic situations if the country of origin and the source country are not identical, as is often the case (ibid., 142). For those genetic resources that have an evident country of origin, and where the country of origin no longer possesses the genetic resource, there is actually no legal entity with sovereign rights over that resource (Henne 1998; Wolfrum and Stoll 1996).

To identify countries with sovereignty over PGRFA, we have to search for the countries of origin, and here our next problem arises. Often it is simply not possible to determine the country of origin of PGRFA (Fowler 2001; Andersen 2001). And in cases where it is possible, there is normally more than one country, often a whole region. For non-domesticated resources it is usually not so difficult to identify one or a few countries of origin – the main problem concerns domesticated resources. Agricultural biodiversity has evolved through the selection, growing and exchange of seeds and plants by farmers over short and long distances for thousands of years – in addition to the processes of natural selection. Even if it is possible to determine when a distinctive property took shape, there will often be several parents to one variety, which have developed their traits based on their parents – and so forth. In theory, these developments could for some varieties be traced back to the dawn of agriculture. In practice, we might be able to distinguish a few recent distinctive turning points in the evolution of the various parent varieties, but for most crops the pedigrees would soon become immensely complicated.

It is important to understand the principle of 'country of origin' in the context of the overall objectives of the Convention (Art. 1), that is, its intentions. We may infer that the CBD is based on the assumption that 'countries of origin of genetic resources' are the adequate Parties to decide over benefit sharing arrangements, and thus the ones that should be responsible for the facilitation of access. However, when the country of origin of a genetic resource cannot be identified, there is also no Party that can be assigned to facilitate access to that resource under the provisions of the CBD – and this runs counter to the basic principles of the Convention (Henne 1998; Fowler 2001; Andersen 2001). In other words, the CBD is not conducive to facilitating access to genetic resources with unclear origins, as is mostly the case with PGRFA. In addition, the bilateral approach to facilitating access to PGRFA implies that breeders of PGRFA, who need access to large amounts of genetic resources for the development of one new plant variety, are often faced with bureaucratic and time-consuming regulations which act to limit actual access (Brush 2004; 2005). And so the development of new varieties based on such access has become difficult.

In seeking to encompass all biodiversity, the CBD negotiators obviously failed to take sufficient account of the particular features of PGRFA in the short time at their disposal for finalizing the Convention text. So far this problem has been bridged at the international level by the Consultative Group on International Agricultural Research (CGIAR), which stores between 20 and 50 per cent of the diversity of major crops conserved in gene banks in trust for the FAO (FAO 1998, 280), facilitating free access according to a material transfer agreement. This has been an intermediate solution in the negotiation period, pending the establishment of an international legally binding agreement on PGRFA, which has now been established with the ITPGRFA. The 2002 Bonn Guidelines on Access and Benefit Sharing confirm the principles of the CBD with regard to access, but also state that they are without prejudice to the access and benefit-sharing provisions of the ITPGRFA.

At the domestic level, however, countries have started to implement the provisions of the CBD on access and benefit sharing as applicable to all types of genetic resources. How this has affected PGRFA management will be further examined in the case studies and in the concluding part of this book.

Secondary Norm: Respect for and Maintenance of Traditional Knowledge

Parties to the CBD have with Article 8j agreed to respect, preserve and maintain traditional knowledge, innovation and practices, as far as possible, as appropriate, and subject to national legislation. Whereas this norm has received substantial emphasis at COP meetings, signifying its importance as a separate norm, the formulations indicate that this must be seen as a secondary norm according to our definition. This norm is important for our analysis because it is the only one to indicate expectations as to how the Contracting Parties should recognize, and act towards, the traditional custodians of biodiversity (implicitly: farmers), and also the sole norm to indicate expectations as to the domestic distributional aspects of benefit sharing.

Reference is made to indigenous and local communities embodying traditional lifestyles relevant for the conservation and sustainable use of biological diversity. The Parties also agree to promote the wider application of such knowledge, innovation and practices. No precise definition is given of the indigenous and local communities referred to. However, it seems evident that traditional farmers belong in that category, even though they were not mentioned until COP VII in 2002, when the links to farmers' rights under the ITPGRFA were highlighted. Nevertheless the Parties were encouraged at COP III in 1996 to build capacity among indigenous and local communities for the *in situ* management of agricultural biological diversity (Decision III/11, Paragraph 15f).

Regulatory Rules of the CBD

Here we will focus on the regulatory rules for implementing the CBD as they relate to PGRFA management – and at the national level. These rules are presented according to the norms highlighted above, with mention of whether they are considered core or secondary rules.

Conservation and sustainable use of biodiversity A core rule is that the Contracting Parties to the CBD, in accordance with their particular conditions and capabilities, shall *develop national strategies, plans or programs* for the conservation and sustainable use of biological diversity and *integrate such concerns* as far as possible into relevant sectoral or cross-sectoral plans, programs and policies (Art. 6). Furthermore the Contracting Parties shall as far as possible and appropriate (that is, as a secondary rule), *identify components of biological diversity* that are important for conservation and sustainable use, and shall *monitor their development* (Art. 7).

For the purpose of *in situ* conservation, various rules are listed that the Parties are expected to implement as far as possible and as appropriate. These are secondary rules according to our definition, and include *establishing a system for protected areas* with particular management requirements, *promoting the protection of ecosystems*, restoring degraded ecosystems, and preventing threats to such ecosystems (Art. 8).

For *ex situ* conservation, various measures are listed that the Parties shall implement as far as possible and as appropriate – again, secondary rules according to our definition. These include the *establishment and maintenance of* ex situ *facilities, preferably in the countries of origin* of components of biological diversity, the *collection of biological resources* from *in situ* condition to these facilities, activities to recover and *rehabilitate threatened species* and to co-operate in *providing financial support* for these activities (Art. 9).

Also when it comes to the sustainable use of components of biological diversity, the rules are secondary according to our definition. The Contracting Parties shall as far as possible and appropriate *integrate such consideration in their mainstream national policy making* (Art. 10), *adopt relevant measures* for the purpose, *protect and encourage customary use* of biological resources when they are sustainable, and *support local initiatives* as well as *co-operation with the private sector* in developing methods for sustainable use of biological resources. For the purpose of the conservation of biological diversity and the sustainable use of its components, the Parties shall also as far as possible and appropriate develop *incentive measures* (Art. 11), and p*romote relevant research and training* (Art. 12).

Finally, the Contracting Parties are to promote and encourage understanding of the importance of the conservation of biological diversity, and of the measures necessary for that purpose, and ensure that it is propagated through the media and in educational programs (Art. 13). This qualifies as a core rule. The Parties shall work together, as appropriate, with other states and international organizations to develop educational and public awareness programs on the issue.

Fair and equitable benefit sharing The Contracting Parties shall take legislative, administrative or policy measures as appropriate for the purpose of the sharing of the results of research and the benefits from the commercial use of genetic resources with the Contracting Party providing such resources (Art. 15). Furthermore, access to and transfer of technology shall be provided to developing countries under fair and most favorable conditions (Art. 16). Each Contracting Party is to take legislative, administrative or policy measures, as appropriate, to ensure this, including measures ensuring that the private sector participates in the transfer. In this context, the Contracting Parties shall facilitate exchange of relevant information (Art. 17), and promote technical and scientific co-operation (Art. 18). All these rules are formulated as mandatory, and are thus core rules according to our definition.

Access to genetic resources Each Contracting Party shall *create conditions to facilitate access* to genetic resources for which they are the country of origin (Art. 15, see discussion above). Although this is formulated as mandatory, that is, as a core rule, the actual *facilitation of access* is a secondary rule, as it is optional to the countries (Para. 15.4).

Respect for and maintenance of traditional knowledge The Contracting Parties shall, as far as possible and as appropriate, and subject to national legislation, respect, preserve, and maintain relevant traditional knowledge and promote their wider application with the approval and involvement of the holders of such knowledge – and encourage the equitable sharing of the benefits arising from the utilization of such knowledge. Due to the introductory formulations, these must be seen as secondary rules according to our definition.

Overview over CBD Norms and Rules Pertaining to PGRFA

Table 6.2 provides an overview of the norms and rules of the CBD as they relate to the management of PGRFA.

Table 6.2 Overview of norms and rules of the CBD pertaining to PGRFA

	Convention on Biological Diversity
Core norms	• Conservation of biological diversity and sustainable use of its components • Fair and equitable sharing of benefits arising from the use of genetic resources
Secondary norms	• Access to genetic resources • Respect for and maintenance of traditional knowledge
Core rules to be implemented by governments	• National strategies, plans or programs for the conservation of biological diversity and sustainable use of its components, and the integration of such concerns in cross-sectoral strategies • Legislative and policy measures to ensure benefit sharing and technology transfer, including measures for private sector participation, exchange of relevant information, the promotion of technical and scientific co-operation • Access to genetic resources is subject to national legislation. Contracting Parties shall create *conditions* for facilitated access to resources for which they are countries of origin, or which they have received in accordance with the CBD • Promote and encourage understanding of the importance of the conservation of biological diversity and develop educational and public awareness programs on the issue
Secondary rules to be implemented by governments	• Establishing a system for protected areas and promote the protection of ecosystems • Establishment and maintenance of *ex situ* facilities, preferably in the countries of origin of components of biological diversity • Activities to recover and rehabilitate threatened species and to co-operate in providing financial support for such activities • Integrate considerations of sustainable use of genetic resources in mainstream national policy-making, and adopt relevant measures for the purpose, including incentive measures and the promotion of local initiatives, relevant research and training • Respect, preserve, and maintain relevant traditional knowledge, promote its wider application and encourage the equitable sharing of the benefits arising from its utilization

Chapter 7

The Agreement on Trade-Related Aspects of Intellectual Property Rights and the Convention for the Protection of New Varieties of Plants

The Agreement on Trade-Related Aspects of Intellectual Property Rights (TRIPS) was adopted on 14 April 1994 as one of the three basic agreements on which the World Trade Organization (WTO) was built.[1] The Agreement Establishing the World Trade Organization entered into force on 1 January 1995, and the TRIPS Agreement came into effect one year later, on 1 January 1996 (Art. 65.1). However, developing countries were allowed to extend their implementation of the TRIPS Agreement until 1 January 2000, and the least developed countries were granted a 10-year extension, up to 1 January 2006 (Art. 65.2), or – if they were unable to meet this deadline – a further extension (Art. 66.1).

The purpose of the TRIPS Agreement is, as stated in its preamble, to promote the effective and adequate protection of intellectual property rights as a means to reduce distortions and impediments to international trade. This is intended to contribute to the promotion of technological innovation and to the transfer and dissemination of technology, to the mutual advantage of producers and users, balancing rights and obligations (objectives in Art. 7). It provides minimum standards for the protection of intellectual property rights in member states, covering such rights as copyrights, trademarks, geographical indications, industrial design, and patents. The provisions on patents (Art. 27) relevant to PGRFA management[2] cover any inventions, whether products or processes, in all fields of technology, provided that they are new, involve

1 The other two were the General Agreement on Tariffs and Trade (GATT), which pertains to goods; and the General Agreement on Trade in Services (GATS). In addition there is the agreement establishing the WTO, and several additional agreements and annexes dealing with the particular requirements of specific sectors and issues.

2 Also the provisions on geographical indication could to some extent be relevant to PGRFA management. Geographical indications 'identify a good as originating in the territory, or a region or locality in that territory, where a given quality, reputation or other characteristic of the good is essentially attributable to its geographical origin' (Art. 22.1). So far this provision has been applied for beverages, for example wines from the Champagne region in France, and to some extent for foodstuffs and other articles; there are also a few examples pertaining to plant genetic resources (see also Dutfield 2000, 87).

an inventive step and are capable of industrial application (Art. 27.1).[3] There are possible exceptions from patentability: (1) inventions that are detrimental to the *ordre public* or morality (Art. 27.2); (2) diagnostic, therapeutic and surgical methods for the treatment of humans or animals (Art. 27. 3[a]); and (3) plants and animals other than micro-organisms, and biological processes essential to the production of plants or animals other than non-biological and microbiological processes. The latter can be excluded from patentability on the condition that the Members 'provide for the protection of plant varieties, either by patents or by an effective *sui generis* system or by any combination thereof' (Art. 27. 3[b]).

Article 27.3 (b) is the focus for this analysis – with several other provisions of the TRIPS Agreement and of other WTO Agreements establishing the context for its implementation.

The scope of interpretations for Article 27.3(b) is related to the term '*sui generis* system' (which means a system of its own kind) and the word 'effective'. The limits for a *sui generis* system and the meaning of an 'effective' *sui generis* system are not explicitly defined in the text. The International Union for the Protection of New Varieties of Plants (UPOV) has held that the most effective way to comply with the provision of an effective *sui generis* system is to follow the model of the UPOV Convention (to be explained in detail below). There are several proponents of this stand (see Helfer 2002, 31). Some of these advocate compliance with the 1978 version of the UPOV Convention, whereas others promote the 1991 version of the same convention, which is the stricter model. Those in favor of the 1991 Act emphasize that this version provides the most extensive protection for plant breeders, whereas those endorsing the 1978 Act maintain that this was the version of UPOV in force when the TRIPS Agreement was adopted. (It was closed for new accessions in 1998.) The International Association of Plant Breeders for the Protection of Plant Varieties (ASSINSEL) in 1999 held an international congress with representatives from more than 1000 seed companies, where they recommended that developing countries should adopt a *sui generis* system based on the 1991 Act (Crucible II Group 2000, 92).

Since the UPOV Convention has been advocated as **the** *sui generis* system for the intellectual property rights protection of plant varieties, and this stand is increasingly being followed by developing countries – including the Philippines – UPOV will be further examined in this chapter, as it relates to the TRIPS Agreement.

The Convention for the Protection of New Varieties of Plants (UPOV Convention) was adopted in Paris in 1961 to ensure that member states acknowledge the achievements of breeders of new plant varieties, by making available to them exclusive property rights for a given period of time. The UPOV Convention provides uniform and clearly defined principles for the protection of plant breeders' rights over plant varieties that are novel, distinct, genetically uniform and stable (to be further elaborated below). The UPOV organization was established through the Convention as the institutional framework for its implementation. The UPOV Convention entered into force in 1968; it was revised in Geneva in 1972, 1978 and 1991, with

3 'Inventive step' and 'capable of industrial application' may be deemed by Members to be synonymous with 'non-obvious' and 'useful', according to the text.

the 1978 Act entering into force on 8 November 1981 and the 1991 Act on 24 April 1998. With each revision, the protection of plant breeders' rights over new varieties of plants that they have developed has been strengthened, as explained below.

An important feature of the UPOV Convention, as compared to TRIPS, is that the former is about intellectual property rights pertaining solely to plant varieties, whereas the latter concerns all kinds of intellectual property. The reason why our focus in this book is on TRIPS, however, is that UPOV was originally established as an organization for OECD countries, and has become relevant for developing countries only through the TRIPS Agreement.[4]

Historical Context: From Business Initiators to Ambivalent Developing Countries

Many NGO activists have maintained that the developing countries have fought against the TRIPS Agreement ever since it came on the table. However, the picture is not that black and white, as we shall see. In determining the development stages of the TRIPS Agreement in terms of Article 27.3(b), it is not difficult to identify the approximate start of the agenda-setting phase, and the precise start of the negotiating phase. However, when it comes to implementation, a different terminology is used in the TRIPS Agreement: it refers to the period when the regulations shall be transformed to national policies as the 'transitional phase'. We will label this period 'implementation period'. The subsequent phase, called the 'review phase' in TRIPS terminology, corresponds to our 'review/evaluation phase'.

Agenda-Setting Phase

The agenda-setting phase, which was to result in the TRIPS Agreement many years later, was introduced in the 1973–1979 Tokyo Round of GATT,[5] when a group of trademark-holding firms organized the Anti-Counterfeiting Coalition which lobbied for the inclusion of an anti-counterfeiting code in the Tokyo Round (Doremus 1996). The initiative attracted the interest of the US and the European Economic Community to draft such a code. This they did, and the draft, which was known as the 'counterfeit code' and aimed at discouraging the import of counterfeit goods, was introduced in the Tokyo Round (Yusuf 1998, 6). However, their initiative did not succeed, and the code was not included in the agenda of the Tokyo Round. Nevertheless, it remained a topic for GATT discussion until the launching of the Uruguay Round in 1986.

---

4 Only South Africa became a member of UPOV before the adoption of the TRIPS Agreement. It joined during the time of apartheid rule, in 1977.

5 The General Agreement on Tariffs and Trade (GATT) was adopted in 1947 in Geneva and entered into force in 1948. Until the entry into force of the WTO it had a provisional organization. Its rules and regulations were developed during trade negotiation rounds, of which the Tokyo Round and the Uruguay Round were the last ones before the adoption of the WTO Agreement. The current Doha Round was introduced in November 2001 as the first trade negotiation round under the WTO.

Table 7.1 Development stages of the TRIPS Agreement

1973–1986	**Agenda-setting phase** introduced with an initiative of the Anti-Counterfeiting Coalition of trademark-holding firms, leading to a proposal from the United States and the European Community on a counterfeit code as part of the Tokyo Round of the GATT trade negotiations. This proposal did not succeed. After further mobilization among business corporations, the US initiated an expert group in 1984 to examine the possibilities of including intellectual property rights in the agenda of the Uruguay Round, and this was finally accepted in 1986.
1986–1994	**Negotiation phase** introduced by the Punta del Este Declaration and marking the start of the Uruguay Round, which included intellectual property rights. Developing countries opposed the idea of a TRIPS Agreement up to 1989. In 1990 they sought for a breakthrough in the negotiations, with a separate draft proposal.
1994–1996	**Interim period**, from the adoption of the WTO agreements in 1994 and to the entry into force of the TRIPS agreement in 1996. During this period the institutional framework for implementing the TRIPS Agreement was established.
1996–2006	**Implementation phase**, in which the developed countries were to have adjusted their legislation to the requirements of Article 27.3(b) of the TRIPS Agreement by 1996, the developing countries by 2000 and the least developed countries by 2006.
1999–	**Review/evaluation phase**, where the TRIPS Council is to review the efforts of the member countries to adjust their national legislation to Article 27.3(b), and if necessary, propose amendments to the General Council of the WTO.

Before that, US business corporations from the copyright, patent and semiconductor industries decided to follow up on the initiative of the trademark-holding firms. During the early 1980s they started lobbying for effective intellectual property rights protection in overseas markets as a trade-related issue. A Multilateral Trade Negotiation Coalition was established in the US, which brought together large financial institutions, manufacturing firms, agribusiness corporations (including seed corporations) and trade organizations (Peng 1990, 209, referred to in Fowler 1994, 176). One of the trade organizations was the Intellectual Property Committee, a coalition of 13 large companies in the fields of pharmaceuticals, seed production and biotechnology (Weissman 1991, referred to in Fowler 1994, 176). With a concerted strategy shared by so many stakeholders from the business sector, the pressure on Washington to take action increased dramatically. The driving forces behind what was to become the TRIPS Agreement had been formed.

Traditionally, the WIPO was the international institution in charge of co-operation on intellectual property rights. However, US business actors were not satisfied with WIPO performance in enforcing such rights or with the patent reform in general (Fowler 1994, 175; Evans and Walsh 1994, 38). The new GATT, as it was seen to emerge during the next trade negotiation round, seemed to provide better enforcement mechanisms. Therefore, the Chamber of Commerce and other actors in the United States urged Washington to press for patent reform in the next round of negotiations (Fowler 1994, 175). In 1984 the US established an expert group to look into the possibility of including intellectual property rights in the upcoming round of trade negotiations (see for example Evans and Walsh 1994, 38–9).

When the negotiators gathered in Punta del Este to launch the Uruguay Round in 1986, the issue of intellectual property rights was brought onto the table. However, it was heavily opposed by several developing countries, and did not make it onto the agenda until the conclusion of the meeting (ibid., 39).

The Punta del Este Declaration marked the inception of the negotiations of the Uruguay Round, and thereby also the negotiation phase that was to result in the TRIPS Agreement:[6]

> In order to reduce the distortions and impediments to international trade, and taking into account the need to promote effective and adequate protection of intellectual property rights, and to ensure that measures and procedures to enforce intellectual property rights do not themselves become barriers to legitimate trade, the negotiations shall aim to clarify GATT provisions and elaborate as appropriate new rules and disciplines. Negotiations shall aim to develop a multilateral framework of principles, rules and disciplines dealing with international trade in counterfeit goods, taking into account work already underway in GATT.

Negotiation Phase

At the beginning of the Uruguay Round, negotiators agreed on some 'General Principles Governing Negotiations', in which it was noted that industrialized countries would not expect developing countries 'to make contributions which are inconsistent with their individual development, financial and trade needs (Section B, Paragraph v). Therefore, developed Contracting Parties should not seek, neither should less-developed Contracting Parties be required to make, concessions that are inconsistent with the development, financial or trade needs of the latter' (Fowler 1994, 176).

During the first years of negotiations, a TRIPS Agreement was strongly opposed by several developing countries. Indeed, the 1988 Montreal mid-term review of the Uruguay Round found TRIPS to be insoluble (Evans and Walsh 1994, 39). Several developing countries continued to insist that 'social clauses' must be inserted in the agreement, and were ambivalent as to whether such an agreement should be incorporated into GATT at all.

Then in the course of 1989, the opposing developing countries changed their positions, and dropped their earlier resistance to a TRIPS Agreement. This radical shift must be understood against the background of the international trade situation, and the pressure for the introduction of intellectual property rights (Yusuf 1998, 9):

6 The Punta del Este Declaration (1986), Section D: Subjects for Negotiation.

- The US government started to introduce effective intellectual property protection in developing countries as a precondition for access to its market under the Generalized System of Preferences (GSP) for developing countries.
- On this background, the US and later the European Economic Community were able to induce significant changes in the legislation on intellectual property rights in many developing countries.
- As more and more developing countries moved towards free market policies, the enactment of effective legislation on intellectual property rights became equated with a certificate of 'good conduct'.
- The developing countries themselves came to perceive a multilateral framework as a lesser evil than bilateral concessions.

In addition, there were the strategic arguments for developing countries to make concessions, since the TRIPS Agreement was a part of the full package of agreements that were to result in the WTO. By yielding on TRIPS, it was felt, gains could be achieved in other important areas of the Uruguay negotiations, like agriculture, textiles and tropical products. TRIPS was regarded as one of the few bargaining cards available to the developing countries.

Thus, in May 1990 a group of 13 developing countries[7] submitted a detailed proposal on a TRIPS Agreement. The proposal was divided into two parts, where the first contained draft rules for procedures to discourage international trade in counterfeit and pirated goods, and the second part set out detailed standards and principles concerning the availability, scope and use of intellectual property rights. This submission is regarded as marking the breakthrough in negotiations on the TRIPS Agreement, as it was taken to indicate that the developing countries had accepted a GATT-based standard-setting approach to intellectual property rights (Yusuf 1998, 9). The proposal also provides insight into the limits for such standards, as seen from the developing countries, and is as such a valuable document for our analysis.

The draft must be understood against the background of all the submissions from other negotiating parties that had already set the agenda for the negotiations. Its proposed provisions on patents are introduced with a range of social objectives, but maintain that patent protection should be available for inventions in all fields of technology that are new, involve an inventive step, and are industrially applicable (proposed Art. 4.1). However, several items are exempted from patentability, among them plant or animal varieties or essentially biological processes for the production of plants or animals.

By dividing the proposed text of a TRIPS agreement into these two parts, the proposing developing countries signaled their determination to emphasize the part dealing with trade in counterfeit goods, while minimizing the part relating to intellectual property rights (Yusuf 1998, 10). At the same time they highlighted the importance of public policy objectives as a basis for regulation of intellectual

7 Argentina, Brazil, Chile, China, Colombia, Cuba, Egypt, India, Nigeria, Pakistan, Peru, Tanzania and Uruguay.

property rights at the national as well as the international level. The way the proposal is formulated also gives weight to the need to respect and safeguard national legal systems and traditions pertaining to intellectual property rights, not least in the perspective of differing levels of economic development.

This proposal provided substantive content to the final TRIPS Agreement with regard to objectives and principles, and helped to form the basis for the final negotiations that led to the adoption of the TRIPS agreement.

Throughout most of the negotiation phase, intellectual property rights over plant varieties were a non-issue. As early as in 1987, the European Community had emphasized that patents should be available for inventions in all fields of technology, except in cases like plant or animal varieties or essentially biological processes for the production of plants or animals.[8] Several other states followed up on this stand. For example, in 1989 Canada maintained that patents should generally be available in all fields of technology, but that it would not be reasonable to oblige all governments to extend patents to an area such as multi-cellular life forms.[9] Here considerably more technical study would be required both domestically and internationally concerning the most appropriate form of protection and the conditions under which it should be accorded. When the US in 1990 submitted to the negotiating group its communication, which was a draft TRIPS Agreement, it mentioned no exemptions at all from patentability.[10] Shortly after, the proposal of the group of 13 developing countries was submitted, where plants and animals were to be exempted from the obligation of patentability – and this meant a breakthrough for the negotiations.

This story tells how reducing the trade in counterfeit goods provided the original motivation for an agreement in this area, and how the issue was extended to cover more and more topics, under pressure from business corporations. Finally, plant varieties were included. The developing countries opposed the idea of a TRIPS Agreement, until they saw the possibility of using TRIPS as a bargaining chip in the trade negotiations. By that time they had also come under heavy pressure to introduce intellectual property rights in their national legislation anyway.

The Adoption of the Agreement and its Implementation

Due to the WTO single undertaking approach, all WTO agreements were adopted as a package in 1994. The provisions directly relevant to PGRFA are covered by Article 27.3(b), but, as we will see later, various other provisions in the TRIPS Agreement and other WTO Agreements must be taken into account in discussing the contents of Article 27.3 [b], which is formulated as follows:

> Members may also exclude from patentability plants and animals other than micro-organisms, and essentially biological processes for the production of plants or animals other than non-biological and microbiological processes. However, Members shall provide for the protection of plant varieties either by patents of by an effective sui generis system

8 GATT *Guidelines* MTN.GNG/ NGll/W/26 (1987).

9 GATT *Standards* MTN.GNG/NGll/W/47 (1989).

10 GATT *Draft Agreement* MTN.GNG/NG11/W/ 70 (1990).

> or by any combination thereof. The provisions of this subparagraph shall be reviewed four years after the date of entry into force of the WTO Agreement.

In transforming the TRIPS agreement into policies and legislation, WTO member countries could, and can, seek WIPO assistance. The WTO and WIPO made an agreement on co-operation on 22 December 1995, which entered into force a few days later, on 1 January 1996.[11] This agreement provides for the exchange of information on, for example, legislation in WTO and WIPO member countries, making available such legislation for developing countries, if necessary translated into a desired language, and making available to developing countries the necessary legal-technical assistance to implement the TRIPS Agreement.

The Committee on Trade and Environment (CTE) is of special relevance in our context. It was established by the General Council of the WTO in 1995 as a discussion forum (not a rule-making body), *inter alia* to consider relevant provisions of the TRIPS Agreement as these relate to environmental concerns. Discussions continued during the implementation phase, and focused on the relation between the TRIPS Agreement and the CBD.

Review Phase

The TRIPS Agreement provides for two types of review. The first concerns the whole agreement, and is aimed at assessing its implementation (Art. 71). It was supposed to be conducted two years after the expiration of the implementation period (in WTO terminology: the transitional period): in 2002 for developing countries and in 2008 for the least developed countries. After these, reviews are meant to be carried out at regular intervals. Any amendments to the text of the TRIPS Agreement following from the reviews shall merely serve the purpose of adjusting to higher levels of protection of intellectual property rights achieved (Para. 71.2). Such proposals must be adopted by consensus by the TRIPS Council and then filed with the Ministerial Council for decision, also by consensus.

The other category pertains solely to Article 27.3(b), which is the only article providing for a review of a limited component in the TRIPS Agreement. This was meant to be carried out in 1999. According to Article 27.3(b), the objects for review are *the provisions of that sub-paragraph.*[12] This marks a difference to the formulation in Article 71, where the review of the TRIPS Agreement as a whole is aimed at evaluating *implementation*. Opinion varies as to how to interpret the formulation on review in Article 27.3(b). Does it provide for a review leading to amendments of that sub-paragraph, or merely a review of implementation up to 1999?

The TRIPS Council agreed at its meeting in December 1998 to initiate the review of Article 27.3(b) through an information-gathering exercise. Member countries that should already have implemented these provisions (that is, industrialized

11 WTO-WIPO cooperation agreement, see <www.wto.org/english/tratop_e/trips_e/wtowip_e.htm>.

12 'The provisions of this subparagraph shall be reviewed four years after the date of entry into force of the WTO Agreement.'

countries) were invited to provide information on how the provisions were treated under national law, whereas other member countries were invited to provide such information on a best-endeavour basis. An illustrative list of question was prepared for the purpose. Responses were received from 23 member countries, three of which were categorized as developing countries (Thailand, South Africa and Zambia), and the EU.[13] Of the three developing countries, two had legislation on plant breeders' rights (Thailand and South Africa), whereas Zambia was in the process of developing such legislation. Most countries with plant breeders' rights legislation provided for exemptions for farmers, permitting them to save and reuse seeds from protected varieties. No information was gathered as to whether farmers were allowed to exchange seeds from their harvest, or to sell them as seeds.

The formal TRIPS reviews of those developing member countries that were to implement the TRIPS by 1 January 2000 commenced in 2000, and included implementation of Article 27.3(b). The procedures provide for written questions from WTO members and replies from the 'reviewee' prior to the review meeting, with follow-up questions and replies during the course of the meeting. At subsequent meetings of the Council, any points emerging from the review session that delegations consider have not been adequately addressed may be raised for further consideration.

By 1 January 2000, 69 developing-country members of the WTO were to have implemented Article 27.3(b) of the TRIPS Agreement, but only 30 per cent had actually done so (See GRAIN 2000, 5). Resistance against this provision was documented. As early as in 1999, developing countries had voiced demands to re-open the text of Article 27.3(b) for negotiations. The collapse of the Millennium Round in Seattle resulted in an abrupt end to the review negotiations on Article 27.3(b). Review meetings were postponed, and a new set of meetings was scheduled for 2000 and 2001. Prior to the Doha Ministerial Meeting in November 2001, several topics were raised in the TRIPS Council for further discussions, mainly by developing countries. These include:[14]

- the meaning of effective protection for new varieties of plant;
- how to handle moral and ethical issues, for example to what extent invented life forms should be eligible for protection;
- how to deal with traditional knowledge and genetic material, and the rights of the communities where these originate;
- whether there is a conflict between the TRIPS Agreement and the CBD.

Demands to re-open the text of Article 27.3(b) for negotiations did not succeed at the Doha Ministerial Conference. However, the Ministerial Declaration from the Conference states that the development dimension shall be taken fully into account

13 WTO *Illustrative list of questions* IP/C/W/273/Rev.1 (2003d).

14 Summarized by the WTO at <www.wto.org/english/tratop_e/trips_e/art27b_background_e.htm>.

in the review process, and that the objectives and principles set out in Articles 7 and 8 of the TRIPS Agreement shall guide the reviews.[15]

Since then, various proposals[16] have been submitted to the TRIPS Council on topics raised prior to Doha, of which the following should be mentioned:

- From the EU: a proposal to examine a requirement that patent applicants disclose the origin of genetic material with legal consequences outside the scope of patent law.[17] Disclosure of the origin of genetic resources would be an important step to ensure fair and equitable benefit sharing, as it would enable clarification of whether the genetic resources were acquired in accordance with the legislation in the respective country.[18]
- From Switzerland: a proposal to amend the WIPO Patent Cooperation Treaty so that national laws require patent applicants to disclose the origins of genetic resources and traditional knowledge. Failure to disclose could hold up a patent grant, or affect its validity.[19] This proposal would put more force behind the demand for disclosing the origin or genetic resources, as it would be a condition for issuing the patent.
- From Brazil, Cuba, Ecuador, India, Peru, Thailand and Venezuela: a proposal based on previous proposals on disclosure of the origins of biological resources and traditional knowledge, 'prior informed consent' for exploitation, and equitable benefit sharing. The group wants the TRIPS Agreement to be amended to make disclosure an obligation; their proposal also examines the weaknesses of alternative methods such as contracts.[20]
- From the African Group: a proposal, including a draft decision, on traditional knowledge designed to prevent 'misappropriation'. The African Group wants all life forms (plants, animals and micro-organisms) to be exempted from patentability and wants *sui generis* protection for plant varieties to preserve farmers' rights to use and share harvested seeds. It proposes requirements on disclosure similar to those in the proposal from Brazil and others above.[21]

Whereas the last two proposals require re-opening of the text for negotiations, the first two may be negotiated without that. This is an important difference: If the texts should be re-opened for negotiations, the output might instead result in higher levels of protection, with reference to Para. 71.2 of the TRIPS Agreement. The two groups of developing countries have shown that they are ready to take the risk, in an effort

---

15 WTO *Ministerial Declaration* WT/MIN(01)/DEC/W/1 (2001), Section 19.

16 See for example WTO *Review* IP/C/W/369 (2002a).

17 WTO IP/C/W/383 (2002b).

18 The Government of Norway has proposed to the General Council of the WTO and the TRIPS Council that the TRIPS Agreement be amended to include a mandatory obligation to disclose the origin of genetic resources and traditional knowledge in patent applications (Communication from Norway, 13 June 2006. Available at <http://docsonline.wto.org/DDFDocuments/t/ip/c/W473.doc>.)

19 WTO IP/C/W/400 (2003a).

20 WTO IP/C/W/403 (2003b).

21 WTO IP/C/W/404 (2003c).

to reduce the standards of the TRIPS Agreement and make them more equitable in a North–South perspective.

These considerations lead us to a key question that has emerged in the wake of the increasing number of proposals. The TRIPS Council considers whether discussions on these subjects have developed far enough for them to be handled immediately in the WTO.[22] Most developing countries support that, as reflected in the proposals from the two groups of developing countries. Several industrialized countries, among them Canada and the US, are inclined to consider such discussions technical and refer them to WIPO. This is also reflected in the proposals from the EU and Switzerland.

Thus we see that the WTO is moving from being the arena for some few powerful states to push through with their policies, to an arena where poor countries are also increasingly managing to voice their concerns. As a result, the US, Canada and some other countries are searching for an additional arena to negotiate outstanding issues, and have chosen WIPO for this purpose. WIPO was originally the international organization for such issues, and has gained renewed relevance since the debates in the TRIPS Council got stuck. This choice can be well understood in the light of 'forum shopping', as an example of how some actors seek forums for particular concerns according to their own interests and bargaining power.

Overview of the Development of the UPOV Convention

In order to join UPOV, a country has to adapt its legislation to the UPOV Convention, as formulated in one of its Acts (since 1998 only UPOV 1991 is open for new accessions) and then apply for membership.[23] UPOV will examine the legislation, and when it has confirmed that it is in accordance with the UPOV Convention, the country will be entitled to be a member of the Union.

As of November 2006, UPOV had altogether 63 members, and 18 of these were developing countries.[24] Of these developing countries 14 memberships were based on the 1978 Act, whereas 4 memberships were based on the 1991 Act (Jordan, Morocco, Tunisia and Vietnam). Several developing countries are in the process of adjusting their legislation to UPOV 1991.

The history of UPOV dates back to the 1950s and the initiative taken by an organization of commercial plant breeders, an organization promoting industrial patents and the International Chamber of Commerce (Fowler 1994, 104). This led to preparatory meetings in 1957, hosted by France. Twelve Western European countries were invited, and six of these decided to found UPOV. This represented an important step in the development of intellectual property rights, since it took

22 As stated at their web-page <www.wto.org/english/tratop_e/trips_e/art27b_background_e.htm>.

23 Countries that are already members of UPOV 1978 may apply for membership based on the 1991 Act when they have adjusted their legislation to that Act. Among the OECD countries the trend has been to do so. Norway has, however, decided to remain with UPOV 1978, with reference to farmers' rights.

24 Countries with economies in transition are not counted in this category.

plant variety protection out of the realm of patent law by creating its own distinct system, a *sui generis* system. With this, two things were achieved: recognition of plant breeders' rights, and at the same time the access of plant breeders to all kinds of plant varieties, whether protected by plant breeders' rights or not. The system did not affect traditional farming, since farmers could continue using their harvest for propagation and exchange, also if it originated from protected crops. This was a kind of open source system of intellectual property rights. However, as protection has become strengthened over the years, the system has lost more and more of that openness. The most significant change was introduced with the 1991 revision. Table 7.2 provides a comparative overview over the main provisions of UPOV 1978, UPOV 1991 and the requirements for patents under OECD patent laws.

Table 7.2 UPOV 1978, UPOV 1991 and OECD patent laws on selected topics

Provisions	**UPOV 1978**	**UPOV 1991**	**OECD patent laws**
Protection coverage	Plant varieties of nationally defined species or genera.	Plant varieties of all genera and species.	Inventions.
	Novelty, distinctness, uniformity, stability and variety denomination.	Novelty, distinctness, uniformity, stability and variety denomination.	Novelty, inventive step, industrial application, enabling disclosure.
Protection period	Minimum 15 years.	Minimum 20 years.	20 years.
Protection scope	Producing for purposes of commercial marketing, offering for sale and marketing of propagating material of the variety.	Producing, conditioning, offering for sale, selling or other marketing, exporting, importing, stocking for above purposes of propagating materials of the variety. If harvested materials are obtained through the unauthorized use of propagating material, certain acts are prohibited if the breeder has had no reasonable opportunity to exercise his right in relation to the propagating material.	In respect of a product: making, importing, offering for sale, selling and using the product; stocking for purposes of offering for sale, etc. In respect of a process: using the process; doing any of the above-mentioned in respect of a product obtained directly by means of the process.

Breeders' exemption	Yes, breeders are free to use a protected variety to develop a new variety. However, repeated use of the protected variety for the commercial production of another variety is not exempted.	Yes. However, in addition to the 1978 provision, essentially derived varieties and varieties that are not distinguishable from the protected variety are not included in the breeders‘ exemption.	No.
'Farmers' privilege'	Yes, farmers are implicitly free to use their harvested material for any purpose, also when it stems from a protected variety.	National governments are entitled to decide whether farmers shall be allowed – within reasonable limits and safeguarding the legitimate interests of the rights holder – to reuse the harvest of protected varieties on their own land holdings without the authorization of the rights holder.	No.
Prohibition of double protection	Yes. Any species eligible for plant breeders' rights cannot be patented.	No.	Up to national laws.

Source: Based on Dutfield (2000, 30), in turn based on Van Wijk et al. (1993), Helfer (2002, 30), and the 1978 and 1991 Acts of UPOV.

According to UPOV 1978, a member country must provide plant breeders' rights for at least five genera or species by the date of the entry into force of the Convention in its territory (Art. 4). The number of genera or species included should increase until at least 24 within eight years after the entry into force. UPOV 1991 includes all genera and species (Art. 3). New members are required to have such a system in place for at least 15 genera or species at the date of its entry into force in their domestic territory. Within 10 years, all genera and species must be included.

Patent systems are generally aimed at inventions, so the question of whether single plants, or new varieties of plants, can be regarded as 'inventions' has been heavily debated for years. Here we will simply note that inventions pertaining to new plants are recognized as patentable in many OECD countries, whereas plant varieties are as yet normally not patentable – the patent systems in the US and Japan constituting important exceptions.

The requirements for plant variety protection (so-called NDUS criteria: Novelty, Distinctness, Uniformity, Stability) are similar in the two versions of UPOV, but UPOV 1991 is more explicit and somewhat stricter than UPOV 1978. The variety must be new, that is, not sold to or disposed by others with the consent of the breeder (*novelty* criterion) (Art. 6 in both UPOV versions). It must be clearly distinguishable from other plant varieties (*distinctness* criterion) (Art. 7 in the 1991 Act; Art. 6 in the 1978 Act). The variety must be genetically uniform in its relevant characteristics (*uniformity* criterion) (Art. 8 in the 1991 Act; Art. 6 in the 1978 Act). These characteristics must remain unchanged after repeated propagation (*stability* criterion) (Art. 9 in the 1991 version; Art. 6 in the 1978 version). Finally, the varieties shall be designated by a denomination to enable their identification and ensure their registration; this denomination is to be used whenever a protected variety is offered for sale, also in other UPOV member countries (Art. 20 in the 1991 Act; Art. 13 in the 1978 Act).

Comparing with OECD patent law, we find certain parallels. The invention has to be 'new' (similar to the novelty criterion), represent an 'inventive step' (similar, but not identical, to the distinctness criterion) and be 'capable of industrial application' (similar to the uniformity and stability criteria). In addition, disclosure of the invention must be provided for, as in the UPOV system (in terms of breeders' exception, further explained below).

The protection period was minimum 15 years in the 1978 Act (Art. 8) for most plant varieties (18 years for trees and wines); this has been increased to minimum 20 years under the 1991 Act (Art. 19) for the same plants (25 years for trees and wines) and is similar to patent law.

An important difference between UPOV 1978 and UPOV 1991 lies in the scope of protection. For both acts, the effect of the right granted to the breeder is that prior authorization shall be required for the production of the protected variety for purposes of commercial marketing, the offering for sale and marketing as such. The breeder decides the conditions for authorization. A new element in the 1991 Act (Art. 14) is that any production and reproduction requires authorization from the rights holder, regardless of whether the intention is commercialization. Also new is that the rights holder's authorization is required for export and import, as well as stocking for any of the above-mentioned purposes. In addition, these provisions apply to harvested material, if the rights holder has not had reasonable opportunity to exercise their right. Finally, the provisions are extended to also cover varieties that are essentially derived from the protected variety, varieties that are not clearly distinguishable from the protected variety, and varieties whose production requires the repeated use of the protected variety. Here the 1991 provisions represent a huge step in the direction of the patent system, as can be seen from Table 7.2.

The important aspects in our context are the exceptions, since they indicate the scope for accessibility and utilization for breeders as well as farmers. First, there is the breeders' exemption. It allows any breeder to use a protected variety for further breeding and to market the new variety without requiring permission from the rights holder. The 1978 version explicitly stated that authorization from the breeder is not required for the utilization of the variety as an initial source of variation, for the purpose of creating other varieties or for the marketing of such varieties (Art. 5.3).

When the repeated use of the protected variety is necessary for the production of another variety, however, such authorization would be required. In the 1991 Act (Art. 15), the breeders' exemption is still valid, but it is further limited. If a new variety is so close to a protected variety as to be 'essentially derived', authorization from the rights holder is required; likewise if a variety is not clearly distinguishable from the protected variety. In patent law there are no such exemptions at all.

The second exemption applies to farmers, and here there is a great difference between the 1978 and the 1991 Acts.[25] According to the 1978 Act a farmer was free to use their harvest for planting or exchange with other farmers, also when it was the harvest of a protected variety. Whereas this was absolute in the 1978 Act, the 1991 Act (Art. 15) leaves it to national governments to define in law whether farmers shall be entitled to use their harvests without the permission of the rights holder, and if so, the conditions under which they are entitled to do so. This optional exception must be implemented within reasonable limits, and in such a way as to safeguard the interests of breeders. If farmers are permitted to use the product of the harvest for propagating purposes under national law, this must be limited to their own land holdings. Exchange of seeds from protected varieties is not allowed. This applies also if the variety is essentially derived from a protected variety, or if it is not clearly distinguishable from that variety. In this way, UPOV 1991 restricts the traditional practices of farmers to exchange seeds, when these seeds stem from protected varieties, or for example 'creole' varieties, if these are essentially derived from a protected variety or are similar to such.

The final important difference between the 1978 and the 1991 Acts concerns double protection. The 1978 Act (Art. 2.1) states explicitly that member states must choose between patent protection and plant variety protection if their national law allows both forms of protection for plant varieties. A plant variety is not to be given 'double protection'. This provision was removed in the 1991 Act, making it possible to apply for double protection of plant varieties. With this decision, the UPOV system is no longer such a clear alternative to patent law, the *sui generis* system envisaged in its early days. It can be used as an additional protection to patents. If a new international patent law treaty is adopted which includes PGRFA, this will further undermine the status of the UPOV system as a *sui generis* system.

From this comparison we see that UPOV has moved from being a distinct *sui generis* system for intellectual property rights, to a system closer to the patent systems. Whether UPOV 1991 can be termed a 'patent-like system', can be discussed. There are some significant differences, of which the most important is probably the breeders' exemption enabling further development of plant varieties for food and agriculture without the rights holders' authorization as long as they are not 'very

25 This is often referred to as the 'farmers' privilege'. In this book the term is avoided, since it is considered to have political connotations. Over the past 10,000 years it has been the customary right of farmers to reuse and exchange their seeds. In the UPOV terminology this right has been reduced to a privilege, which connotes an exemption provided to them by breeders. One could also turn this the other way around, saying that the breeders are the ones who have a privilege, as they have freely and without remuneration been allowed to use farmers' seeds to develop – over time – the cultivars used today.

similar' to a protected variety. However, the strengthening of breeders' rights under UPOV 1991 has clear parallels to patent law, so we may say that the UPOV system is getting closer to the patent system.

UPOV member states and the UPOV Secretariat maintain contact with and provide legal, administrative and technical assistance to the governments of a growing number of states, particularly developing countries that are considering membership in UPOV. Regional seminars are conducted where representatives from the countries in the region are invited to participate. This can make it difficult to distinguish between advocacy for UPOV membership, and the assistance applied for by the countries.

A Note on the WIPO System

As noted above, the TRIPS Council and the UPOV are interlinked with WIPO in various ways. The origins of WIPO can be traced back to the Paris and Berne Conventions adopted in 1883 and 1886, respectively (Dutfield 2000, 95).[26] The Convention Establishing the World Intellectual Property Organization was signed in Stockholm in 1967, and entered into force in 1970. WIPO became a UN Specialized Agency in 1974. The WIPO Convention has been revised twice, in 1979 and 2003. WIPO is headquartered in Geneva, and works to promote the development of measures for the protection of intellectual property rights and to harmonize national legislation in this field, perform the administrative tasks of the various unions, offer legal-technical assistance, and encourage the conclusion of international agreements designed to promote the protection of intellectual property (Art. 4). Today, most international conventions related to intellectual property rights – a total of 23 treaties, including the WIPO Convention – are administered by WIPO. As of March 2007, WIPO had 184 member states.

Within WIPO, the process with the potentially most comprehensive future impacts for developing countries is the Draft Substantive Patent Law Treaty (SPLT), being negotiated in the Standing Committee on the Law of Patents (SCP).[27] This is one of several processes introduced after the Director General of WIPO announced the new WIPO Patent Agenda in August 2001.[28] As drafted,[29] this treaty will contribute to the further harmonization of international patent law, in terms of *inter alia* uniform substantive patent law standards on prior art, novelty, utility and inventiveness, requirements relating to sufficient disclosure and the drafting of claims.

---

26 The Paris Convention pertains to the protection of industrial property, whereas the Berne Convention concerns the protection of literary and artistic works. Further conventions administered by WIPO are the Madrid Agreement Concerning the International Registration of Trademarks (1891), the Lisbon Agreement for the Protection of Appellations of Origin and their International Registration (1958), the Patent Cooperation Treaty (1970), and the Budapest Treaty on the International Recognition of the Deposit of Micro-organisms for the Purpose of Patent Procedure (1977). All these treaties have been revised or amended at least once.

27 For a thorough analysis of these processes, see Tvedt (2005).

28 WIPO Memorandum A/36/14 (2001).

29 WIPO Draft Treaty SCP/10/4 (2004).

The WIPO Intergovernmental Committee on Intellectual Property and Genetic Resources, Traditional Knowledge and Folklore (IGC) was established by the WIPO General Assembly in October 2000.[30] It is an international forum for debate and dialogue concerning the interplay between intellectual property, and traditional knowledge, genetic resources, and traditional cultural expressions (folklore). The IGC takes into account, and seeks to work complementarily to, related activities undertaken by other international institutions, such as the bodies of the CBD and the FAO. We return to the IGC in connection with our discussions in Chapter 8 on regime overlap.

Institutional Context: Clustered in the WTO

The TRIPS Agreement is one of the three main pillars on which the WTO is founded. As such, it is clustered with the other two pillars: the General Agreement on Tariffs and Trade (GATT) and the General Agreement on Trade in Services (GATS). All three agreements were adopted in 1994 as a package, together with the Agreement Establishing the World Trade Organization and various agreements on specific topics. The clustered regimes are thereby embedded in the WTO.

The WTO Cluster

According to the Preamble of the Agreement Establishing the World Trade Organization, the overall objective is to raise standards of living among the Parties, ensure full employment and a large and steadily growing volume of real income. The WTO seeks to expand the production of and trade in goods and services, based on utilization of resources in line with the objective of sustainable development, seeking to protect and preserve the environment. This is to be done in a manner consistent with the needs and concerns at the various levels of economic development. The developing countries, and especially the least developed among them, should be ensured a share in the growth of international trade commensurate with the particular needs of their economic development.

The Preamble goes on to state the desire to reduce tariffs and other barriers to trade, and eliminate discriminatory treatment in international trade relations. In other words, the immediate objective is to provide full competitive opportunities of trade on equal terms among member countries. The core question in debates on the WTO in civil society organizations as well as academe is whether its social and environmental goals can be achieved with full trade liberalization, as envisaged in the WTO agreements, or whether that is in fact detrimental to these goals. We will not delve into that discussion here, but focus on the TRIPS and how it is clustered and embedded in the WTO, in terms of its Article 27.3(b).

The WTO Agreement establishes a single institutional framework for all agreements covered by the WTO. The Ministerial Conference of the WTO, which meets at least biannually, is the highest decision-making body. A General Council

30 WIPO document WO/GA/26/6 (2000).

supervises the operation of the agreement and the ministerial decisions on a regular basis. This General Council has established subsidiary bodies for the three pillars of the WTO: the Goods Council, the Services Council and the TRIPS Council. The General Council also acts as a Dispute Settlement Body and a Trade Policy Review Mechanism.

The Dispute Settlement Mechanism is a cornerstone of the WTO. It was established under GATT long before the WTO came into being. In 1988 it was revised and strengthened, and has remained integrated in GATT. All dispute settlement relating to the TRIPS Agreement, as well as to other agreements, is to take place there. The General Council acts as the Dispute Settlement Body. The dispute settlement procedures are comprehensive and include panel reports and appellate reports to be adopted by the Dispute Settlement Body. Once these are adopted, the respective country must implement the recommendations within a given time frame. There are further rules on compensation to the offended party by the offending party. If the latter does not comply with the decisions made by the Dispute Settlement Body, it can be suspended from the WTO, which means that it would have no protection against trade sanctions or other disadvantaging trade practices. The Dispute Settlement Body can also decide to invoke trade sanctions towards an offending member country, in order to force the country to change its policies. With these enforcement mechanisms, the compliance procedures of the WTO are far stronger than those of the IU, the ITPGRFA or the CBD.

A member can propose an amendment to any of the provisions in any agreement to the Ministerial Conference, but the decision to submit the proposed amendment to the members for acceptance shall be taken by consensus (Art. 10 of the WTO Agreement). In practice, it is very difficult to make any changes in the texts of WTO agreements.

Decision-making in the WTO is based on consensus. The procedures that have emerged for arriving at consensus have been heavily criticized from many sides (see for example Narlikar and Wilkinson 2004) for their lack of transparency and integration of member countries. Green Room Meetings are frequently used. These are small group meetings called on the initiative of the Director General. The average number of participating countries is approximately 20, and almost always includes the 'QUAD' (Canada, Japan, the EU and the US). Little information about these meetings is normally provided to those not invited. Another prevalent form is the Friends of the Chair procedure, whereby the Chair appoints facilitators to organize discussions in specific areas. Several developing countries have complained about the lack of transparency in these processes, and the lack of geographical representation (ibid., 450).

When we consider the number of items negotiated, the number of more or less official meetings conducted, and the lack of transparency, it becomes evident that countries need huge delegations to keep abreast of developments. Also here we find great discrepancies between the developing and developed countries. Whereas the US and EU delegations are estimated to have had approximately 800 members each, the Nigerian delegation counted 12, the Central African delegation counted 3, and Barbados had 8 members at the Cancun meeting of the Ministerial Conference (ibid., 452).

Due to the deep controversies about the WTO, and the lack of democracy in its decision-making structures, considerable tension always attends the ministerial meetings, and large demonstrations often take place outside the buildings. However, if the WTO fails to progress in terms of implementation and new trade negotiations, strong actors like the US will concentrate on bilateral negotiations with individual developing countries – and that would mean a much tougher deal for the latter, as their bargaining position is strongest in coalitions. Such bilateral agreements are already emerging as a new factor in this complex picture.[31] Therefore it is important also for developing countries to keep the WTO negotiations on track.

Since the TRIPS Agreement opens for the possibility of *sui generis* systems of intellectual property rights over plant varieties, and the TRIPS Council refers to the UPOV Convention in this context, we will briefly examine the institutional context of UPOV.

Links to UPOV and WIPO

At its inception in 1961, the UPOV Convention was a separate regime and the Union an independent institution, but was later linked to WIPO in terms of administrative matters. Since then, the Director General of WIPO is the Secretary-General of UPOV, and is assisted by a Vice Secretary-General for UPOV. The UPOV Office is the secretariat of the Union and is located at WIPO headquarters in Geneva. It has a small international staff. The UPOV Council is the decision-making body. Each member country has one vote in the Council.

With regard to the WTO, WIPO helps to implement the TRIPS Agreement in terms of information sharing, and provision of technical assistance – for example, in preparing legislation, training, institution building, and in modernizing intellectual property systems as well as enforcement,[32] particularly for developing countries. WIPO is thus relevant in our context, not only because it contributes to administering the UPOV Convention, but also because it has a co-operation agreement with the WTO and works to strengthen intellectual property protection in the world.

Potential Effects of the Institutional Structure

The TRIPS is thus deeply embedded in the WTO, and closely clustered with the other agreements within the WTO umbrella. International implementation of Article 27.3(b) will be determined by this whole system with its strong enforcement procedures. In addition it is interlinked with the UPOV and the WIPO system, both of which promote the strengthening of intellectual property rights. The way this link has taken shape might contribute to raising the minimum standards for the protection of intellectual property rights over PGRFA to a higher level than that envisaged in Article 27.3(b).

---

31 For more information on bilateral agreements, see the section titled 'A Note on 'TRIPS-Plus' Agreements', at the end of the chapter.

32 Joint WTO-WIPO Press release, 21 July 1998.

Resulting Norms and Rules: Intellectual Property Rights

Before turning to the norms of the TRIPS Agreements and the UPOV Convention, we should stress an important difference between the IU/ITPGRFA and CBD on the one side, and the TRIPS Agreement and the UPOV Convention on the other: as we have seen, enforcement procedures are much stronger in the latter category than in the former.

Core Norm Pertaining to PGRFA: Effective Sui Generis Protection for Plant Varieties as a Minimum Standard

The TRIPS Agreement includes a long list of norms and rules, but here we focus on relevance for PGRFA. As noted, countries belonging to the WTO are, according the TRIPS Agreement, free to choose whether to include plants in their patent systems or not. Under the TRIPS Agreement, the core norm pertaining to PGRFA is that plant varieties must be protected by a type of system that is effective.

The wording of the core norm provides flexibility with regard to forms of protection. It is possible to apply a patent system, it is possible to apply an effective *sui generis* system, and it is possible to provide double protection with patents and a *sui generis* system combined. Since TRIPS provides minimum standards for the protection of intellectual property rights, the crucial question is what the TRIPS Council will accept as a minimum standard for an effective *sui generis* system. In this regard, the TRIPS Agreement gives some guidance, which we will discuss in light of ongoing review negotiations in the TRIPS Council and the foregoing process leading to the adoption of the TRIPS Agreement.

The formulation in Article 27.3(b) reflects the lack of consensus among the negotiating parties on the form of protection to be set as a minimum standard in the TRIPS Agreement (Correa 1998, 197). At the time of the negotiations, plant varieties were protectable in terms of plant breeders' rights; they were not patentable in Europe,[33] whereas in the United States and in Japan they were patentable, and still are. Additionally, in the US it was, and still is, possible to protect a variety with plant breeders' rights and patents, that is, a combination of both ('double protection'). In the vast majority of developing countries, plant varieties were neither patentable nor protectable in terms of plant breeders' rights or other intellectual property rights. The text of Article 27.3(b) is in its formulation a compromise between the European system on one side and the US and the Japanese system on the other. The minimum standard is set much higher than common practice in developing countries.

It should be noted, though, that the text does not explicitly refer to plant breeders' rights as a *sui generis* system. Subsequently, there is no reference to the international convention in force with regard to the intellectual property protection for plant varieties, the UPOV Convention. If member countries develop other systems, which are considered effective by the TRIPS Council, these systems would also have to

33 They are still not patentable, according to the EU patent directive. However, there has been heated debate as to whether new legislation on patents in the EU countries will in practice enable the patentability of plant varieties.

be accepted as *sui generis* systems. Many civil society organizations and scholars regard this possibility as an option for developing countries (for example CIPR 2002; Helfer 2002; Correa 1998; GRAIN 1998 and 1997; Leskien and Flitner 1997), as their farming systems are so different from those of developed countries, and because most of them have no tradition of plant breeders' rights or patents in the field of plant varieties.

So far, however, the UPOV model appears most prevalent in implementing the TRIPS Agreement in developing countries when it comes to Article 27.3(b). The most likely reasons for that seem to be:

- that such a model appears to have the best prospects for being accepted by the TRIPS Council;
- that it is advocated by the UPOV and some of its member countries;
- that UPOV and WIPO offer technical and administrative assistance for developing such legislation;
- that it is regarded as a shortcut to compliance with the TRIPS Agreement, as compared to the development of a *sui generis* system in countries with scarce legal and administrative resources.

In order to determine, as precisely as possible under the current conditions, what the norm of an effective *sui generis* system means as a minimum standard of the TRIPS Agreement, we will first identify the principles to be followed in developing a *sui generis* system. Thereafter we discuss the UPOV model as one option for complying with Article 27.3(b), followed by a discussion of various possibilities of other potential *sui generis* systems.

Basic principles for a minimum protection of intellectual property rights over plant varieties The TRIPS Agreement provides some guidance in provisions other than Article 27.3(b) for the framing of a *sui generis* system for the protection of plant varieties. These provisions must be considered the absolute minimum requirements for an effective *sui generis* system. Leskien and Flitner (1997) and Helfer (2002) have summarized these minimum requirements for domestic *sui generis* legislation as follows:

First of all, TRIPS has no definition of 'plant variety', so it does not allow any limitation as to number of varieties or species of plants. Thus it applies to all plant varieties in all species and botanical genera, and this will have to be reflected in national legislation. If UPOV 1978, which provides for a minimum number of species or genera to be protected, is used as a model, domestic legislation may have to be changed on this point, depending on how the limitation on species or genera is formulated.

Second, a *sui generis* system has to be a system of intellectual property rights. As a point of departure, the whole Agreement is about intellectual property rights. Furthermore, Article 1(2) of the TRIPS Agreement specifies that the term 'intellectual property' is to refer to 'all categories of intellectual property that are the subject of Section 1 through 7 of Part II'. As Article 27.3(b) is one of the components in this part of the TRIPS Agreement, it clearly refers to intellectual property rights.

Finally, Article 68 on the TRIPS Council provides for monitoring and consultations on intellectual property rights. Obviously, it is not possible to review legislation unless notification has been given. Since the TRIPS Council has explicitly requested its member countries to notify of their *sui generis* laws for the protection of plant varieties, with reference to Article 63(2) on notifications of law, it has already indicated that it regards the *sui generis* option as a system for intellectual property protection (Leskien and Flitner 1997, 28). This conclusion implies that governments must ensure that rights holders are granted the right to exclude all third parties from engaging in specified activities concerning the protected variety – or at least the right to receive equitable remuneration for such activities.[34]

A further specification is that a *sui generis* system must be a system of intellectual property rights additional to those already dealt with in the TRIPS Agreement. Members cannot use the trademark system as a *sui generis* system, since their obligations in this regard are already covered under Article 15. The same goes for geographical indications, as their obligations are covered under Article 22; likewise for trade secrets, covered by Article 39. A *sui generis* system must literally be a system 'of its own kind'.

Third, national treatment has to be granted to all applicants for intellectual property rights under any *sui generis* system, whether they are nationals or citizens of any other WTO member country. This goes back to Article 3 of the TRIPS Agreement, which provides for national treatment. Countries using the UPOV model as a *sui generis* system will have to extend the principles of reciprocity from applying to citizens of all UPOV member countries to applying to citizens of all WTO member countries.

Fourth, account must be taken of the Most-Favored-Nation treatment provided for in Article 4 of the TRIPS Agreement. This implies that national laws must ensure that any privileges or immunity granted to any other country must be accorded immediately and unconditionally to all other WTO member countries. However, Leskien and Flitner (1997, 31) maintain that this obligation is of minor practical relevance: only in exceptional cases would countries grant better conditions to foreigners than to their own nationals.

Finally, fifth, any *sui generis* system must include the regulation of enforcement of rights by private parties. This is provided for in Articles 42–49 of the TRIPS Agreement. Leskien and Flitner (1997, 32) argue that this is most likely the meaning of 'effective' as it pertains to *sui generis* systems, since this term is mostly applied in Articles 42–49, except for in Article 27.3(b). Any *sui generis* law must contain procedures that enable breeders to enforce the rights granted to them. Thus it is the rights holders that are responsible for uncovering any infringements of their rights, and for taking appropriate action through the judicial system.

Helfer (2002, 34) asks whether there are any additional requirements for an effective *sui generis* system other than those provided here. He maintains that international tribunals have concluded that rights granted in a treaty must be

34 For further discussion on the question of whether *sui generis* systems have to be systems of intellectual property protection, see Leskien and Flitner (1997, 28) and Helfer (2002, 32).

interpreted to make them effective, if not, they are considered illusory (Helfer 1998, 403). On this basis, it is reasonable to assume that states that have implemented a *sui generis* system that follows the five principles dealt with above are unlikely to have their laws challenged successfully by the TRIPS Council.

The UPOV model as a sui generis option for intellectual property rights over plant varieties As noted above, the UPOV model is actually two models – one based on the 1978 Act and one based on the 1991 Act. The core norm in the UPOV 1978 Act is that plant breeders' rights must be granted for plant varieties that are novel, distinct, uniform and stable, as exclusive rights to control production for commercial marketing, offering for sale and the marketing of the protected variety. Thereby breeders are free to use the material for further breeding, and farmers are free to use their harvest of the protected variety for any purpose. A secondary norm is that double protection with plant breeders' right and patents is prohibited.

The core norm in the UPOV 1991 Act is that plant breeders' rights have to be granted for plant varieties that are novel, distinct, uniform and stable, as exclusive rights to control production for commercial marketing, offering for sale and the marketing of the protected variety, including the harvest thereof. Breeders are free to use the material for further breeding under certain conditions – but farmers are no longer entitled to freely use the harvest of a protected variety. They may be entitled to reuse the seeds from their harvest on their own land holdings within reasonable limits if the national governments decide so, but are not allowed to offer them for exchange or sale. National governments may also decide that farmers are not entitled to reuse the seeds from their harvest, or that they must pay royalties to the rights holder if they do so.

Other sui generis options for developing countries From the five minimum requirements listed above, there are many options available to developing countries – such as combining the system with obligations from other international regimes like the IU/ITPGRFA, for example in terms of farmers' rights, and the CBD, in terms of the protection of traditional knowledge. Several countries (among them India as well as Zambia, Thailand, Costa Rica and Bhutan) have sought such solutions. Instead of going into these solutions here, we will briefly outline the spectrum within which a *sui generis* system is possible (based on Leskien and Flitner 1997).

There is no requirement that plant varieties must be novel, distinct, uniform, and stable to be eligible for protection. A *sui generis* system could extend protection also to genetically heterogeneous varieties, enabling farmers to apply for intellectual property protection as well. Instead of uniformity, stability and distinctness, the requirement could be, for example, that varieties must be identifiable and useful, with the meanings of those criteria carefully defined.

Countries developing a *sui generis* system are free to decide what material is to be protected, as long as the system covers all plant species. For example, they can choose whether to protect only propagating material, or also the harvested material stemming from a protected variety. They may distinguish between reproductive and vegetative propagating material, and in addition they can determine a shorter or longer duration for the protection period.

Finally, there is a great scope for variety in defining the acts requiring prior authorization from the rights holder. For example, the breeders' exemption can be maintained and farmers can be granted the right to use their harvest as they deem, as long as they do not engage in the commercial marketing and sale of a protected variety.

In addition, it is possible to add a requirement as to disclosure of origin of the genetic material used in developing the variety, as well as a requirement for prior informed consent from those who provided it according to their national legislation.

This discussion has shown that the core norm of the TRIPS Agreement with regard to intellectual property rights to plant varieties is a standard considerably lower than those currently being implemented with reference to UPOV. The TRIPS Agreement, as it is formulated, permits a variety of solutions based on the minimum standards given. Whether such other solutions would be accepted by the TRIPS Council remains an open question.

Secondary Norm: Patenting of Animals and/or Plants is Optional

Countries may decide whether or not to include plants (and animals) in their patent systems. Most developing countries have chosen not to include plants. There is also the possibility of including components of plants, such as genes, in patent systems. This is a contentious issue as such patents could have effects also for the use of plant varieties. Legislation varies in developing countries in this regard. It has been considered beyond the scope of this book to go further into the highly complex discussion on patents, due to space limitations.[35] The emphasis of this book is on plant variety protection as this is currently the most urgent intellectual property rights issue in developing countries with regard to PGRFA management.

Core Regulatory Rules

The regulatory rules of the TRIPS Agreement concerning intellectual property rights to PGRFA are all core rules: there are no rules that are optional in this regard. As noted, the TRIPS Agreement obliges WTO member developing countries to implement its provisions.

Implementation of Article 27.3(b) must include *enforcement procedures*, as required in the general enforcement obligations set out in the TRIPS Agreement. Such procedures must permit effective action against any act of infringement of intellectual property rights (Art. 41.1) and be fair, equitable and not unnecessarily

35 Developing countries may be affected by patents in Northern countries, limiting their export opportunities to the countries where such patents are held, or – if patents are claimed in developing countries – stakeholders in these countries may be affected. However, there have been extremely few cases, each of which is highly complex. Among the more well known examples is the US yellow bean patent, based on a Mexican bean variety, which has provided serious barriers to bean exports to the United States from Mexican farmers (Andersen and Tvedt 2002).

complicated, costly or time-consuming (Art. 41.2). The judicial authorities shall have the authority to require *infringers to pay damages* to compensate the rights holder for injuries suffered due to infringements, including attorney's fees (Art. 45). Member governments are required to provide for *criminal procedures and penalties,* at least in cases of willful trademark counterfeiting or copyright piracy on a commercial scale (Art. 61). Reactions may include imprisonment and/or monetary fines. Such reactions may also be applied in other cases of IPR infringement if this is done 'willfully and on a commercial scale'.

The TRIPS Agreement provides for *technical co-operation* (Art. 67). It specifies that developed country members shall provide technical and financial assistance to developing countries and least developed countries, upon request and on mutually agreed terms. Such co-operation is to include assistance in the preparation of laws and regulations on the protection and enforcement of intellectual property rights. They are also to include support for the establishment of domestic offices and agencies relevant to these matters and to the training of personnel.

Countries that opt to implement Article 27.3(b) of the TRIPS Agreement in accordance with the *UPOV Convention* are free to choose whether they will apply for membership in UPOV. If they do so, they will be required to implement many regulatory rules, which by far exceed the minimum standards of the TRIPS Agreements. These include comprehensive legal remedies, an authority entrusted with the task, and systems for public information. Furthermore they involve detailed and specified criteria on conditions for protection and requirements for applications and examination thereof, and a series of specified rules for the scope and duration of breeders' rights as well as procedures for variety denomination. Since many countries choose this path in implementing Article 27.3(b) it is listed in the overview in the next section. However, it should be borne in mind that this is only one of several options for complying with Article 27.3(b) of the TRIPS Agreement.

Overview of Norms and Rules of the TRIPS Agreement and the UPOV Convention

Table 7.3 provides an overview of the norms and rules of the TRIPS Agreement pertaining to the management of PGRFA and of the UPOV Convention, since the latter is the most frequently used model for *sui generis* systems of TRIPS-related intellectual property rights to plant varieties.

Table 7.3 Overview of norms and rules of the TRIPS Agreement and the UPOV Convention pertaining to PGRFA

	TRIPS Agreement	UPOV Convention (1991 Act)
Core norms	• An effective *sui generis* system for the protection of plant varieties, as a minimum standard.	• Protection of breeders' rights over plant varieties which are new, distinct, uniform and stable.
Secondary norms	• Member countries may exclude from patentability plants and animals and essential biological processes for their production. Micro-organisms may, however, not be excluded.	• Within reasonable limits and safeguarding the interests of breeders, member countries may permit farmers to use products of their harvest that originate from protected varieties, for propagating purposes on their own lands.
Core rules to be implemented by governments **(no secondary rules)**	• Developing countries to implement Article 27.3(b) in terms of legislation and institutional arrangements by 2000, and least developed countries by 2006. • Various specified procedures to be introduced at the national level to ensure legal enforcement of the legislation.	• Contracting Parties to adopt all measures necessary, including legal remedies, an authority entrusted with the task, and systems for public information. • Detailed criteria on conditions for protection and requirements for applications and examination thereof. • Detailed rules for the scope and duration of breeders' rights and detailed procedures for variety denomination.

A Note on 'TRIPS-plus' Agreements

Parallel to the negotiations in the WTO and the implementation and revision of the TRIPS Agreement, another process has been taking shape. Industrial countries and groups of such countries often negotiate bilateral and regional trade and investment agreements with individual or groups of developing countries. Under such constellations, the developing countries have a considerably weaker bargaining position. Increasingly, bilateral and regional agreements between industrialized and developing countries have come to include provisions on intellectual property rights. In general, developing countries find that they have to introduce intellectual property rights at higher standards than those provided for under the TRIPS Agreement, and they are often obliged to join UPOV 1991 by a certain deadline – which is why these agreements are often called 'TRIPS-plus' agreements (Vivas-Eugui 2003). The US, the European Free Trade Association and the European Union are among the active proponents and initiators of bilateral and regional agreements, and have established such agreements with many developing countries.

In the long run, this may pave the way for more stringent international regimes on intellectual property rights, for example in the form of the new Substantive Patent Law Treaty currently being negotiated under WIPO (see Musungu and Dutfield 2003; Corea 2004; and Tvedt 2007).

Chapter 8

Regime Overlap, Interaction and Resulting Constellations

Drawing on the preceding three chapters, we will now focus on regime overlap, interaction and the resulting regime constellations with which developing countries are faced. We start out with an overview of the driving forces behind the international regimes and their mechanisms of influence, as a basis for the analysis. With this as a background, we will turn to the overlap and interaction between the international regimes in greater detail, focusing on two and two regimes. First we will analyze the overlaps and interaction between the IU/ITPGRFA and the CBD, then we will turn to the CBD and TRIPS/UPOV, and finally to TRIPS/UPOV and the IU/ITPGRFA. For each set of regimes, we will begin with an overview of the interaction between the two regimes in a time perspective, seeking to identify synergies, barriers and strategic opportunities. Against this background, we then identify overlaps between the two regimes, and proceed to systematically analyze interaction between them with regard to each of the identified overlap-items. Finally, we draw together our findings on the resulting overall regime constellations with regard to PGRFA management and derive assumptions on what can be expected to be the aggregate effects for developing countries.

Regime Interaction: Driving Forces and Mechanisms of Influence

From the history of regime formation and interaction regarding PGRFA management, it appears that developments have been dominated by two main fronts, with some bridge-builders and issue-specific factions in-between. On the one side, there were a few powerful industrialized countries, among them the US, Japan, Australia and Canada; on the other, a large majority of developing countries.

In the case of the TRIPS Agreement, national and multinational corporations were the driving forces behind the group of industrialized countries that worked for the inclusion of an agreement on intellectual property rights in the WTO, and more specifically to include provisions on PGRFA. These forces possessed power resources, particularly coercive force, as became visible when the developing countries finally accepted the TRIPS Agreement, as described in Chapter 7. Structural power is seen as the main mechanism of influence: the constellations of driving forces, their use of power resources and how this shaped and determined the structures of the global political economy within which developing countries, their political institutions, their economic enterprises and their professional people have to operate – all indicate structural power at work.

NGOs were decisive in initiating and pushing for negotiations towards the CBD, and were as such the initial driving forces. When formal negotiations started, certain coalitions had been formed between NGOs and responsible government entities, particularly around the issue of conservation (largely Northern countries) and around benefit-sharing (largely Southern countries). Policy core beliefs about the value of conservation, or the justice of benefit sharing, formed the glue that held these coalitions together. Some country delegations acted as brokers – the Nordic countries, for example – all of this indicating that CBD negotiations had clear patterns of advocacy coalitions. The negotiation focus changed from a conservationist approach to a more holistic approach that included sustainable development and equity aspects, due largely to a process of learning and norm diffusion. Learning and norm diffusion are seen as the main mechanisms of influence here.

There were also NGOs behind the IU and the ITPGRFA, but FAO member states seem to have been the main driving forces. Initially there were two main groups, but the actor landscape diversified during the negotiations that led to the ITPGRFA. Also the interests of the various groups changed along the way. There are no clear patterns of any specific main mechanism of influence: several mechanisms seem to have been at work. Financially strong industrialized countries were first skeptical to the IU, and demanded that intellectual property rights be recognized – which indicates elements of coercive force. But there were also elements of learning and norm diffusion, as with the *State of the PGRFA report* (FAO 1998), which helped to create an understanding of the critical situation for PGRFA. In addition, international implementation of the IU, as well as negotiations on the ITPGRFA (and the follow-up after its adoption) struggled due to scarce institutional capacity as to funds as well as changing delegations and general lack of political clout behind the negotiations. We can therefore see patterns of all three mechanisms of influence indicated in Chapter 3.

The groups of countries used the negotiation forums in various ways. Here we can detect clear elements of 'forum shopping'. Initially, the developing countries fought for recognition of PGRFA as a common heritage of mankind, accessible to all, and to which no intellectual property rights should be possible, and used the IU negotiations as a forum to voice their demand. The most economically powerful Northern countries were less pleased about this principle, as it would not be in line with their regulations on intellectual property rights. However, many years later they were to be the ones who advocated accessibility of PGRFA, using the ITPGRFA negotiations as a forum – after developing countries had started introducing CBD-derived access legislation that reduced such accessibility. Now the fronts had reversed fully, and the developing countries were the ones skeptical to free access – as long as there were not sufficient provisions to ensure benefit sharing. These concerns were voiced first in the CBD negotiations and later in the negotiations leading to the ITPGRFA.

Initially, developing countries saw intellectual property rights to PGRFA as a new form of imperialism to which breeders in their countries would have little or no access; the North would be able to exploit the genetic resources of the South for profit, without giving anything in return. When, in the late 1980s and early 1990s, the developing countries lost in the battle against intellectual property rights in WTO negotiations, the demand for fair and equitable sharing of the benefits from the use of genetic resources arose as a response voiced in the CBD and later in the ITPGRFA. Developing countries demanded that accessibility should be made dependent on benefit-sharing arrangements – on a bilateral basis – and in practice discarded the 'common heritage of mankind' principle they had fought for in the IU.

As seen in Chapter 7, industrialized countries have sought to address PGRFA-related intellectual property rights, discussed under the TRIPS Agreement, also in the WIPO, since little progress has been made on Article 27.3(b) in the WTO, due to sharp criticism from developing countries. At WIPO, negotiations of the draft Substantive Patent Law Treaty may lead to new constellations regarding intellectual property rights over PGRFA.

It follows that different countries have opted for different forums to voice their interests. Such choices are normally made on the basis of their considered basic game power as well as negotiating power.[1] Economically powerful industrialized countries were involved in all negotiations, but chose TRIPS as their main arena, and are increasingly involved in the WIPO. Developing countries have also been involved in all processes, but were particularly successful in connection with the adoption of the IU and the CBD. In recent years, they have struggled to keep the debate on PGRFA-related intellectual property rights on the table in the TRIPS Council, while industrialized countries have been trying to get it moved to WIPO.

An additional element not taken into account by the forum-shopping perspective is how the composition of delegations affects negotiations. At the FAO, delegations are dominated by representatives of ministries of agriculture; at the CBD-COP, delegations are dominated by representatives from ministries of the environment; and the WTO is dominated by delegations from ministries of foreign affairs and of trade. How harmonized the various delegations from a country are depends on the extent to which their countries have developed policies and mandates on the issue – and how precise these are. Due to scarce institutional capacity in developing countries, it is not self-evident that all delegations from a country will speak with one voice. In this picture also the leadership[2] of certain delegation heads may be decisive in shaping negotiations and creating breakthroughs.[3] Differences between delegations from the same countries and their composition in terms of leadership may contribute to explaining the differing rationales behind the regimes and the political clout behind the negotiations – and thus also regime interaction.

1 Basic game power refers to the parties' control over the resources in question as well as their economic and technological capacity to make use of these resources, whereas negotiation power refers to capabilities based on strength in numbers, coalition and leadership (Underdal 1997, 17). As this aspect has been analyzed in detail in Rosendal (2000), it will not be further pursued here.

2 Young (1991) distinguishes between various types of leadership, such as entrepreneurial and intellectual leadership, as factors that help to explain regime formation.

3 As an example, Batta Bjørnstad (2004, 64) describes how Cathleen Enwright from the US delegation sought a breakthrough of the ITPGRFA negotiations with regard to farmers' rights, stating 'For proposing this, I will probably lose my job.'

Overlaps and Interaction between the IU/ITPGRFA and the CBD

We will now proceed to the interaction of, and functional overlaps between, two and two international agreements. We will start out with overviews before going into the details of the single items of regime overlap.

Interaction of the IU/ITPGRFA and the CBD in a Time Perspective

Figure 8.1 summarizes and compares the development stages set out in the respective sub-chapters on the historical contexts of the two regimes.

<table>
<tr><td rowspan="2">IU/ ITPGRFA</td><td rowspan="2">Agenda setting</td><td rowspan="2">Nego- tiation</td><td colspan="10">Implementation phase for the IU</td><td>A[a)]/ interim period[b)] ITPGRFA</td></tr>
<tr><td colspan="3">A[c)]/Implementation</td><td colspan="2">Agenda setting</td><td colspan="6">Negotiations on revision of the IU</td></tr>
<tr><td>Timeline</td><td>79-80</td><td>81-82</td><td>83-84</td><td>85-86</td><td>87-88</td><td>89-90</td><td>91-92</td><td>93-94</td><td>95-96</td><td>97-98</td><td>99-00</td><td>01-02</td><td>03-04</td></tr>
<tr><td>CBD</td><td></td><td colspan="4">Agenda setting</td><td colspan="2">Negotiations</td><td>A[d)]</td><td colspan="5">Implementation phase for the CBD</td></tr>
</table>

Figure 8.1 Comparing development stages: IU/ITPGRFA and CBD
a) Adoption of the International Treaty on Plant Genetic Resources for Food and Agriculture
b) The ITPGRFA entered into force 29 June 2004, and thereby also the implementing phase
c) Adoption of the International Undertaking on Plant Genetic Resources
d) Adoption of the CBD and ratification period

Three main features stand out with regard to interaction between the two regimes during these development stages:

- *1979–1992*: Until the adoption of the CBD, there was little interaction between the two regimes. They developed from different angles, and PGRFA did not become an issue with regard to the CBD until the formal negotiations started.
- *1992–2001*: Through Resolution 3 of the Nairobi Final Act on the adoption of the CBD in 1992, the negotiating parties initiated a new round of negotiations on the IU. In a way the CBD overruled the IU. On the other hand, it provided the impulse needed to renegotiate the IU in the form of a legally binding regime on PGRFA, and to resolve the outstanding issues highlighted in the Nairobi Final Act.
- *2001–2004*: Adopted in 2001, the resultant ITPGRFA represents a solution to several of the challenges raised with the CBD, operationalizing its goals with regard to PGRFA management and leading to a division of labor between the two regimes, which has develop further in the years thereafter.

Generally, the interaction between the two regimes in the period between 1992 and 2001 was characterized by some synergy, but was dominated by serious barriers. The period after 2001, in turn, has largely been characterized by synergy, but with some remaining barriers – which are, however, not insurmountable.

'Interaction with barriers' refers to the challenges involved in harmonizing an international agreement for agriculture with an international agreement for the environment with emphasis on non-domesticated biodiversity. As we have seen, core provisions of the CBD did not fit with the particular conditions and needs for PGRFA management. Therefore negotiations took time, attention and resources. The ITPGRFA was not adopted until almost 10 years after the adoption of the Nairobi Resolution. Due to these lengthy negotiations, the IU remained formally in force but was not followed up – and little happened to realize the CBD objectives with regard to PGRFA management. This vacuum situation meant that the erosion of PGRFA could continue unabated, without proper international regulation, for twelve years.[4]

Great uncertainty arose with regard to the international regulation of PGRFA, due to an unclear division of labor during this period and expectations that a new international agreement would soon be in place. The division of labor between the entry into force of the CBD and of the ITPGRFA was characterized as follows:

- The CBD related to gene-bank material collected after its entry into force, whereas the bulk of the material collected prior to that sorted under the FAO.
- Farmers' rights were the purview of the FAO, whereas *in situ* conservation, such as plant varieties in farmers' fields, was covered by the CBD.

In legal terms this division of labor between the CBD and the IU might have seemed clear, but in practical terms this was not the case. It was difficult to determine whether a plant variety or a genetic trait belonged to the 'before CBD' or the 'after CBD' category. Among the reasons for that were the differing information routines in gene banks, including a lack of information in many cases. Moreover, accessions collected prior to the CBD could still be grown in *in situ* conditions, and would thereby apply to both regimes. An important aspect of farmers' rights is that these are related to genetic material grown in their fields. Thus, farmers' rights and *in situ* conservation were and always will be closely linked. Therefore it was highly uncertain what the CBD actually covered in the period until the ITPGRFA entered into force, and consequently in what cases CBD provisions like access and benefit-sharing rules would apply.[5]

In 1995, the Conference of the Parties to the CBD – after some uncertainty – affirmed that agricultural biodiversity was covered by the objectives and provisions of the

4 The ITPGRFA entered into force in June 2004, but it took more than 12 years to get a new international agreement on PGRFA in place.

5 The IARCs solved these problems by developing an Interim Standard Material Transfer Agreement regulating the transfer between their gene banks and PGRFA recipients; see below.

Convention.[6] Since then, the COP-CBD has followed the negotiations under FAO, and sought to encourage their finalization. Now that the ITPGRFA is in place, this initiative by the CBD negotiators, and the COP follow-up, must be regarded as interaction for mutual benefit. The FAO got a binding treaty on PGRFA, and the parties to the CBD got a treaty that operationalized its own principles with regard to PGRFA.

With the ITPGRFA, there are new possibilities for interaction between the two regimes. As we will see below, most of the barriers have been solved. However, there are still remaining hurdles at the multilateral as well as domestic level. The main problems at the multilateral level pertain to the division of labor resulting from the Multilateral System of the ITPGRFA, and crops and other relevant plants not covered by that system (see Chapter 5). The main obstacles at the domestic level relate to legislation which has already been introduced to implement the CBD, and which may need to be changed with regard to PGRFA. Especially in developing countries, legislative processes are demanding in terms of scarce financial and institutional resources, and a new round on these acts and regulations so soon after their adoption may not receive top priority.

In a time perspective, we see that the ITPGRFA, which was in an early stage of development (agenda setting and negotiations), was influenced by the CBD, which was in a later stage of development (adoption and implementation). The interaction with barriers was a result of a conflict between the interests related to the management of non-domesticated and domesticated genetic resources, which were difficult to unite under the framework of joint norms. Thereby the established regime, the CBD, determined the direction of the newcomer, the ITPGRFA. The synergetic interaction was the result of a learning situation where the diffusion of norms and ideas from one regime (the CBD) influenced the material content of another (the ITPGRFA) so that they were adapted to the particular conditions of the other regime. This process was challenging, and it was not certain that the negotiators would succeed. Because they did succeed, we can look back to a synergetic interaction after overcoming most of the barriers.

Resulting Overlaps between the IU/ITPGRFA and the CBD: Overview

As a result of the dynamics described above, there is comprehensive overlap between the functional scope of the IU/ITPGRFA and the CBD. In fact, all aspects of the functional scope of the IU and the ITPGRFA are totally covered by that of the CBD, since both agreements concern PGRFA management, including conservation, access and benefit sharing.[7] However, the functional scope of the CBD is larger: it applies to all other components of biological diversity as well (except for human biodiversity), whereas the functional scope of the IU and the ITPGRFA is more detailed in terms of regulatory rules.

---

6 Secretariat of the Convention on Biological Diversity (2001), 313, Decision I/9/1995, Annex, points 5.9 and 6.3. This was also decided for coastal and marine biological diversity (ibid., point 5.3).

7 The IU did not originally address benefit sharing, but aspects of benefit sharing are dealt with in the agreed interpretations of the IU in 1989 (FAO Conference Resolution 4/89, paras 4 and 5, and FAO Conference Resolution 5/89).

Table 8.1 Overlapping norms and rules: IU/ITPGRFA and CBD

CBD → **IU/ITPGRFA ↓**	**Core norms:** *– Conservation, sustainable use* *– Benefit sharing*	**Secondary norms:** *– Access* *– Recognition of traditional knowledge*	**Regulatory rules:** *– Domestic policy and legal measures* *– GEF (international)*
Core norms: *– Conservation, sustainable use* *– Access*	1.1 Compatibility with regard to conservation and sustainable use (both) 1.2 Divergence with regard to access (IU/ITPGRFA) and benefit sharing (CBD)	1.3 Divergence with regard to access (both) 1.4 Compatibility with regard to traditional knowledge (CBD), and conservation and sustainable use (IU/ ITPGRFA)	1.10 Compatibility with regard to conservation and sustainable use and GEF 1.11 Divergence with regard to regulatory rules for *ex situ* conservation
Secondary norms: *– Benefit sharing* *– Farmers' rights* *– Limited intellectual property rights (no interaction with the CBD)*	1.5 Compatibility with regard to benefit sharing (both) 1.6 Compatibility with regard to benefit sharing (CBD) and farmers' rights (IU/ ITPGRFA) 1.7 Neutral overlap on conservation (CBD) and farmers' rights (IU/ITPGRFA) but with possibility of barriers	1.8 Compatibility with regard to farmers' rights (IU/ITPGRFA) and traditional knowledge (CBD) 1.9 Compatibility with regard to benefit sharing (IU/ ITPGRFA) and access (CBD)	1.12 Divergence with regard to division of labor on benefit sharing before the ITPGRFA, compatibility after 1.13 No binding regulatory rules on farmers' rights and little interaction between the two regimes hitherto
Regulatory rules: *– Domestic policy and legal measures* *– International networks for* ex situ *conservation* *– Multilateral system (ITPGRFA)* *– Funding strategy*	1.14 Compatibility with regard to norms and rules for conservation and sustainable use 1.15 Compatibility with regard to norms and rules for *ex situ* conservation 1.16 Compatibility and divergence with regard to the multilateral system and the norm of benefit sharing	1.17 Compatibility and divergence with regard to norms and rules on access regulation 1.18 Limited compatibility between traditional knowledge and IU/ITPGRFA regulatory rules, since there are no binding rules on farmers' rights	1.19 Co-ordination of domestic policy and legal measures is not provided for 1.20 Compatibility between the funding strategy and the GEF provisions of the CBD

The CBD and the IU/ITPGRFA have more or less the same countries as parties/signatories.[8] They also have basically the same constellations of parties/signatories with regard to interests. However, the two regimes have emerged from different backgrounds, rationales and interests, as explained in Chapters 5 and 6. Whereas the IU emerged from an agricultural rationale, the point of departure for the CBD was nature conservation. This means that the IU and the ITPGRFA are basically agricultural agreements, whereas the CBD is an environmental convention that takes into account aspects of sustainable development – in terms of requirements for sustainable use of resources and for the fair and equitable sharing of benefits from such use. Therefore, the basic interests behind the two regimes differ. The vital concern behind the IU and the ITPGRFA is to make PGRFA available for present and future agriculture (with benefit sharing as a secondary norm), whereas the vital concern behind the CBD is to maintain biological diversity and provide for the fair and equitable sharing of benefits arising from the use of its components.

Table 8.1 provides an overview of overlaps in functional scope between the IU/ITPGRFA and the CBD, where twenty items of overlap were identified. In the subsequent sections, these items of overlaps will be explored in greater detail and the interaction between the two regimes in these regards will be analyzed.

Detailed Analysis of Overlap and Interaction between IU/ITPGRFA and the CBD

In this section we will analyze item for item, as identified in Table 8.1 with regard to overlaps and interaction. For the sake of simplicity, we will focus on items 1.1–1.12, which are considered the most important in our context.[9]

Item 1.1: Compatibility with regard to conservation and sustainable use Both agreements provide for the conservation and sustainable use of PGRFA as a core norm. The CBD does not refer explicitly to PGRFA, but to biological diversity in general, of which PGRFA is a component. Both agreements address *in situ* as well as *ex situ* conservation in addition to sustainable use. The way these issues are addressed in the three agreements is compatible, also when a main emphasis of the CBD in terms of *in situ* conservation implicitly concerns non-domesticated resources in general and protected areas in particular. Basically, there is no conflict between the norm of conserving biological diversity in protected areas and the norm of conserving and utilizing PGRFA in a sustainable way. There is a potential for conflict if people living within such protected areas are not allowed to continue

8 Several countries that voted for the ITPGRFA at its adoption in the FAO in 2001 have not yet signed or ratified. Most of these were signatories to the IU, and they will most likely sign/ratify sooner or later. The CBD has generally more Parties/Signatories than the IU/ITPGRFA. China and Japan are the largest countries in terms of population and economy that are Parties to the CBD but neither Parties nor Signatories to the IU or the ITPGRFA. Both are members of the FAO Commission Plant Genetic Resources for Food and Agriculture (CGRFA). Japan was one of the two countries to abstain from the vote on the adoption of the ITPGRFA at the FAO in 2001 (the other was the US, which nevertheless signed the Treaty later on, thus becoming a Signatory to the ITPGRA, as it is to the CBD).

9 An analysis of the remaining items is found in Andersen (2007).

their traditional agricultural practices, or to collect and use other plants traditionally employed for food, medicines, clothes and shelter – but this is more a question of rules and modes of implementation than of the *norms* as such.

As described in Chapters 5 and 6, there has been interaction between the two regimes regarding conservation and sustainable use of PGRFA. Most of this interaction has had the character of the COP to the CBD supporting FAO efforts, and of information gathering instigated by the COP. Since 2002, the Global Strategy on Plant Conservation has been the main CBD instrument concerning PGRFA, aimed at monitoring efforts underway in the FAO and elsewhere towards specified objectives.

We may conclude that interaction between the bodies of the CBD and the IU/ITPGRFA on this item has been synergetic, and that the scope for synergies has improved after the adoption of the ITPGRFA in 2001 and the Global Strategy on Plant Conservation in 2002.

Item 1.2: Divergence with regard to access and benefit sharing As noted in Chapter 6, a core norm of the CBD is the sharing of benefits derived from the utilization of components of biological diversity, and this is also envisaged as a basic incentive for the conservation of biological diversity. In the IU/ITPGRFA, access to PGRFA is the basic motivation towards which all other efforts are directed. Accessibility of PGRFA diversity, now and in the future, depends on proper conservation, sustainable use, availability of information and the facilitation of such access. The ultimate goal is to make these resources available for present and future production needs and food security: any other form of benefit sharing is subordinate to that. Benefit sharing, except for mutual accessibility, was not addressed in the IU, although aspects of it were formulated in the agreed interpretations in 1989.[10] Not until the harmonization of the IU with the CBD, leading to the ITPGRFA, was benefit sharing addressed as a norm pertaining to PGRFA management. This norm became operationalized in the form of a multilateral system quite unlike the bilateral system of benefit sharing envisaged under the CBD.

The problem with the bilateral approach to benefit sharing is that it was not conducive to facilitating access to PGRFA – partly because it is difficult to identify the legitimate country which is supposed to provide access, and partly because bilateral access procedures thus far have proven overly restrictive and burdensome. That is why the two core norms are considered as diverging.

However, that does not mean that benefit sharing is unimportant when it comes to PGRFA: indeed, this is clearly stated as an objective in the ITPGRFA. But it is vital that it does not hamper access to PGRFA, since that is crucial not only to further breeding but also to the very existence of these resources. That is why the negotiating parties established the Multilateral System of Access and Benefit Sharing, which better meets the particular challenges of benefit sharing regarding PGRFA. In this way, they solved the divergence between the CBD and the FAO regime on this issue. However, the solution was not complete, as not all crops could be included. As noted

10 FAO Conference Resolutions 4/89 and 5/89 (1989a; 1989b).

in Chapter 5, the bilateral approach remains dominant for non-Annex 1 crops and forage plants.[11]

The interaction between the two regimes here has aspects of barriers as well as synergy. The ITPGRFA negotiators were challenged by the CBD to find an appropriate arrangement for access and benefit sharing appropriate for PGRFA. Since the CBD approach was not suitable, it posed difficult barriers that took years to overcome. Nevertheless the result would probably not have seen the light of the day, had it not been for the initiative taken by the negotiators of, and the Conference of the Parties to, the CBD. Therefore it seems appropriate to see this as a synergetic effect of the interaction between the two regimes.

Item 1.3: Divergence with regard to access Access to PGRFA is a core norm under the IU as well as the ITPGRFA, but is a secondary norm under the CBD, where it concerns the components of biological diversity (and not only PGRFA). Under the IU, the norm was based on the universal principle of PGRFA as the common heritage of mankind. In 1991, this principle was interpreted as subject to national sovereignty. Thereby the terms upon which access should be provided appeared uncertain, and the FAO Conference stated that further clarification was needed.[12] With the Convention came a new principle according to which 'countries of origin' were to decide over access and benefit-sharing arrangements, according to their jurisdiction, and to which the FAO had to harmonize the IU. This approach was not at all adequate for facilitating access to PGRFA: not only would it be extremely difficult to identify such countries – even if this could be done, there would normally be more than one country, often many. If implemented according to the literal requirements of the CBD, this approach would in fact hamper access to PGRFA. This overlap represented a serious divergence between the two agreements.

Basically, we can note little interaction between the two regimes on this issue. The COP-CBD has continued implementing the Convention as envisaged in the text, and has deepened this understanding in the Bonn Guidelines on Access and Benefit Sharing.[13] Several countries have introduced access legislation and institutional structures and procedures based on the CBD principles pertaining to all components of biological diversity, including PGRFA. Thereby they have created a situation that may present hurdles for implementing the ITPGRFA in terms of access. During the years up to the entry into force of the ITPGRFA, the situation concerning access to PGRFA was highly uncertain, and many countries were restrictive in providing access (see for example ten Kate and Laird 1999, 32, and further documentation in Chapter 11). The IARCs of the CGIAR operated with an intermediate Standard Material Transfer Agreement, which enabled them to provide for access. However, this standard agreement could not solve all problems arising from the bilateral

11 Except for genetic resources stored by the IARCs, as explained in Chapter 5.

12 FAO Conference Resolution 3/91 (1991).

13 The current negotiations of an international regime on access to genetic resources and benefit-sharing are aimed at adopting an instrument to effectively implement the provisions in Article 15 and Article 8(j) of the Convention and the three objectives of the Convention (UNEP 2004, UNEP/CBD/COP Decision VII/19).

approach to access and benefit sharing.[14] Generally, the international transfer of genetic resources has decreased considerably since 1992.[15] Of the various reasons for this decline, the access regulations derived from, or inspired by, the CBD are considered particularly important (ibid., and further documentation in this book). With the Multilateral System on Access and Benefit Sharing, the ITPGRFA represents a solution to these problems – but only for those plant species and varieties covered by the system (including those of the IARCs), and under the management and control of the parties and in the public domain.

Against this background, we may say that interaction between the two regimes has hit many barriers when it comes to regulating access to PGRFA.

Item 1.4: Compatibility with regard to traditional knowledge, and conservation and sustainable use The provisions of the International Undertaking and the ITPGRFA on conservation and sustainable use (core norm) depend heavily on co-operation with local communities with traditional knowledge (secondary norm of the CBD). In collecting PGRFA from farmers' fields or other *in situ* conditions, it is necessary to obtain all information available on the growing conditions, features and uses. Without such passport data, the seeds as such can be of little use. The holders of this (often traditional) knowledge are normally local communities, including indigenous peoples. Thus we find full compatibility between the two norms.

As we saw in Chapters 5 and 6, there were no indications on interaction between the two regimes on this issue up to 2002. In 2002, the COP-CBD requested the CBD Executive Secretary to examine, in collaboration with the FAO, the implications of the ITPGRFA on the issues under CBD Article 8(j) and related provisions. In 2004, the Executive Secretary presented a note to the Seventh Meeting of the COP.[16] Many similarities were identified, especially between Article 8(j) of the CBD and Article 9 of the ITPGRFA. Therefore governments would need to ensure coherence when framing legislation, policies, strategies and action plans for the conservation and sustainable use of biological diversity in general, and for PGRFA in particular. Also the further work of the Governing Body of the ITPGRFA was considered of great relevance for the issues raised in those two Articles.

Item 1.5: Compatibility with regard to benefit sharing Benefit sharing is a core norm under the CBD and a secondary norm under the ITPGRFA. It was not addressed

14 According to Dr Michael Jackson, then Head of the Genetic Resources Center at the International Rice Research Institute (IRRI) in the Philippines, several developing countries demanded more control over the plant varieties collected in international gene banks after the entry into force of the CBD (interview with the author, 3 May 2000). These demands resulted in greater bureaucracy and problems in facilitating access, but did not lead to any benefit sharing with regard to PGRFA accessions stored in international gene banks.

15 See the database of the CGIAR System-wide Information Network for Genetic Resources (SINGER) at <www.singer.cgiar.org/SINGER/Overview/Distribution/User/user.htm>, and enter '*transfer*' by '*user type and year*'. See also ten Kate and Laird (1999, 312).

16 UNEP/CBD/COP/VII /Inf. 18 (2004).

in the IU, although aspects of it were formulated in the agreed interpretations in 1989.[17]

This indicates that benefit sharing was always a secondary concern under the FAO regime, whereas it was a central concern to the CBD negotiators. The formulations in the ITPGRFA as well as the process leading up to it show that a learning process has taken place, where the norms from the CBD were adapted to the specific conditions pertaining to PGRFA. Benefit sharing is, however, organized differently under the two regimes, as we noted in Chapters 5 and 6:

- The most important benefit under the IU/ITPGRFA is continued access to PGRFA to ensure present and future food production, in addition to exchange of information, the access to and transfer of technology, capacity building and monetary benefits. Under the CBD, there is a greater focus on monetary benefits.
- The potential recipients with whom benefits should be shared differ. Under the CBD, the 'countries of origin' of the accessed genetic resources are to receive the benefits and to encourage equitable benefit sharing with their indigenous and local communities. This may work well for non-domesticated resources, but not for domesticated resources. By contrast, with the ITPGRFA, benefit sharing is facilitated through the Multilateral System for access to genetic resources from the Annex 1 crops that are under the control and management of the Contracting Parties and in the public domain. The Governing Body ensures that the benefits channeled to the multilateral system flow directly or indirectly to farmers in all countries who conserve and sustainably use PGRFA, and particularly farmers in developing countries and countries in transition (Art. 13.3) – regardless of whether they were the ones who provided the resources from which benefits were derived.

In addition, the parties to the ITPGRFA as well as the CBD agree to make available information, provide access to and transfer of technology and strengthen capacity building, particularly from developed to developing countries and countries in transition on a general basis – without reference to access to specific genetic resources.

The norms between the two regimes on this item are compatible, but the rules have been developed differently to meet different challenges. We find the greatest difference between the multilateral system of the ITPGRFA, and the bilateral approach of the CBD. Due to the result of this interaction, we can conclude that there have been synergies.

Item 1.6: Compatibility with regard to benefit sharing and farmers' rights Farmers' rights as approached under the IU (agreed interpretations in 1989 and 1991) and the ITPGRFA contain aspects of benefit sharing. FAO Conference Resolution 5/1989 provided for farmers to participate fully in the benefits derived from the improved use of PGRFA, through plant breeding or other scientific methods. FAO Conference

17 FAO Conference Resolutions 4/89 and 5/89 (1989a; 1989b).

Resolution 3/1991 provided for an international fund to support conservation and sustainable use of PGRFA, particularly in developing countries, which can also be seen as a benefit-sharing measure. Article 9 of the ITPGRFA suggests that contracting parties take measures to ensure the right of farmers to participate equitably in sharing benefits arising from the use of PGRFA (section 9.2.b). These provisions are compatible with the provisions in the CBD on benefit sharing, in particular with the Bonn Guidelines on Access and Benefit Sharing. The Bonn Guidelines advise Contracting Parties to support measures, as appropriate, to enhance the capacities of indigenous and local communities to represent their interests fully at negotiations on benefit sharing (para. 16.a.vii). It also advises users to ensure fair and equitable benefit sharing in conformity with the mutually agreed terms they establish with the indigenous and local communities or stakeholders involved (Art. 16.b.ix).

Despite these clear overlaps, it has not been possible to determine any specific interaction between the two regimes with regard to benefit sharing (core norm under the CBD) and farmers' rights (secondary norm under the IU/ITPGRFA). We may note that there are potentials for synergies between the two.

Item 1.7: Neutral overlap on conservation and farmers' rights but with possibility of barriers Conservation of biodiversity is a core norm under the CBD, whereas recognition of farmers' rights is a secondary norm under the IU/ITPGRFA. The norm on farmers' rights has several components; for the IU, these are expressed in the non-binding agreed interpretations referred to as FAO Conference Resolution 5/1989. The component most relevant here concerns the responsibility of the international community to ensure the continuation of farmers' contributions in conserving, improving and making available plant genetic resources. This formulation is strengthened in the non-binding provision of the ITPGRFA (Art. 9.2.c), which advises that farmers should be allowed to participate in decision-making on PGRFA. The overlap situation with the CBD pertains particularly to its provision on protected areas (Art. 8a–8c). For countries implementing these provisions, farmers would have to be entitled to participate in decisions concerning protected areas of relevance for PGRFA. The situation here has parallels with that regarding item 1 on the norms of the two regimes pertaining to conservation and sustainable use above. Much depends on how these norms are realized. There is the possibility of divergence if farming is categorically prohibited in protected areas. On the other hand, this is to some extent counterbalanced by Article 8j of the CBD, on respect for indigenous peoples and local communities.

Depending on interpretation, this latent divergence may result in interaction with barriers. As noted in section 6.3.1, the core norms of the CBD on conservation of biological diversity focus on *in situ* management of non-domesticated resources, particularly in protected areas, where farming is often forbidden or severely limited, because it is considered to have negative impacts on biodiversity. Whether traditional agricultural practices are sustainable and compatible with the objectives of nature conservation in protected areas is often a question of population density, since that determines how long an area is allowed to recover after having been used for agriculture. Slash-and-burn agriculture is often seen as a threat to forests, since it can be a source of devastating forest fires. In historical perspective, however, it has normally been a sustainable practice, as long as the forests were allowed to recover fully after being burned down and used for agricultural purposes, and as long as enough pristine forest surrounded the agricultural sites. In many traditional farming systems, forest or bush clearing is an important component that comes into basic conflict with the norms of nature conservation in protected areas. Whether the two can be reconciled depends on the extent of the problem, and the potentials for regulation. There are no easy shortcuts to conflicts on farming and protected areas, and specific solutions will need to be found for each case. Generally, nature protection depends on the co-operation of the people living in the area in question, who are often also good safeguards against illegal external interventions there.[18] It is also normally in the interest of farmers that their natural surroundings are maintained, since these provide ecosystem services such as food, medicines, fire and shelter.

There has not been any direct interaction between the two regimes at the international level with regard to the conservation of biodiversity in protected areas and farmers' rights. However, since 2002 we may note interaction with regard to farmers' rights (Art. 9 under the ITPGRFA) and traditional knowledge (Art. 8j under the CBD), as further elaborated under item 1.8. The work of the COP-CBD on implementing Article 8j on traditional knowledge shows that there is scope for solutions. On this background we may conclude that the interaction between these two norms has potentials for synergies as well as conflicts, depending on which strategies the implementing agencies choose, and the level of interaction.

Item 1.8: Compatibility with regard to farmers' rights and traditional knowledge As the Executive Secretary of the CBD expressed in his note to the COP of the CBD in 2004,[19] several similarities and parallels can be identified between the norms on farmers' rights under the ITPGRFA (Art. 9) and those of indigenous and local communities under the CBD (Art. 8j). There are no conflicts between these two secondary norms as norms,[20] so we will consider them to be compatible. Also the earlier resolutions on farmers' rights adopted by the FAO Conference as agreed interpretations of the IU are compatible with Article 8j of the CBD. Both regimes address the respect for and recognition of the knowledge, innovation and practices of indigenous or local communities or, in the case of the FAO regimes, of farmers. Both provide for the maintenance of these practices and knowledge and encourage their wider application.[21] They also both address benefit sharing.

Despite these clear parallels, there was no interaction on this issue between the two regimes at the international level until 2002. Then the COP-CBD requested the CBD Executive Secretary to examine, in collaboration with FAO, the implications

18 This is the general experience of, for example, the Rainforest Foundation Norway: see <www.rainforest.no>.

19 UNEP/CBD/COP/VII /Inf. 18 (2004a).

20 As the norms are not very detailed, although quite comprehensive, the picture might change in situations where they are realized – depending on what components are emphasized, and how.

21 For the ITPGRFA, this is covered in Article 6.

of the ITPGRFA on the issues under CBD Article 8(j) and related provisions. This resulted in the note by the Executive Secretary in 2004, referred to above.

There is great scope for synergies with regard to implementing the two norms at the international as well as domestic levels. For example, the introduction of legislation on the protection of traditional knowledge can pave the way for an understanding of the legitimacy of farmers' rights. So far these potentials do not appear to have been utilized, but the initiative by the COP-CBD, resulting in the said note, indicates that this may change in the future.

Item 1.9: Compatibility with regard to benefit sharing and access Benefit sharing was a secondary concern under the IU from 1989 on (due to FAO Resolution 5/89) and was not fully developed until the adoption of the ITPGRFA. Access to the components of biological diversity is considered a secondary norm under the CBD. Until the ITPGRFA was adopted, the provisions of the CBD constituted the sole international regulation that provided systematically for access to PGRFA and other genetic resources combined with the sharing of benefits arising from their use. Access regulation was the point of departure for sharing of benefits. Although the CBD provisions on access, based on a bilateral approach, did not fit with the conditions pertaining to PGRFA management, they represented a norm that could serve as the point of departure for formulating the ITPGRFA. Up to the adoption of the ITPGRFA, there was interaction with barriers, due to the problems of developing a system for access-based benefit sharing concerning domesticated genetic resources in harmony with the CBD. As these barriers were overcome with the Multilateral System, we may say that the interaction between the two regimes has been synergetic here.

Item 1.10: Compatibility with regard to conservation and sustainable use and GEF Whereas conservation and sustainable use is a core norm under the IU/ITPGRFA, the provisions of the Global Environment Facility (GEF) as a financial mechanism are regulatory rules under the CBD. The GEF provides financial support to developing countries and countries in transition, *inter alia* for the conservation of biological diversity and sustainable use of its components, including PGRFA. Thus we see full compatibility between the two.

However, the focus of the GEF with regard to biological diversity has been mainly on non-domesticated resources. There is scope for greater synergies as to the conservation and sustainable use of PGRFA if GEF supports more projects of this kind.

Item 1.11: Divergence with regard to regulatory rules for ex situ conservation Both regimes have regulatory rules on *ex situ* conservation. Whereas their norms are compatible in this respect, their regulatory rules are not, but these differ with regard to strategy. According to the CBD, each contracting party is to establish and maintain facilities for *ex situ* conservation and research (Art. 9.b) and receive financial support for this work (Art. 9.e), whereas the IU provides for international co-operation towards this end (Art. 6). According to ITPGRFA, the Contracting Parties are to co-operate to promote the development of an efficient and sustainable system of *ex situ* conservation (Art. 5.1.e). There is a basic difference between these approaches. The CBD approach requires that each country maintains its own gene bank(s), whereas the IU/ITPGRFA requires that countries work together for this purpose. As developing countries often have great problems with *ex situ* conservation, due to financial, institutional and infrastructure constraints, many accessions are in danger of losing their fertile vigor; indeed, many accessions have probably already been lost (FAO 1998). Therefore, it was considered important in terms of the objectives of the IU/ITPGRFA to ensure conservation through international co-operation. From the point of view of the CBD, control with genetic resources and support for domestic efforts towards conservation were important aspects and reasons for the formulation of Art 9.b. Spending scarce resources to build new gene banks under fragile conditions, instead of using already established *ex situ* facilities, could, however, prove counterproductive to the norm of conservation.[22] Therefore the overlap between these regulatory rules is considered diverging.

It has not been possible to identify interaction between the two regimes on this point. However, the Global Strategy for Plant Conservation, adopted at COP 6 in 2002 (Decision VI/9), makes no mention of the requirement that each country should have its own gene bank. Instead it refers to the FAO, the CGIAR and international networks. This may be interpreted as an indication that some learning has taken place. Against this backdrop, we may say that these regulatory rules contained the potential for conflicts when implemented at the domestic level, but that there is scope for synergy with the COP to the CBD learning from FAO experience, in that all institutions work towards the same goals. The Global Strategy for Plant Conservation may also be interpreted as giving preference to the FAO regulation of *ex situ* conservation.

Item 1.12: Divergence and compatibility with regard to benefit-sharing rules Whereas the norms on benefit sharing were and are compatible (see item 1.5), the regulatory rules differ. Interaction between the two regimes in this regard was characterized by discussions among the negotiating parties to the ITPGRFA on how to harmonize the new agreement with the CBD, while also applying a different approach with regard to regulatory rules. It became necessary to identify other regulatory rules to fulfill the intentions of the CBD on benefit sharing. After the adoption of the ITPGRFA, a division of labor was established between the two regimes. The ITPGRFA regulates all genetic resources included in the multilateral system, whereas the CBD regulates genetic resources collected after the entry into force of the CBD and not included in that system. The resources not covered by these two regimes are not covered by any international regulations. This division of labor indicates compatibility between the two regimes in this respect.

22 Whether it would be counterproductive to the norm of benefit-sharing is another question. Some sort of control by the provider of the resources is necessary to enable benefit-sharing arrangements. However, if the varieties are lost as a result of poor storage facilities in national gene banks, there will be no benefits to share. The efforts of the FAO to enable countries to exercise control over accessions in international gene banks are one way of dealing with the problem, the Multilateral System is another.

After years beset with barriers and difficulties in getting the question of benefit sharing moved under the negotiations leading to the ITPGRFA, a breakthrough was finally achieved and an adequate regulatory approach was found to the norms on benefit sharing under the CBD. Therefore we can conclude that, although there were barriers between the two regimes on this issue prior to the adoption of the ITPGRFA, the later division of labor solved these problems.

Overlap and Interaction between the CBD and the TRIPS/UPOV

We will now turn to the interaction of, and functional overlaps between, the CBD and the TRIPS/UPOV. Again, we will first focus on overviews before going into the details of the single items of regime overlap.

Interaction of the CBD and TRIPS/UPOV in a Time Perspective

The development stages of the TRIPS Agreement with regard to Article 27.3 (b), and of the CBD, as illustrated in Figure 8.2, have three characteristics. First, the agenda-setting and negotiation phases lasted considerably longer for the TRIPS Agreement than for the CBD. There are several reasons, the most important probably being that the TRIPS Agreement was negotiated as part of the larger WTO 'package deal', so the process required greater co-ordination. Second, negotiations leading to the CBD started after the negotiation towards the TRIPS Agreement had begun, and were finalized long before the adoption of that Agreement. This is important for understanding how the material contents of the TRIPS negotiations could influence the negotiations towards the CBD. Third, the implementation phases of both agreements ran basically parallel.

TRIPS (27.3.b)	Agenda setting			Negotiations				A[a] + Interim period	Implementation phase for TRIPS: (Impl.)	Review/evaluation			
Timeline	**73–80**	**81–82**	**83–84**	**85–86**	**87–88**	**89–90**	**91–92**	**93–94**	**95–96**	**97–98**	**99–00**	**01–02**	**03–04**
CBD		Agenda setting			Negotiations		A[b]	Implementation phase for the CBD					

Figure 8.2 Comparing development stages: TRIPS and CBD
a) Adoption of the TRIPS Agreement
b) Adoption of the CBD and ratification period

The most important features of interaction between the two regimes in this time perspective concern the negotiation and implementation of the two agreements:

1981–1994 The emerging systems of intellectual property rights had a decisive influence on the positions of the negotiators to the CBD, and resulted in special emphasis on the norm and regulatory rules on benefit sharing during the formation of the regime. In her study of the CBD, Rosendal (2000, xi and 110) found that the developing countries won through with their demands in the CBD negotiations. As the current study shows, the industrialized countries won through with their claims in the TRIPS negotiations. For some countries, there was an obvious conflict between the two, which is why they acceded to only one of the two agreements. Most notably, the US refused to ratify the CBD.[23] An important reason was the wording of the CBD Article 16(5), that intellectual property rights systems should not run counter to the objectives of the CBD.[24]

1994–2004 The degree to which patents or other intellectual property rights systems derived from the TRIPS Agreement limit access to, or the use of, plant varieties is decisive to considerations as to whether they conflict with the overall objectives of the CBD and its provisions on access. If developing countries introduce intellectual property rights systems that restrict access to PGRFA in order to comply with the TRIPS Agreement Article 27.3(b), there is an open conflict. As we will see, there is considerable evidence that such systems are being established as implementing measures. This indicates that the TRIPS Agreement in its domestic implementation tends to constitute barriers to CBD implementation regarding PGRFA access and use.

The time dimension alone is not sufficient to explain the interplay between two regimes: it has to be supplemented with the factor of power. The two regimes have had more or less simultaneous development stages, but the TRIPS Agreement is considerably stronger than the CBD in terms of enforcement mechanisms as well as driving forces. The effects have been 'interaction with barriers', in that the more powerful regime (TRIPS) has posed some central hurdles to domestic implementation of the less powerful regime, the CBD, as we will come back to below.

Resulting Overlaps between the CBD and the TRIPS/UPOV: Overview

The overall policy objectives and principles of the TRIPS Agreement pertaining to PGRFA differ from those of the CBD and the ITPGRFA in that they concern the issue of intellectual property rights exclusively. We will compare a part of the TRIPS Agreement (Art. 27.3[b]) with the whole of the two other agreements.

The Secretariat of the TRIPS Council has noted that intellectual property rights proved to be a sensitive issue under the negotiation of the CBD as well as in later COP meetings.[25] At its third meeting in 1996, the COP-CBD recognized that 'further work is required to help develop a common appreciation of the relationship between intellectual property rights and the relevant provisions of the TRIPS Agreement and the CBD' (Decision III/17). Special reference was made to 'issues relating to technology transfer and conservation and sustainable use of biological diversity and the fair and equitable sharing of benefits arising out of the use of genetic resources, including the protection of knowledge, innovations and practices of indigenous and

23 The US only signed the CBD.

24 Expert panel IV, cited in Rosendal 2000, 149.

25 See WTO Note IP/C/W/216 (2000).

local communities embodying traditional lifestyles relevant for the conservation and sustainable use of biological diversity.' Most CBD norms are considered relevant with regard to intellectual property rights.

The CBD entails two provisions on intellectual property rights that neutralize each other. They may be regarded as an attempt to seek a balance between existing agreements on such rights and to ensure that future international and national laws on intellectual property rights will not run counter to the Convention (Andersen 2002):

- In the case of technology subject to patents and other intellectual property rights, access and transfer shall be provided on terms consistent with the adequate and effective protection of intellectual property rights (Art. 16.2).
- Patents and other intellectual property rights may influence the implementation of the CBD; thus the parties are obliged to co-operate in order to ensure that such rights are supportive of and do not run counter to its objectives (Art. 16.5). Such efforts are expected at the level of international laws as well as in national legislation.

There is no conflict between the provisions of the CBD and the TRIPS Agreement here. Since the CBD points in a complementary as well as a diverging direction on this issue, it is not possible to derive a clear norm from the Convention in this regard. That is why intellectual property rights are not covered by the identified norms and regulatory rules in Chapter 6 above – which in turn is the reason why this issue is not further dealt with in terms of CBD norms/rules below.

Table 8.2 provides an overview of overlaps in functional scope between the CBD and the TRIPS/UPOV, where 12 items were identified.

Detailed Analysis of Overlap and Interaction between the CBD and TRIPS/UPOV

In this section we will analyze item for item, as identified in Table 8.2 with regard to overlaps and interaction. For the sake of simplicity, we will focus on items 2.1–2.3, which are considered the most important in our context.[26]

Item 2.1: Neutral overlap with regard to benefit sharing and effective sui generis protection, but resulting in interaction with barriers Fair and equitable sharing of the benefits arising from the utilization of genetic resources is a core norm under the CBD, whereas an effective *sui generis* system for the protection of intellectual property rights is the core norm under the TRIPS Agreement. Critics of the WTO, particularly among civil society organizations, maintain that there is a basic conflict between these two norms, but the evidence does not provide support for this. After all, there is no clause in the TRIPS Agreement specifying that it is not permissible to share the benefits arising from the income from intellectual property rights to genetic resources. Independent of whether patents, plant breeders' rights or other forms of intellectual property rights are used, the possibilities to share potential benefits are basically the same, as long as this aspect is not explicitly regulated.

26 An analysis of the remaining items is found in Andersen (2007).

Table 8.2 Overlapping norms and rules: CBD and TRIPS

CBD → **TRIPS 27.3(b) and related provisions ↓**	**Core norms:** *– Conservation, sustainable use (no direct overlap)* *– Benefit sharing*	**Secondary norms:** *– Access* *– Recognition of traditional knowledge*	**Regulatory rules:** *– Domestic policy and legal measures* *– GEF (international) (no direct overlap)*
Core norm: – Effective *sui generis* protection for plant varieties as a minimum standard	2.1. Neutral overlap with regard to benefit sharing and effective *sui generis* protection, but resulting in interaction with barriers	2.2. Neutral overlap with regard to access and effective *sui generis* protection, but resulting in interaction with barriers 2.3. Neutral overlap with regard to the recognition of traditional knowledge and effective *sui generis* protection, but resulting in interaction with barriers	2.7 Compatibility between regulatory rules on domestic policy and legal measures under the CBD and the norm of an effective *sui generis* protection
Secondary norms: – Patenting of animals and/or plants is optional	2.4 Limited compatibility with regard to benefit sharing and optional patenting	2.5 Marginal overlap with regard to access and optional patenting 2.6 Marginal overlap with regard to the recognition of traditional knowledge and optional patenting	2.8 Marginal overlap with regard to domestic policy and legal measures derived from the CBD and optional patenting
Regulatory rules: – Developing countries to implement Article 27.3(b) by 2000, least developed countries by 2006 (no overlap) – Strong domestic procedures to ensure legal enforcement	2.9 Strong regulatory rules from the TRIPS Agreement reinforce legislation derived from this agreement as compared to the CBD regulations on benefit sharing	2.10 Strong regulatory rules from the TRIPS Agreement reinforce legislation derived from this agreement as compared to the CBD regulations on access 2.11 Strong regulatory rules from the TRIPS Agreement reinforce legislation derived from this agreement as compared to the CBD regulations on traditional knowledge	2.12 In general the regulatory rules of the TRIPS Agreement are stronger than those of the CBD, enabling it to override the implementation of the CBD in potential cases of conflict

When, however, benefit sharing is not accounted for in domestic legislation on intellectual property rights to genetic resources, and the holders of such rights do not voluntarily engage in fair and equitable benefit-sharing arrangements, the result is a conflict between the two norms. Countries providing the genetic resources directly, via international gene banks or other channels, are not granted any remuneration for their contribution – not even access on favorable terms to the protected variety. Particularly for developing countries that lack the financial and institutional capacity to make use of established systems of intellectual property rights, such practices are often regarded as provocative.

While the TRIPS Agreement does not prohibit the sharing of benefits from intellectual property rights, it does not provide for such sharing, either. One possible measure could be disclosure of source countries of PGRFA in patent applications (see for example Correa 2003), or disclosure of legal provenance or compliance with national law, as has been discussed more recently. This could be followed up in various ways, and would enable control of whether the rules of the CBD (and, since 2004, the ITPGRFA) on benefit sharing have been followed. However, there exists no such rule, so the overlap between the two regimes must be regarded as neutral in this respect.

On the other hand, the CBD does also not oblige countries with systems for intellectual property rights over genetic resources to ensure benefit sharing with the countries that provide these resources. Under the CBD, the burden of proof rests with the provider countries of genetic resources, not with the recipients and users of these resources. Provider countries are to undertake legislative and policy measures to ensure benefit sharing with those who receive PGRFA from them. As PGRFA-provider countries generally have scant control over these resources,[27] their possibilities to develop benefit-sharing arrangements are highly limited.

This is the situation that has emerged through the implementation of the TRIPS Agreement and the CBD so far. The TRIPS Agreement has been implemented largely without any provisions regulating benefit sharing with provider countries. Many developing countries have responded with more or less restrictive legislation on access to genetic resources, in order to capture some of the benefits from the use of their resources. This means that interaction between the two regimes poses barriers to benefit sharing under the CBD, and results in restricted access to genetic resources.

Item 2.2: Neutral overlap with regard to access and effective sui generis protection, but resulting in interaction with barriers Effective *sui generis* protection is, as a minimum standard, the core norm of the TRIPS pertaining to genetic resources; under the CBD, access to such resources is a secondary norm. The TRIPS Agreement does not explicitly state that *sui generis* systems on intellectual property rights must restrict access to protected varieties, but is neutral in this regard. For example UPOV 1978, which has been used as a *sui generis* system by several countries,[28] enables

27 Due to the collection of PGRFA from these countries in international gene banks, by multinational corporations and the historical exchange between countries over the centuries.

28 The number is decreasing as steadily more of these countries convert to UPOV 1991.

farmers to use the harvest from protected varieties for any purpose, and it allows other breeders to use protected varieties for the development of new varieties. Such a system does not pose any threat to the norm of access to genetic resources. The text of the TRIPS Agreement is neutral with regard to regulation of access to genetic resources. It is through the implementation of the TRIPS Agreement that conflicts may arise.

Since UPOV 1978 is closed for new members, most countries currently developing *sui generis* systems of intellectual property rights follow the 1991 Act of the UPOV Convention. As noted in Chapter 7, this act leaves it to the national governments to decide whether farmers are to be entitled to reuse the seeds from protected varieties. If they are allowed to reuse their seeds, this is to take place only on their own landholdings and within reasonable limits, thereby safeguarding the rights of the plant breeder. Farmers are no longer allowed to exchange seeds without the permission of the rights holder if the seeds stem from protected varieties – which reduces their accessibility. Whether governments allow farmers to reuse protected seeds or not, these regulatory rules represent a restriction on access, as compared to the 1978 Act of the UPOV Convention.

A frequent misunderstanding with regard to UPOV 1991 is that it generally prohibits farmers from reusing their seeds. In fact, any legislation based on UPOV applies solely to the use of *protected* varieties. Farmers' varieties remain unregulated, as long as these are not protected by any other legislation.[29]

Proponents of UPOV 1991 as a *sui generis* system for implementation of the TRIPS maintain that plant breeders' rights do not limit the access to genetic resources, because they are applicable only to new varieties of plants, whereas all the old varieties remain freely accessible. Furthermore, they hold that the protected varieties are accessible upon remuneration to the plant breeder, which is necessary as an incentive to plant breeding. Also the new varieties represent a contribution to the diversity of PGRFA; and, after 20–25 years, they too will be accessible without plant breeders' rights, as that is the duration period for such rights. Seen isolated from the context, these arguments are all valid.[30] However, when we look at the context in which plant breeders' rights are applied, we find that the impacts may be detrimental to access. When several farmers in an area start to use protected varieties and thereby lose their possibility to exchange seeds with other farmers, those who

29 Seed laws are increasingly regulating the exchange and sales of farmers' varieties in many countries. In Norway, for example, it is prohibited for farmers to exchange seeds from traditional varieties that are not protected with intellectual property rights. It is even prohibited to give them away. These regulations are currently being discussed with a view to increasing farmers' legal space to maintain genetic diversity in agriculture.

30 However, it is not granted that plant varieties as such will be available after the protection period. In Germany, the owner of the plant breeders' rights over the potato variety 'Linda', Europlant, sought to withdraw the variety from the market towards the end of its protection period (Andersen 2005c). This was possible by withdrawing certification for 'Linda' with arguments of plant health. Potato farmers mobilized and protested, arguing that this was merely a tactic for Europlant to get its new potato variety 'Elana' into the market. The case shows that there is no guarantee that plant varieties will be made freely available after the end of a period of plant breeders' rights protection.

have not converted to protected varieties can obtain seeds only in exchange with the remaining farmers. When the number of farmers among which seed exchange is possible decreases, also the number of available plant varieties for these farmers decreases, thereby reducing their access to PGRFA. Since these are often poor farmers, they are not able to purchase protected seeds and the related technology (such as fertilizers and pesticides). They may also grow their crops on marginal lands not suitable for the protected varieties because of poor soil quality, difficult climatic conditions or for other reasons. Poor traditional farmers are therefore indirectly the losers in areas where protected varieties are introduced.

A major study conducted in middle-income countries (Van Wijk and Jaffe 1995, referred to in CIPR 2002, 61) supports this interpretation. It concluded that commercial farmers together with the seed industry emerged as the principal beneficiaries from plant variety protection systems, whereas poor farmers had not benefited directly from protection. Potentially they could be adversely affected by restrictions on seed saving and exchange in the future, through reduced access.

An additional challenge is the fact that the genetic diversity decreases in an area where protected varieties are introduced, since the protected varieties are by definition genetically homogeneous, whereas farmers' varieties are normally genetically heterogeneous. Often the providers of protected varieties are multinational corporations; with effective marketing strategies, they can take over an extensive share of the seed market within a short time, thereby comprehensively reducing the genetic diversity in that seed market. In Brazil, for example (CIPR 2002, 65), Monsanto entered the seed marked for maize in 1997 when the country introduced legislation on plant breeders' rights and joined UPOV (1978 Act). Within two years, Monsanto's market share had soared from 0 per cent to 60 per cent. By 1999, only one Brazilian-owned firm remained in the market, with a share of 5%. Whereas the new varieties of maize may have meant increased yields and better quality, they have also replaced traditional varieties and thereby reduced the prospects for access to diverse maize varieties in Brazil.

Summing up our discussion here, the overlap between the TRIPS Agreement and the CBD is neutral with regard to access to genetic resources and effective *sui generis* systems. However, the interaction between them is one of barriers, because UPOV 1991 has emerged as a model for implementing the TRIPS Agreement.

Item 2.3: Neutral overlap with regard to the recognition of traditional knowledge and effective sui generis protection, but resulting in interaction with barriers Effective *sui generis* protection is, as a minimum standard, the core norm of the TRIPS regarding genetic resources, whereas the CBD provides for respect for and maintenance of traditional knowledge as a secondary norm. This latter norm includes 'the equitable sharing of the benefits arising from the utilization of such knowledge, innovations and practices' (Art. 8j). As the overlap between these two norms concerns the component of benefit sharing, and the overlap between benefit sharing (CBD) and effective *sui generis* protection has been thoroughly discussed under item 2.1 above, we refer the reader to that discussion.

Overlap and Interaction between the TRIPS/UPOV and IU/ITPGRFA

Finally will turn to the TRIPS/UPOV and the IU/ITPGRFA. Once again we will first focus on overviews, before going into the details.

Interaction of TRIPS/UPOV and the IU/ITPGRFA in a Time Perspective

The interaction between the TRIPS/UPOV and the IU/ITPGRFA is illustrated in Figure 8.3, which compares development stages.

<table>
<tr><td rowspan="2">IU/ ITPGRFA</td><td rowspan="2">Agenda setting</td><td rowspan="2">Nego-tiation</td><td colspan="9">Implementation phase for the IU</td><td colspan="2" rowspan="2">A[a)]/interim period [b)] ITPGRFA</td></tr>
<tr><td colspan="3">A[c)]/Implementation</td><td colspan="2">Agenda setting</td><td colspan="4">Negotiations on revision of the IU</td></tr>
<tr><td>Timeline</td><td>79–80</td><td>81–82</td><td>83–84</td><td>85–86</td><td>87–88</td><td>89–90</td><td>91–92</td><td>93–94</td><td>95–96</td><td>97–98</td><td>99–00</td><td>01–02</td><td>03–04</td></tr>
<tr><td rowspan="2">TRIPS (27.3.b)</td><td colspan="4" rowspan="2">Agenda setting (from 1973 and onwards)</td><td colspan="4" rowspan="2">Negotiations</td><td rowspan="2">A[d)] + Interim period</td><td colspan="4">Implementation phase for TRIPS</td></tr>
<tr><td>(Impl.)</td><td colspan="3">Review/evaluation</td></tr>
</table>

Figure 8.3 Comparing development stages: IU/ITPGRFA and TRIPS

Referring to the presentations in Chapters 5 and 7, this figure indicates four phases of interaction between the two regimes.

1979–1986 The agenda-setting phase for the IU started six years after the first initiatives to include intellectual property rights in the GATT negotiations. However, during the agenda-setting and negotiation phases of the IU, there were no indications that an agreement on intellectual property rights would result from the GATT negotiations, or that this would address plants and plant varieties. Shortly after the adoption of the IU, the US initiated the expert group on intellectual property rights in 1984; and in 1986, the issue was successfully included in the agenda for the Uruguay Round, as explained in Chapter 7. There were still no indications as to the inclusion or non-inclusion of PGRFA in the negotiations, but several developing countries were alerted by the fact that this issue was included in the trade agenda.

1987–1991 In 1987, the issue was brought to the table by the EU, which proposed that patents should be available for all inventions – with some exceptions, among them plant varieties, as explained earlier. New proposals came from other countries in the following years, under heated debate. This was the time when FAO members felt the need to formulate the agreed interpretations of the IU on plant breeders' rights in order to attract more signatories. The debates were heated, and, following three resolutions by the FAO Conference, the IU norms regarding access to PGRFA were considered so ambiguous that the conference deemed them to be in need of clarification. The dynamics of the GATT negotiations on intellectual property rights were among the most influential factors for these amendments to the IU (on this, see for example Fowler 1994, 201).

1992–2001 While the TRIPS Agreement was still in the making, the initiative was taken for new negotiations on revising the IU to bring it in harmony with the CBD. These negotiations had just started when the TRIPS Agreement, together with the whole WTO package, was adopted. This led to a new situation. Now the question became how to design the new agreement on PGRFA in a compatible way, while also taking measures to hinder potential negative effects of that agreement for the objectives of the FAO agreement. Negotiations were complicated and heated. The overlap with the TRIPS Agreement was an additional reason why they took so long.

2001–2004 The solution was found in the provisions referred to above, prohibiting the application of intellectual property rights on material from the Multilateral System in the form it is received, and ensuring benefit sharing in cases where material is further developed and protected by intellectual property rights.

Whether these provisions of the ITPGRFA will affect the interaction between the two regimes depends on the interpretation of both agreements, and their implementation. Seen from the WTO perspective, the prohibition of intellectual property rights on materials from the Multilateral System in the form it is received – particularly if interpreted broadly – and the requirements for benefit sharing could constitute barriers to effective implementation of the relevant TRIPS provisions. The WTO system, with its economic sanctions, is considerably stronger than the CBD and the ITPGRFA. However, this does not mean that the WTO will necessarily overrule the provisions of the ITPGRFA pertaining to intellectual property rights. The outcome of the interplay between these two regimes remains to be seen. In general we may say that the potential of the parties to the ITPGRFA to modify the effects of the TRIPS agreement will depend on their political will to pursue this possibility.

It is interesting to highlight the illustrative character of this development for the potentials of a regime in an early development stage to modify the effects of a regime in a later development stage. The negotiating parties to the regime under formation, the ITPGRFA, had a strategic opportunity to modify the outcome of the established regime (TRIPS Agreement, Art. 27.3.b), and they made use of this in the formulation of the Treaty.

Resulting Overlaps between TRIPS/UPOV and IU/ITPGRFA: Overview

Our analysis of the TRIPS Agreement focuses on Article 27.3(b), with related provisions in the various WTO Agreements, and not the TRIPS Agreement as such. As with the overlap between the CBD and the TRIPS Agreement, there are also two categories involving the relation between the IU/ITPGRFA and the TRIPS Agreement: *direct overlap* in provisions on intellectual property rights, and *indirect overlap* in terms of the effect on other issues. Both forms are dealt with in detail under the items of overlap and interaction below. Generally we find items of compatibility and of divergence in both categories.

Table 8.3 Overlapping norms and rules: IU/ITPGRFA and TRIPS

TRIPS 27.3(b) and related provisions → **IU/ITPGRFA ↓**	**Core norm:** – Effective *sui generis* protection for plant varieties as a minimum standard	**Secondary norms:** – Patenting of animals and/or plants is optional	**Regulatory rules:** – Implementation deadlines (no overlap) – Strong domestic procedures to ensure legal enforcement
Core norms: – Conservation, sustainable use (no direct overlap) – Access	3.1 Divergence between the core norms of access to PGRFA and effective *sui generis* protection, resulting in interaction with barriers (IU) and strategic opportunities (ITPGRFA)	3.2 Marginal overlap with regard to norms on access (IU/ITPGRFA) and optional patenting (TRIPS)	3.9 Divergence between the strict rules for implementation of the TRIPS Agreement and the norms on access under the IU/ ITPGRFA
Secondary norms: – Benefit sharing – Farmers' rights – Limited intellectual property rights (ITPGRFA)	Neutral overlap between the norms on effective *sui generis* protection and – 3.3 benefit sharing and – 3.4 farmers' rights, resulting in interaction with barriers (IU) and with strategic opportunities (ITPGRFA) 3.5 Divergence between effective *sui generis* protection and limited intellect-ual property rights (ITPGRFA)	3.6 Marginal overlap between benefit sharing and optional patenting 3.7 Marginal overlap between farmers' rights and optional patenting 3.8 Marginal overlap between norms on limited intellectual property rights and optional patenting, but with the possibility of barriers (TRIPS) and of strategic opportunities (ITPGRFA)	Scope for divergence as well as compatibility in the relation between strict procedures for implementation of the TRIPS Agreement and: – 3.10 benefit sharing – 3.11 farmers' rights Scope for divergence with regard to: – 3.12 limited intellectual property rights (ITPGRFA)
Regulatory rules: – Domestic policy and legal measures – Ex situ networks (no direct overlap) – Multilateral System ITPGRFA) – Funding strategy	3.13 Compatibility with regard to the norm of an effective *sui generis* system and the regulatory rules on domestic policy and legal measures 3.14 Compatibility with regard to funding strategy (ITPGRFA) and the norm of an effective *sui generis* system	3.15 Marginal overlap between optional patenting and domestic policy and legal measures 3.16 Divergence and compatibility with regard to the norm on optional patenting and the regulatory rules of the Multilateral System	3.17 Compatibility with regard to domestic policy and legal measures, and strong domestic procedures for intellectual property rights 3.18 Divergence between the Multilateral System and strict domestic procedures for intellectual property rights

The constellations of signatories and parties to the CBD and the IU/ITPGRFA are similar, with a few exceptions. This means that also the constellations between the CBD and the TRIPS Agreement and the IU/ITPGRFA and the TRIPS Agreement are generally similar. As of January 2007, the WTO had 150 member countries and is still growing.[31] As noted, two of the strongest proponents of intellectual property rights for plants and plant varieties – the United States and Japan – were initially hesitant to the new Treaty. The US has in the meanwhile signed, and is considering ratification by 2009.[32]

As detailed in Chapter 5, the IU and the ITPGRFA in particular have a range of provisions pertaining directly to intellectual property rights. This is the result of intensive interaction between the two regimes ever since the IU was initiated, and has shaped the IU/ITPGRFA in various ways during the development stages of the two regimes.

Table 8.3 provides an overview of overlaps in functional scope between TRIPS/UPOV and the IU/ITPGRFA, where 18 items were identified.

Detailed Analysis of Overlap and Interaction between TRIPS/UPOV and the IU/ITPGRFA

In this section we will analyze item for item, as identified in Table 8.3 with regard to overlaps and interaction. For the sake of simplicity, we will focus on items 3.1–3.5, which are considered the most important in our context.[33]

Item 3.1: Divergence between the core norms of access to PGRFA and effective sui generis protection, resulting in interaction with barriers and strategic opportunities A core norm of the IU was the availability of PGRFA, originally free of charge, on the basis of mutual exchange and on mutually agreed terms to all IU signatories. After the agreed IU interpretations in 1989 (see Chapter 5), minimum restrictions and charges were to be imposed on the transfer of PGRFA; and plant breeders' rights as provided under the UPOV Convention were seen as compatible with the Undertaking. This happened two years before the adoption of the 1991 Act of the UPOV Convention, and only after considerable controversy between North and South.

When the TRIPS Agreement was adopted in 1994, it did not refer to any particular *sui generis system*, and was thus neutral in terms of overlap with the IU. However, as UPOV 1991 was increasingly advocated as the model for *sui generis* systems, and countries in North and South started to follow that path, barriers arose in interaction between the two regimes. The agreed interpretation of IU in 1989 referred to the 1978 Act of the UPOV Convention, but since 1991 the new UPOV Act has imposed restrictions on the availability of PGRFA for farmers, indirectly affecting their access to these resources, as explained above. This obvious conflict

31 WTO (2003e), *Understanding the WTO*, p. 2.

32 Stated by the delegate from the United States at the Second Session of the Governing Body of the ITPGRFA (29 October–2 November) on 2 November 2007.

33 An analysis of the remaining items is found in Andersen (2007).

was not addressed with regard to the IU, but instead discussed under the negotiations that led up to the ITPGRFA.

In negotiating the ITPGRFA, the parties had not only to harmonize the provisions with the CBD, but also ensure that there was no conflict with the TRIPS Agreement. As a response to that Agreement, the negotiators therefore developed formulations that prohibit intellectual property rights on material from the multilateral system or their genetic parts and components in the form they are received. This was a compromise between those wishing a stricter provision and those who did not want any such provision. The underlying intention is clear: to ensure broadest possible access to PGRFA. How it will be interpreted remains open: A narrow interpretation will allow intellectual property rights on the material after only minor alterations, whereas a broad interpretation will limit this scope. For example, the formulations of UPOV 1991 on essentially derived varieties as well as certain other varieties (Art. 14.5) could be used in interpreting the formulation 'in the form received'. That would mean that a variety that was either essentially derived from a variety received from the Multilateral System, not clearly distinguishable from such a variety, or whose production required the repeated use of such a variety, would not be eligible for intellectual property rights. Thus we see that the ITPGRFA provisions on intellectual property rights represent a strategic opportunity for shaping the outcome of the TRIPS Agreement in order to overcome the barriers between the two regimes and ensure access to PGRFA – at least for the genetic resources of Annex 1 plants.

Item 3.2–3.5: Various items of overlap Items 3.2 to 3.5 are closely interrelated with item 3.1. The overlap with regard to the core norm of access under the IU/ITPGRFA and the secondary norm of optional patenting under the TRIPS Agreement (item 3.2) is marginal, because the IU was (and the ITPGRFA is) focused mainly on plant varieties. However, the formulation in the ITPGRFA on prohibition of intellectual property rights over PGRFA in the form received from the Multilateral System includes their parts and components. In this sense there is overlap between the TRIPS Agreement and the ITPGRFA. Referring to item 3.1 above, we can define this overlap as neutral and the resulting interaction as one of strategic opportunities for implementation of the ITGRFA.

Items 3.3 and 3.4 concern the overlap between effective *sui generis* protection and benefit sharing (3.3) and farmers' rights (3.4). This overlap is neutral, since TRIPS has no provisions for or against benefit sharing, and because the norms on benefit sharing and farmers' rights do not exclude *sui generis* protection. Farmers' rights and benefit sharing are closely related in this context. The overlap between farmers' rights and effective *sui generis* protection has mainly to do with the right of farmers to participate on an equitable basis in the sharing of benefits arising from PGRFA use. For that reason, these two overlap constellations are discussed under benefit sharing below.

The norm of benefit sharing was first fully developed under the Multilateral System of the ITPGRFA, but was also addressed in the agreed interpretations of the IU in 1989. The barriers that developed between the IU and the TRIPS Agreement in this regard were similar to those described under item 2.1 above: The intellectual property rights introduced in many countries following from the TRIPS Agreement did not provide for benefit-sharing arrangements, and voluntary arrangements for benefit sharing were not usual with regard to PGRFA. At the same time access to genetic resources decreased, as a result of the developments described under item 2.2 above. A new problem has come with the regimes that have been developing in many countries to ensure benefit sharing by regulating access to *inter alia* PGRFA on a bilateral basis, thereby creating new obstacles to access to these resources, as described in Chapter 6.

The ITPGRFA represented a solution to several of the problems described above, among other things as a response to the TRIPS Agreement. In addition to limiting intellectual property protection, it provides for the sharing of benefits arising from the commercialization of PGRFA incorporating material from that system (para.13.2.d.ii): If the product is available without restrictions to others for further research and breeding, the recipient who commercializes it shall be encouraged to make a payment. If it is not available without restrictions, the recipient will be obliged to make a payment, to be determined by the Governing Body. The latter option refers to intellectual property rights, which restrict access for further research and breeding.

In responding to the TRIPS and its implementation, the negotiators of the ITPGRFA developed strategic opportunities to ensure the sharing of benefits from the commercialization of PGRFA through intellectual property rights. Here priority is given to farmers in the South, who are caretakers of the diversity of PGRFA (para. 18.3. and 18.5). This is why the interaction between the ITPGRFA and the TRIPS Agreement can be described as one with strategic opportunities both with regard to benefit sharing and with regard to farmers' rights.

Item 3.5 describes the overlap between effective *sui generis* protection under the TRIPS Agreement and limited intellectual property rights under the ITPGRFA as potentially diverging. How ITPGRFA limits intellectual property rights is described under items 3.3 and 3.4 above. In prohibiting intellectual property rights on these resources, it provides a barrier to effective *sui generis* protection of intellectual property rights to plant varieties. It could be argued that the provisions of the two agreements are not divergent, as the ITPGRFA provisions apply to material in the form that has been received from the multilateral system and that is therefore not novel, which is a criterion for patents as well as plant breeders' rights. However, how to interpret the formulation 'in the form received' is not yet clear. A narrow interpretation (as indicated above) would mean no real divergence between the two. Any broader interpretation, however, would affect implementation of the TRIPS Agreement with regard to intellectual property rights to plant varieties. That is why we designate this overlap as potentially diverging. As indicated under item 3.1 above, there is some scope for strategic opportunities for the ITPGRFA in this context, in that it may ensure better access to PGRFA through these regulations. On the other hand, this may in turn affect implementation of the TRIPS Agreement as barriers to effective *sui generis* protection.

Resulting Regime Constellations

In this section, the essence of the analyses in Part III will be derived in terms of resulting regime constellations and assumed aggregate regime effects in developing countries. Table 8.4 summarizes the regime constellations regarding PGRFA management.[34]

The contents of Table 8.4 will be explained in the following sub-sections. Here we will also seek to derive assumptions on potential aggregate effects in developing countries. In considering potential aggregate effects of regime constellations at the national level, there are basically three sets of factors to account for: (1) the contents of the regime constellations as they appear to the countries in question, (2) regime interaction pertaining to national implementation, and (3) the specific framework conditions in the countries in question. The first set of factors is about general expectations, without taking the particular conditions in each country into consideration, and will be our focus here. The other factors will be dealt with in Part IV.

Constellations Regarding Conservation, Sustainable Use and Utilization

This is the only one of the PGRFA management components to be directly affected by only two of the three regimes, namely the IU/ITPGRFA and the CBD, although it is also indirectly affected by the TRIPS Agreement. As such, it has already been dealt with under item 1.1. At this point we can simply note that the norms of the two regimes are generally compatible with regard to conservation and sustainable use, but that the rules diverge with regard to *ex situ* conservation. The divergence is the result of the CBD providing for domestic *ex situ* facilities in countries of origin of PGRFA in order to ensure national control – as a response to globally emerging intellectual property rights systems, whereas the IU and the ITPGRFA call for international co-operation. Domestic *ex situ* facilities would provide greater control over the resources for 'countries of origin'. Due to scarce financial and institutional resources, international co-operation could ensure better maintenance of acquired accessions. In addition to this divergence, there was confusion with regard to the division of labor between the CBD and the FAO for PGRFA management. The Global Plan of Action sought to speed up the conservation and management work, but could not solve the confusion about the division of labor. Until 2002, there was only marginal interaction between the two regimes, since the COP-CBD was awaiting the results of the ITPGRFA negotiations. In the meantime the COP-CBD concentrated on the conservation of wild biological diversity, particularly in protected areas, and referred *ex situ* conservation of PGRFA to the FAO. With the ITPGRFA, the division of labor is more clarified, and there is scope for synergies between the two.

34 This simplified overview relates the regime constellations to the PGRFA management components identified in Chapter 3. The component of PGRFA utilization is merged with sustainable use (first item), as the same constellations apply for both. Intellectual property rights, not included in the list of components necessary for PGRFA management in Chapter 3, have been included here for the purpose of overview, as has a further central item: funding. This is just to indicate the funding situation at the international level. It will not be dealt with as a separate component at the national level, as it is a precondition for all other management components.

Table 8.4 Resulting regime constellations for developing countries regarding PGRFA management

Central items	Regimes	Overlap	Interaction	Regime constellations
Conservation/ sustainable use and utilization of PGRFA	IU/ITPGRFA CBD (TRIPS is an indirect factor)	Compatible norms on both items but diverging rules with regard to *ex situ* conservation	Marginal interaction up to 2002. Confusion with regard to division of labor. Since 2002 a CBD strategy on plant conservation is to ensure co-operation with FAO	From 1993 to 2004 no clear international guidelines on *in situ* management of PGRFA. *Ex situ* conservation was primarily to be carried out by national gene banks
Access to PGRFA	IU/ITPGRFA CBD TRIPS	Diverging norms among all three regimes	As a response *i.a.* to TRIPS, the CBD got provisions regulating access. Its bilateral approach made it difficult to harmonize IU with CBD. The result was a Multilateral System under ITPGRFA	From 1993, developing countries were to establish access legislation and from 2000 legislation on intellectual property rights. After 2004, access to specified crops to be ensured through the Multilateral System
Benefit sharing	IU/ITPGRFA CBD TRIPS	Diverging norms between CBD and IU/ITPGRFA, neutral with TRIPS, but scope for conflict	The CBD approach to benefit sharing, spurred by TRIPS Agreement, made harmonization with IU difficult. With the Multilateral System the problem was solved with regard to benefits from specific crops	Since 1993, benefit sharing arrangements to be included in access legislation as appropriate, on a bilateral basis. No specific rules for PGRFA. After 2004, the Multilateral System applies to Annex 1 plants
Farmers' rights (FR)/ traditional knowledge (TK)	IU/ITPGRFA CBD TRIPS	Compatibility between CBD and IU/ITPGRFA, neutral with TRIPS, but scope for conflict	No interaction between CBD and IU, but CBD-COP initiative towards ITPGRFA in 2002. Barriers between CBD and IU due to access rules. Scope for synergies between CBD and ITPGRFA	Pressure to establish policies on TK, but not on FR. Rules introducing access legislation do not account for FR. Rules on the introduction of IPR after 2000 account for neither FR nor TK
Intellectual property rights (IPR)	IU/ITPGRFA CBD TRIPS	Neutral norms between IU + CBD and TRIPS, diverging between TRIPS and ITPGRFA	IPR a central reason for the CBD regulation on access and benefit sharing; they made negotiation of the ITPGRFA difficult, and resulted in provisions limiting IPR	Since 2000, increased pressure for developing countries to establish IPR systems for PGRFA. Heavy pressure to follow UPOV in these efforts, but no legally binding obligation
Funding pertaining to PGRFA management	IU/ITPGRFA CBD TRIPS	Compatible overlap for IU/ITPGRFA and CBD; and TRIPS and ITPGRFA; neutral for IU/ CBD & TRIPS	Scope for synergy between IU/ITPGRFA and CBD, but little interaction up to 2002. Part of ITPGRFA funding strategy since then, which also includes IPR related benefits	Marginal funding available for PGRFA management up to 2004. Since then greater scope but still uncertain how the funding strategy is to ensure sufficient funds

As a result of the interaction between the two regimes, there were no clear international guidelines on *in situ* management of PGRFA from 1992 to 2004. *Ex situ* conservation was primarily to be carried out by national gene banks, according to the legally binding international framework of the CBD.

Against this background, we may expect that one aggregate effect of these regime constellations for domestic implementation in developing countries was scant attention to *in situ* conservation and sustainable use of PGRFA up to 2002. CBD countries would be inclined to follow the COP-CBD: wait for the new agreement on PGRFA, not least since the IU, which was about to be superseded, was not legally binding. Secondly, from 1992 to 2004 we may expect a greater focus on the establishment and maintenance of national gene banks and less attention to international co-operation on *ex situ* conservation, which may have affected conservation activities negatively. From 2004 international co-operation on *ex situ* conservation has received more emphasis, with the entry into force of the ITPGRFA, and with the establishment of the Global Crop Diversity Trust, as an independent fund under international law.[35]

All in all, we can assume that from 1992 to 2004 regime constellations contributed to postponing crucial efforts necessary for the conservation, sustainable use and utilization of PGRFA.

Access to PGRFA

As we have seen, there are diverging norms between all three regimes with regard to access. The CBD provisions on access to PGRFA were motivated by the trend towards intellectual property rights to plants – as a reaction particularly to the emerging TRIPS Agreement. This in turn made negotiations for the new ITPGRFA difficult. The result was a Multilateral System to facilitate access to a number of specified plants.

The resulting constellations were threefold. From 1993, developing countries were to establish access legislation on genetic resources. As the CBD did not distinguish between domesticated and non-domesticated resources, countries establishing such legislation tended to include PGRFA in their general access legislation. From 2000, developing countries were to have established intellectual property rights legislation to plant varieties. There has been massive pressure to follow UPOV in these efforts. Since 2004 they are to include in the Multilateral System the PGRFA of Annex 1 plants under their management and control and in their public domain.

The result of these regime constellations has been reduced access to PGRFA up to 2004. The bilateral system of the CBD was designed with a view to non-domesticated resources and not to PGRFA. As we have seen (and will explore in greater depth in Parts IV and V), the first generation of access legislation led to more bureaucratic and time-consuming procedures and reduced accessibility, without generating the envisaged benefits to the provider countries of PGRFA.

In a counterfactual perspective, we could ask whether the situation would have been different without the CBD. Would countries have introduced access

35 See: <www.croptrust.org/main/index.php>.

legislation anyway? It will never be possible to answer this with certainty. As we will see in the next part, we can be quite confident that actors in the Philippines would have tried to develop some sort of regulation on non-domesticated plants for the pharmaceutical sector, but that such a regulation would not have covered PGRFA. It would also probably not have been that strict. It also seems likely that some efforts would have been undertaken in Latin America to regulate access to wild biodiversity, since financial benefits were expected, particularly with regard to pharmaceutical products. PGRFA would probably not have been covered by such legislation, because biodiversity was generally associated with non-domesticated plants and animals, and not with PGRFA – until the CBD. The CBD provided a framework for defining access regulation, and was also a source of inspiration. A basic assumption is therefore that it made a substantial difference, not least with regard to the inclusion of PGRFA in access regulation. More precisely how the CBD affected access legislation in developing countries will be discussed in the next part of this book.

The TRIPS Agreement affects access from another angle. Most countries follow UPOV 1991 in their implementation of Article 27.3 (b) of the TRIPS Agreement – more or less. Depending on how this is done, it affects access to PGRFA. If farmers are not allowed to exchange or sell farm-saved seeds from protected varieties, as provided for in UPOV 1991, that reduces their access possibilities. If farmers are not allowed to use their own farm-saved seeds for the next harvest (optional in UPOV 1991), it also reduces their access to PGRFA. Similarly, the reduced possibilities for breeders to use protected varieties of plants, as compared to UPOV 1978, lessen accessibility for breeders. Also here a counterfactual perspective is relevant: Would developing countries have followed UPOV without the TRIPS Agreement? Probably they would – as a result of bilateral agreements between individual countries and groups of countries, the TRIPS-plus agreements highlighted in Chapter 7. However, negotiating such agreements on a bilateral basis has taken more time, so the process would probably have been slower.

The interaction among the three agreements was not conducive to access to PGRFA until the ITPGRFA entered into force. Access to Annex 1 plants is likely to be easier, but access to other plants remains a difficult point.

For the period studied in this book, we may expect that the aggregate effect of the international regimes pertaining to PGRFA management was reduced legal access to these resources.[36]

---

36 Whether access is achieved by illegal means is another question, and one that cannot be dealt with here. However, this option should be kept open, since there is not such a long tradition of restrictions on access to PGRFA, and most farmers and local people would only be proud to share their seeds and to feel that these are deemed valuable by visitors. Seeds are also easy to export and import because of their size. Moreover, information about new access legislation may not be distributed sufficiently in developing countries with scarce institutional and financial resources, so that violators of such legislation may not realize they are breaking the law.

Fair and Equitable Sharing of PGRFA-Related Benefits

The norms on benefit sharing are divergent between the CBD and the IU/ITPGRFA, and neutral with the TRIPS Agreement. The divergence between the CBD and the IU/ITPGRFA has two dimensions. First, it relates to the priority of the two norms: Whereas access is more important than benefit sharing under the ITPGRFA, benefit sharing is more important than access under the CBD, as noted in Chapters 5 and 6. Second, and as explained in Chapter 6, the bilateral approach to benefit sharing under the CBD is ill-suited for PGRFA, because it is most difficult to identify countries of origin, and since there is little potential for PGRFA-related benefit sharing on a bilateral basis. For the plants included in the Multilateral System, these barriers were solved with the multilateral approach to access and benefit sharing under the ITPGRFA. The overlap with the TRIPS Agreement on this item is neutral, since TRIPS has no provisions for or against benefit sharing. However, none of the agreements provides for disclosure of origin of PGRFA in applications for intellectual property rights. Therefore the burden of regulation for benefit sharing remains with the provider countries of PGRFA and not with applicants for intellectual property rights. In effect, intellectual property rights systems derived from the TRIPS Agreement and UPOV will therefore not provide for the sharing of benefits derived from the use of PGRFA. In the case of the Multilateral System, the burden of proof remains with the providers of genetic resources and the 'Third Party Beneficiary'. Specific provisions have been included in the Treaty on benefit sharing in the case of intellectual property rights on material developed on the basis of PGRFA received from the Multilateral System. This regulation can be seen as the utilization of a strategic opportunity.

Thus we may expect as an aggregate effect of the three regimes that the most important benefit with regard to PGRFA – access to these vital resources for food production – was reduced from 1992 and up to 2004. Likewise few, if any, other forms of benefit sharing can be expected to result directly from the transfer of genetic resources. However, there might be examples of indirect benefit sharing in terms of development co-operation in PGRFA management. We lack sufficient background data to develop assumptions as to what benefits there will be to share for the Multilateral System.

The Recognition of Farmers' Rights and Traditional Knowledge

When it comes to farmers' rights and the recognition of traditional knowledge, there is compatibility between the CBD and the IU/ITPGRFA, as shown above. Between these two agreements and the TRIPS Agreement the overlap is neutral, since TRIPS has no provisions directly dealing with farmers' rights or with the recognition of traditional knowledge.

Until 2002, there was no interaction between the CBD and the IU with regard to this item, despite obvious parallels between the two. With the ITPGRFA, some scope for synergy has emerged, and there is a joint understanding of potential synergies, as expressed in UNEP/CBD/COP VII/ Inf. 18.

There was pressure to establish policies on traditional knowledge, but not on farmers' rights – until 2004. The rules concerning the introduction of access legislation after 1993 do not cover farmers' rights; and the rules on the introduction of intellectual property rights after 2000 include neither farmers' rights nor traditional knowledge. Both types of legislation may have negative effects on the realization of farmers' rights, perhaps also with the introduction of policies on traditional knowledge.

Access regulation may negatively impinge on the realization of farmers' rights if it makes it difficult to establish community seed banks, back-ups of such seed banks in national or international gene banks, or if it hinders participatory plant breeding. It may also obstruct work on documenting traditional knowledge, as has been the case in Peru, Colombia and several other countries (see for example Brush 2004 and 2005; Andersen 2005b), where the protection of traditional knowledge has been incorporated in access regulation. In fact, it has been reported that scientists working for PGRFA conservation are not even allowed to talk with farmers without prior permission (ibid., see for example Andersen 2005b, 55). Surely this represents a considerable obstacle to conservation work for the benefit of farmers. Such experiences indicate that there may not always be synergies between recognition of traditional knowledge – combined with access legislation – and farmers' rights. On the other hand, it should be possible to overcome these barriers once farmers' rights are recognized and taken into account.

Although the TRIPS Agreement is neutral with regard to farmers' rights and the recognition of traditional knowledge, the way in which it is implemented produces barriers to these issues. Obstacles emerge when farmers' access to PGRFA is restricted through intellectual property rights to protected varieties, when their traditional use of these resources is limited, and when traditional knowledge is used for products or processes for which intellectual property rights have been granted.

We may expect that, even though little happened to realize farmers' rights up to 2002, work for the recognition of traditional knowledge under the CBD progressed during this period. This, combined with access regulation in many countries, may have limited farmers' rights, as explained above. We may also expect that implementation of the TRIPS Agreement, following UPOV 1991 more or less, will lead to a deterioration of farmers' customary rights to save, use, exchange and sell plant varieties. It may be that farmers' rights will not receive high priority when the ITPGRFA is implemented, since its provisions are optional to governments. Realization of farmers' rights will therefore depend on initiatives by the parties to encourage the national and international implementation of these ITPGRFA provisions.

All in all, we may expect deterioration in farmers' rights as the aggregate effect of the regime constellations up to 2004. This may impact on the livelihoods of traditional farmers as well as their ability to maintain PGRFA.

Intellectual Property Rights to PGRFA

The overlaps between the norms of the IU, CBD and TRIPS Agreement on intellectual property rights to PGRFA are neutral. The IU addressed intellectual property rights only in the agreed interpretations of 1989,[37] making clear that there was no

37 FAO Conference Resolution 4/89 (1989a).

contradiction between the IU and plant breeders' rights as known under the UPOV Convention (1978 Act). The CBD provides for the recognition and consistency with intellectual property rights but obliges the parties to ensure that such rights do not run counter to the objectives of the CBD. The overlap between the TRIPS Agreement and the ITPGRFA is divergent, since the ITPGRFA prohibits intellectual property rights on material from the Multilateral System in the form it is received.

It was primarily because of the emerging regime on intellectual property rights to PGRFA that access and benefit sharing received such emphasis in the CBD. In turn, negotiations on the ITPGRFA were made difficult by the combination of the provisions of the TRIPS Agreement on intellectual property rights to PGRFA and the access and benefit-sharing regulation under the CBD. As we saw in chapters 5 to 7 and at the start of this chapter, these constellations resulted from the way in which various driving forces pursued their interests and used forum shopping as a strategy. The different origins and rationales behind the regimes also played a role. These were the main reasons why negotiations became so protracted. However, the end-result was that the ITPGRFA contained provisions that modified the outcome of the TRIPS Agreement – an example of negotiating parties making use of the strategic opportunity of developing the most recent regime.

We may expect that, as aggregate effects of these regime overlaps, most developing countries did not implement the TRIPS Agreement before they were forced to: that most of them started these processes close to 2000 (although, as noted, the least-developed countries have later deadlines), and that they have been following UPOV 1991 more or less, due to substantial pressure – as noted in Chapter 7. Subsequently several developing countries have adopted legislation on this point (as we will explore in greater depth in Part V), but domestic implementation of such legislation is probably only starting in most cases. How the Multilateral System will modify such legislation remains to be seen.

Funding and PGRFA Management

Funding has always been a bottleneck to PGRFA management, not least for the implementation of the IU/ITPGRFA. There is compatible overlap between the provisions of the IU/ITPGRFA and the CBD, in that the GEF is seen as one of the future financial sources for both agreements. However, the GEF will need substantially greater financial resources if it is to make a difference with regard to PGRFA management, since its mandate is already very broad. As for the ITPGFRA and the TRIPS Agreement, we also find compatibility, in that the funding strategy for the ITPGRFA includes benefit sharing from intellectual property rights based on genetic material from the Multilateral System.

The aggregate effect of this situation is expected to be marginal funding for PGRFA management up to 2004, and somewhat better scope for such funding after that. Future funding perspectives remain uncertain.

Summary of Findings from Part III

To summarize the findings in Part III we compare the intentions behind the single regimes, as analyzed in Chapters 5 to 7, with the assumed aggregate regime effects presented in this chapter.[38] Table 8.5 provides an overview of the findings.

Above, we concluded that the regime constellations from 1992 to 2004 can be assumed to have contributed to delaying crucial efforts necessary for the conservation, sustainable use and utilization of PGRFA. However, a core intention behind the IU/ITPGRFA and the CBD has been to strengthen the efforts for the conservation and sustainable use of PGRFA. If these assumptions regarding the aggregate regime effects are supported by our findings, this means that the regimes have not worked according to their intentions with regard to PGRFA conservation and use in the period under study – due to regime interaction.

We also concluded that reduced legal access to PGRFA could be expected as an aggregate effect of the international regimes in the period under study. However, access to genetic resources is a central intention behind the IU/ITPGRFA as well as the CBD. Even the core countries behind the provisions of the TRIPS Agreement on plants and plant varieties advocate free access to PGRFA, which is not protected by intellectual property rights. If our assumptions are supported by our findings, this means that the regimes have worked contrary to their intentions in the period studied with regard to PGRFA – and again, this is due to regime interaction.

Thirdly, we concluded that few if any direct benefits from the use of PGRFA were assumed to be shared as a result of new legislation on access and benefit sharing, as a follow-up to the CBD. Quite the converse: the major benefit with regard to PGRFA – access to these vital resources – would be limited. Thus, a factor crucial for food security – free access to PGRFA – would be sacrificed for few if any benefits. If this assumption is supported by our findings, it means that the interaction between the TRIPS Agreement and the CBD has caused a situation that has worked contrary to the intentions behind the CBD as well as those of the IU/ITPGRFA.

And fourthly, we concluded that recognition of farmers' rights would deteriorate as a result of the interaction among all the regimes studied here. However, recognition of farmers' rights is a norm under the IU/ITPGRFA, and is covered implicitly by the CBD. If our assumption is supported by our findings, it means that the international agreements pertaining to PGRFA management have worked contrary to these norms on farmers' rights.

38 This summary follows the core components of PGRFA management listed in Table 3.1, but lumps together three of them (*in-* and *ex situ* management and utilization) because they are closely interrelated in this context. It also includes intellectual property rights and funding, as it is useful to understand the regime interaction in this regard in order to see how this has affected the management of PGRFA.

Table 8.5 Comparing assumed aggregate regime effects in developing countries with single-regime intentions

<table>
<tr><th>Central items</th><th>Regimes</th><th>Single-regime intentions</th><th>Assumed aggregate effects in developing countries</th></tr>
<tr><td rowspan="2">Conservation sustainable use and utilization of PGRFA</td><td>IU/ITPGRFA</td><td>Domestic in situ management, inter-national co-operation on ex situ management</td><td rowspan="2">Up to 2002, scant attention to in situ management. Since 1992, greater ex situ conservation focus on national gene banks (due partly to indirect effects of TRIPS Agreement)</td></tr>
<tr><td>CBD</td><td>Domestic in situ management, country based ex situ facilities</td></tr>
<tr><td rowspan="3">Access to PGRFA</td><td>IU/ITPGRFA</td><td>Expeditious access to PGRFA, from 2004 facilitated through the Multilateral System</td><td rowspan="3">From 1993, access to PGRFA increasingly limited by legislation on access to genetic resources, and from 2000 by legislation on intellectual property rights. After 2004, access to specified crops ensured through the Multilateral System</td></tr>
<tr><td>CBD</td><td>Access to be facilitated by 'country of origin', upon prior informed consent</td></tr>
<tr><td>TRIPS</td><td>Limited access to PGRFA protected by IPR</td></tr>
<tr><td rowspan="3">Benefit sharing</td><td>IU/ITPGRFA</td><td>Major benefit is access; no regulation of other forms of benefit sharing up to 2004</td><td rowspan="3">The major benefit, access, was reduced until 2002; few (if any) other direct benefits. After 2004, access is ensured to a list of crops, and benefits from specified intellectual property rights to be warranted</td></tr>
<tr><td>CBD</td><td>Subject to legislation, benefit sharing to be arranged upon mutually agreed terms</td></tr>
<tr><td>TRIPS</td><td>No provisions on benefit sharing</td></tr>
<tr><td rowspan="3">Farmers' rights/ traditional knowledge (TK)</td><td>IU/ITPGRFA</td><td>Farmers' rights remain optional to governments and are not expected to receive high priority</td><td rowspan="3">Little has happened to realize farmers' rights so far, more with regard to TK. Access legislation limits farmers' rights. After 2000, farmers' rights may deteriorate if intellectual property rights are introduced without legislation on recognition of farmers' rights</td></tr>
<tr><td>CBD</td><td>Subject to legislation, parties shall as far as possible respect and maintain TK – and ensure benefit sharing</td></tr>
<tr><td>TRIPS</td><td>No provisions on farmers' rights/TK</td></tr>
<tr><td rowspan="3">Intellectual property rights (IPR)</td><td>IU/ITPGRFA</td><td>IU compatible with UPOV 1978. Limitation of IPR under ITPGRFA</td><td rowspan="3">Intellectual property rights cause worry in the South regarding access to PGRFA. Since 2000, increased pressure on developing countries to establish IPR systems pertaining to PGRFA. Follow-up generally as late as possible, based on UPOV</td></tr>
<tr><td>CBD</td><td>IPR compatible with the CBD, but parties shall ensure that they are supportive of the CBD objectives</td></tr>
<tr><td>TRIPS</td><td>Effective sui generis systems for plant varieties. Optional patenting of plants</td></tr>
<tr><td rowspan="3">Funding for PGRFA management</td><td>IU/ITPGRFA</td><td>No funds for domestic implementation through IU, likely with ITPGRFA</td><td rowspan="3">Marginal funding for PGRFA management up to 2004. Since then better scope, but it remains uncertain how the ITPGRFA funding strategy can ensure sufficient funds</td></tr>
<tr><td>CBD</td><td>Marginal funds for domestic PGRFA implementation through GEF</td></tr>
<tr><td>TRIPS</td><td>IPR ensure remuneration for breeders</td></tr>
</table>

All in all, the assumed aggregate regime effects for the period between the entry into force of the CBD (1993) and the entry into force of the ITPGRFA (2004) for PGRFA management are largely negative – with little domestic emphasis on conservation and sustainable use, reduced access, few (if any) other shared benefits, limitation of farmers' rights to PGRFA and marginal funding for PGRFA management. We can note one major achievement for PGRFA management during this period, but that must be seen as an action undertaken in spite of regime interaction and not because of it: the establishment of the *International Network of Ex Situ Collections under the Auspices of the FAO.* This brought the collections of PGRFA in the gene banks of the CGIAR institutions under FAO auspices, and increased their availability.[39] Indeed, this was a positive development – however, it was the result not of regime interaction, but of an initiative taken by the FAO. With the entry into force of the ITPGRFA in 2004, prospects emerged for positive developments concerning PGRFA management as well as regime interaction in this regard.

In the remainder of the book we examine whether our assumptions on the aggregate effects of the international regimes pertaining to PGRFA management have been correct, and also inquire how and why these effects developed as they did. Finally, we will discuss the prospects for change in light of recent developments within – and the interaction between – the various regimes.

39 In addition the CGIAR institutions introduced the System-wide Information Network for Genetic Resources (SINGER), an invaluable database for individual collections in the gene banks, containing all available characteristics and information. Later the Global Crop Diversity Trust substantially improved the situation for *ex situ* conservation.

PART 4
Domestic Responses to the International Regime Constellation Pertaining to Plant Genetic Resources for Food and Agriculture: Cases from the Philippines

In order to analyze how the international regime constellations affect PGRFA management in developing countries, we now turn to the case studies. We begin in Chapter 9 by introducing the Philippines so as to establish a better understanding of the conditions under which the findings from the case studies will have general relevance for other developing countries and as a context for the case studies themselves. On this backdrop the policies pertaining to PGRFA are presented. This forms the foundation for the analysis of domestic policy decisions and goal achievements concerning PGRFA management. We then analyze the state of policy decisions as well as of goal achievements pertaining to the core components of PGRFA management, as identified in Chapter 3. We end by discussing whether the findings can be explained as effects of the international regimes. This question of attribution is further deepened in Chapter 10.

In the two embedded case studies on access and benefit-sharing regulation and plant breeders' rights legislation in Chapter 10, we investigate the processes leading to policy decisions and implementation, with a view to examining the relative effects of the international regimes studied in this book. By focusing on the actors engaged in these processes and how they influenced the legislation processes, we seek to grasp the factors that influenced the positions of these actors and the effects on goal achievement, including the international regimes and their interaction. On this basis it will be possible to determine the extent to which the international regimes affected the policies in question. Central in Chapter 10 is the discussion of how the mechanisms of influence outlined in Chapter 3 may explain our findings, taking the propositions developed in that chapter as points of departure. Chapter 10 ends with conclusions in this regard.

Chapter 9

Effects in the Philippines

The Philippine Context

We begin with an introduction to the Philippines, before highlighting its characteristics as a case country for the study of PGRFA management, with a view to the potentials for generalization to other developing countries.

Basic Data

The Republic of the Philippines is an archipelago of 7,107 islands in tropical Southeast Asia, north of the equator. Some 1,190 of the islands are inhabited by altogether approximately 90 million people. The islands are divided into four groups: Luzon in the north, the Visayas in the central and eastern areas, Mindanao in the south and Palawan in the far west. The topography of the Philippines is characterized by mountainous areas with hill slopes divided by fertile plains and surrounded by the sea and by coral reefs. The climate is tropical, with an average annual temperature of 27°C, characterized by high humidity and abundant rainfall.

Despite the rapid devastation of biological diversity in the Philippines, there is still a wealth of flora and fauna. As for PGRFA, a total of 1,663 species of plants relevant to agriculture have been recorded in the Philippines (DENR and UNEP 1997, 70). Of these, 477 species have food value, 353 species have feed/fodder value, 632 species have medicinal/ herbal value, and 201 have ornamental value. In addition, 35 species are considered as fiber crops. Of the crops, 28 food crops and 15 medicinal plants have been found to be endemic to the Philippines.

Principal farm products are rice, corn, coconut, sugarcane, *abaca* (Manila hemp), tobacco, maguey and pineapple (NEDA 1995, iii; Department of Agriculture 1995, 13). A wide variety of fruits are grown, including banana, mango, citrus fruits and papaya. Vegetables and root crops like mungbean, yardlong bean, cowpea, tomatoes, eggplant, cabbage, mustard, lettuce, sweet pepper, pea, sweet potatoes, sugar beets, cassava and squash are raised for domestic consumption (Department of Agriculture 1995, 13).

Despite rapid urbanization, around 50 percent of the population still live in rural areas, where the vast majority depend on farming for income/livelihood. (Fishing is also important.) More than half of all rural families live below the poverty line, according to the Philippine Bureau of Agricultural Research (2001, 5). The social indicators of development show some progress in recent decades, particularly with regard to literacy,[1] life expectancy, access to safe drinking water, entailing a general decline in child malnutrition and infant mortality rates (World Bank 2001). Two thirds of the 5 million poorest people in the Philippines live in rural areas. Average farm family income is only about half that earned by urban families. Despite the vast potential of agriculture, its per capita output has been low, further declining over the past two decades (ibid., 8). As of 2001, altogether 11.3 million people were directly employed in agriculture – or 37.4 per cent of the work force (ibid., 5).

Upland areas constitute about 60 per cent of the total land area and are predominantly hilly and sloping (Department of Agriculture 1995, 15). Holdings are small and produce is grown for own consumption, to some extent for sale. Each farm normally grows several crops. There is a high diversity of crops and within crops, since farmers rely heavily on own propagating material and the exchange of such. The lowlands are more typically monoculture, particularly in rice, corn, coconuts, sugar and pineapple, and mainly in the irrigated (as opposed to rain-fed) lands.

Colonial Heritage and Roots of Civil Society Movements

In 1565, colonial rule was established on parts of the archipelago by Spain. The colony was named *the Philippines* after King Philip II of Spain. Spain brought profound changes to cultural, economic and political life in the Philippines. The Roman Catholic Church was introduced from the very beginning and became deeply rooted in Philippine culture. Today 83 per cent of the population is Roman Catholic, whereas 9 per cent are Protestants.[2] The Philippines is the only country in Asia with a Christian majority. Muslims make up 5 per cent of the population, and live predominantly in the Mindanao region.

The most important changes in economic life came with the introduction of property rights to land and the increasing acquisition of land holdings for the Church and the Spanish colonists (Zaide 1999, 88). As a result, most of the land in central areas was owned by Spanish nobility, the Roman Catholic Church and a small group of Filipino aristocrats. The peasants shared their harvests with the landowners, normally paying more than half the crop to the landlords.[3] A hacienda system was created, which paved the way for patron–client structures (Gonzales 2001, 288). Along with these structures came the introduction of forced labor, where Filipino males between 16 and 60 years were required to work for the Spanish colonists for up to 40 days a year (Zaide 1999, 88). These developments prepared the ground for feudal structures in the Philippines, as well as for the still ongoing land disputes.

Fired by two novels by the famous Philippine medical doctor José P. Rizal[4] on the abuses of the colonists and priests, the self-educated lower-middle-class clerk

---

1 The share of adult literates was 93.5 per cent in 1995, according to NEDA (1995, iv). It has, however, decreased slightly since then (to 90 per cent in 2000, according the Philippine census).

2 According to most sources; see for example Wikipedia: Demographics of the Philippines at <www.answers.com /topic/demographics-of-the-Philippines>.

3 US Department of State (1950) in Schirmer and Shalom 1987, 70–77.

4 *Noli Me Tangere* ('touch me not') and *El Filibusterismo* ('the subversive').

Andres Bonifacio (Ileto 1979, 79) founded a secret society for the liberation of the Philippines in 1892, the Katipunan.[5] It was based on such traditional Philippine values as the virtue of *lakas loob*, inner strength, and brotherhood (the meaning of *katipunan*). The objective was not only political self-rule but also inner liberation, as was underlined with special initiation rites for all newcomers to the society (Bankoff and Weekley 2002, 101; Ileto 1979, 75–113). This was a spiritual and political movement in one, building on the nation's cultural foundations and seeking to develop them for independence and a new genuinely Philippine society.[6] The movement spread quickly: it had 20,000 members by 1896, when it was discovered and the revolution started. Many battles were fought and labor strikes organized. During the rebellion, Rizal, who did not participate, was captured and executed under dramatic circumstances in Manila in 1896.

However, the Katipunan split in two factions – one following Bonifacio, and one following Emilio Aguinaldo, who proved to be a better military leader. The rivalry ended in fighting between the factions, and the execution of Bonifacio by a squad of Aguinaldo's soldiers. Thereafter much of the force went out of the movement. Following a peace agreement in 1887, Philippine independence was proclaimed on 12 June 1898,[7] under the condition of close ties to Spain, and that Aguinaldo went in exile to Hong Kong (Zaide 1999, 120). On this background, he declared that Katipunan no longer existed (Aguinaldo, quoted in Ileto 1979, 157), thereby overlooking the spiritual dimension of the Katipunan movement as well as the nation-building aspect. The revolution was transformed to a purely political idea (Bankoff and Weekley 2002, 102), and many of the revolutionaries felt that they had not achieved their aim. The lost spiritual dimension of the revolution is an issue among some groups in civil society even today.[8] This history provides a background to understanding the strength of NGOs in the contemporary Philippines.

Following the Spanish-American war in 1898, the United States attacked the Philippines to combat the Spanish, and was welcomed by the Filipino revolutionaries as allied forces. Aguinaldo and his men returned to the Philippines. The revolutionaries established a congress and adopted the first republican constitution in Asia (Bankoff and Weekley 2002, 1). Representatives from the US Artillery participated in the celebrations and co-signed the document. However, the new republic was short-lived indeed. Under the terms of the 10 December 1898 Treaty of Paris between Spain and the United States, the Philippines was handed over from Spain to the United States, becoming a US colony. The Filipinos regarded this as a deep betrayal, and Emilio Aguinaldo organized a massive guerrilla force and declared war against the

5 *Kataastaasan Kagalang-galang na Katipunan ng mga Anak ng Bayan* ('most respected and honorable society of the sons of the country').

6 On Philippine identity, see also Constantino (1974 and 1982) and Jocano (1997).

7 Historians have never agreed on the exact date of Philippine Independence, and many dates are presented in the literature. However, the official date is 12 June 1898.

8 In January 2005, the Philippine civil society activist, author and Right Livelihood Award winner Nicanor Perlas, together with various groups and organizations, conducted a conference which he called Karangalan (having honor/dignity), which – inspired by history – sought to address the spiritual dimension as a prerequisite for a sustainable development along with traditional Philippine values.

new colonists in 1899. The United States mobilized two thirds of its army (126,000 soldiers) for the war in the Philippines. Nevertheless, it took two years to defeat the guerrillas (Dusik 1986, 65; Zaide 1999, 132). Between 200,000 and 600,000 Filipino civilians lost their lives in, or as a result of, the brutal battles in the Philippine–American War until 1901, when Aguinaldo was captured, surrendered and gave his oath of allegiance to the United States.[9] This was seen as a second betrayal by many revolutionaries; some explained it with the termination of the Katipunan in 1898, and the subsequent loss of *lakas loob*, inner strength (Bankoff and Weekley 2002, 102). Fighting continued in parts of the country for several years.

During the Second World War, Japan invaded the Philippines in 1942. The following three years are among the darkest in Philippine history. Japanese rule in the Philippines was characterized by arbitrariness and atrocities, with the 'Death March' marking the height of this brutality; around 62,000 Filipino soldiers and 11,000 US troops were captured upon capitulation and forced to walk 120 kilometers from Bataan to Tarlac without food, water or medicine. Only 56,000 reached Tarlac, where they were interned in Camp O'Donnell, a concentration camp; of these soldiers, 24,000 died in internment. The memory of the Death March is deeply rooted in the Philippine culture, and shared suffering is one of many factors that can explain the close ties that developed between the Philippines and the United States despite the betrayal earlier in the century. Another factor is the US liberation of the Philippines from the Japanese in 1945, in collaboration with a Philippine resistance army, the *Hukbalahap*.[10]

On 4 July 1946, the Philippines celebrated its independence. Manuel A. Roxas was the first elected president of the new republic. In his two years in power before he died, Roxas adopted two agreements with the United States which were to have a wide-ranging impact on the development of the young republic: the Bell Trade Act and the Military Bases Agreement.

The Bell Trade Act[11] was passed in the US Congress and approved by the Philippine Congress only two days before independence (Schirmer and Shalom 1987, 88).[12] Through the Act, the Philippine economy was linked to the US economy in several ways. First, a system of preferential tariffs was established, which undermined the control of the Philippine government over imports and exports. Second, various further restrictions were placed on the control of the Philippine government over its economy. For example, the currency exchange rate was fixed and the Philippine government was obliged not to put any restrictions on financial transfers from the

9 Historians have never agreed on how many people died as a result of the war. The lowest figures (Dusik 1986; Zaide 1999) are around 200,000 civilians, 16,000 Philippine guerilla soldiers and 4,200 US soldiers, whereas a higher estimate is a total of 600,000 (Schirmer and Shalom 1987, 19; Bankoff and Weekley 2002, 64).

10 *Hukbo ng Bayan Laban sa Hapon*, which means 'people fighters against Japan', in short *Hukbalahap*. After the war, the US Army turned against the Hukbalahaps due to their Marxist foundations (US Department of State 1950, in Schirmer and Shalom 1987, 70–77).

11 The Philippine Trade Act of 1946 is known as the Bell Trade Act after its main sponsor, C. Jasper Bell of the US Congress.

12 An Executive Agreement formalizing the agreement between the two countries was signed on Independence Day.

Philippines to the United States (Bell Trade Act, article V).[13] Third, it granted US citizens equal rights with Filipinos to exploit, develop and utilize natural resources, including agriculture, timber, water and mining, as well as to operate public utilities (ibid, article VII). Despite the fact that the Philippine Constitution had reserved the development of natural resources and public utilities to Filipinos, or to corporations that were at least 60 per cent Filipino, US citizens could now own up to 100 per cent of these enterprises. Other foreign nationals could own only up to 40 per cent (Schirmer and Shalom 1987, 87).

The Bell Trade Act provoked Philippine nationalists. Nine years later, in 1955, Washington agreed to replace the act with a new treaty, the Laurel-Langley Agreement (Schirmer and Shalom 1987, 94). The new treaty was in effect similar to the Bell Trade Act. In 1970, US companies accounted for 80 per cent of the foreign-owned equity in the Philippines – a higher percentage than when the Philippines had been a US colony (Schirmer and Shalom 1987, 125). The Laurel-Langley Agreement was in force until 1974, but the structures it had shaped of US economic interests in the Philippines remained.

The Military Bases Agreement secured many military bases as property of the United States. This was originally for 99 years, but according to the revision in 1966, the agreement was to be terminated in 1991, which it was. Two of the bases were the largest US military bases outside the United States: Clark Air Base and Subic Naval Base (Schirmer and Shalom 1987, 96). Altogether, the 16 bases occupied a total area of 240,000 hectares throughout the country (Daroy 1988, 15). The Philippines could not place any restriction on the use of the bases, and the agreement allowed the United States to recruit Filipino volunteers into the US Armed Forces. Finally, it prohibited the Philippines from allowing any other power to have bases in the Philippines without the consent of the United States.

These two agreements were to shape Philippine politics decisively in the first four decades of formal independence. They also provided the foundation for further ties to the USA, which in turn provide an important context for understanding several of the PGRFA-related Philippine policies, particularly the Plant Variety Protection Act. In return, the Philippines received extensive development aid and military assistance, *inter alia* to combat the Hukbalahaps (Schirmer and Shalom 1987, 100).

Ferdinand Marcos, His Heritage, and People's Power

Philippine economy made good progress in the first 20 years of independence. In 1965 Ferdinand Marcos was elected president. He stayed in office for 21 years. Before introducing Martial Law in 1972 (followed by immense human rights abuses), Marcos started a comprehensive infrastructure program of roads, bridges, schoolhouses, government buildings, cultural centers and irrigation systems (Zaide 1999, 165). For these projects he earned great popularity and was the only president to be re-elected after the war (in 1969). However, his success was based on borrowed money. Ferdinand Marcos drew the Philippines into one of the most serious debt crises in Asia.

---

13 Excerpt printed in Schirmer and Shalom 1987, 89.

In 1983, one of his strongest opponents, Benigno (Ninoy) S. Aquino returned to the Philippines from exile in the US, but was assassinated at Manila Airport. More than two million people attended the funeral (Diokno 1988, first photo page after p. 175), which came to mark the beginning of open protest against Marcos in the form of massive demonstrations and political strikes. At the same time, the New People's Army (NPA), the military wing of the 1968-established Communist Party of the Philippines (CPP) had become so strong that it threatened political power (Bello 1987, 2; Diokno 1988, 165; Chapman 1987). Massive protests against Ferdinand Marcos, supported by the Catholic Church and parts of the military together with American pressure, finally made Marcos flee the country in 1986. This victory is often referred to as *People's Power* and as a peaceful revolution.

Ninoy Aquino's widow, Corazon (Cory) Cojuangco Aquino gave her oath as president of the Philippines the same day as Marcos fled. Political prisoners were set free. A new constitution was approved by the people in a free plebiscite in 1987. In the same year a new congress was elected, and local governments were elected the following year. The most important change in the first year of her presidency was probably the creation of democratic space.[14] Several new newspapers emerged, political talk shows were broadcast on television, and even communist leaders who were released from jail were interviewed in the media (Nemenzo 1988, 235).

Corazon Aquino inherited a budget deficit of unmanageable size after Ferdinand Marcos (Nemenzo 1988, 245). In 1972, foreign debt was USD 2.6 billion (Bankoff and Weekley 2002, 100). By the end of Marcos' presidency, the Philippines had a debt of approximately USD 28 billion (Kotte 1988, 226; Nemenzo 1988, 243; Garcia 1988, 35). Interest on debt amounted to more than 45 per cent of export earnings (Garcia 1988, 35). The Philippines has still not recovered from its debt burden, which is one reason why it ranks among the economically poorest in Southeast Asia. In this situation, the country is dependent on the Bretton Woods institutions, which in turn means that trade liberalization is the main direction of the country's economic policy.

One aspect of the debt burden is corruption, which assumed new proportions under the Marcos dictatorship. Marcos became known as *Mr. Ten Percent*, referring to the commissions demanded for every contract for the government's extensive infrastructure projects (Tiglao 1988, 41). Huge sums landed in his personal accounts abroad, and he and his family lived in inconceivable luxury.[15] A system of 'cronies' was established, friends who had or were placed in strategic positions and received great privileges in return for their support to the president. So ingrown was corruption in the political system that it continued under the successors to Marcos: Aquino (1986–1992) and Ramos (1992–1998) – and flourished again under Joseph Estrada

14 The concept of *democratic space* was introduced in the Philippines by Ed Garcia after the assassination of Ninoy Aquino in 1983, first referring to the areas of freedom which Filipinos could shape under dictatorship, through collective acts of defiance, and later to the freedoms which Corazon Aquino promised to restore (Nemenzo 1988, 234).

15 For example, it is commonly known that his wife Imelda possessed more than 3,000 pairs of shoes (Nemenzo 1988, 268, photo page 11).

(1998–2001) (Zaide 1999, supplement 2001, 3).[16] Estrada was himself accused of bribery, graft and corruption, betrayal of public trust, and culpable violation of the constitution (Zaide 1999, supplement 2001, 3), was ousted from office in 2001 by the second peaceful upheaval, known as *People's Power II*, and arrested. His Vice-President, Gloria Macapagal-Arroyo (2001–) was sworn in as President the same day and inherited the political structures from Joseph Estrada, as well as many of his cronies.[17] Corruption remains a substantial problem. President Arroyo was elected for a new presidential period in 2004, but has later been accused of election fraud after a tape with her voice – in dialogue with the highest responsible officer for the election – was made public, indicating that she demanded a certain election result from him.[18] The situation led to massive protests and political disorder followed by human rights abuses in terms of persecution and assassinations of journalists and political activists, still ongoing today.

The Political System

The President of the Republic of the Philippines is both the head of state and head of government. The government has three branches, in theory co-equal (Gonzales 2001, 261):[19]

- the executive, headed by the President;
- the bicameral legislature: the Philippine Congress, with the Senate and House of Representatives, headed by the Senate President and Speaker of the House respectively;
- the judiciary, with the Supreme Court as its highest body, headed by the Chief Justice.

The president, vice-president and senators are elected each sixth year on a nationwide basis. Members of the House of Representatives are elected each third year from 208 legislative districts and from so-called party lists representing labor, peasant, urban, poor, indigenous communities, women, youth and other sectors as stipulated by law (Gonzales 2001, 272). This is a special feature of the Philippine democracy, with roots in People's Power I. The party flora is diverse, with parties of many nuances and in various coalitions.

The 1987 Constitution institutionalized the influence of organizations from civil society on political processes. The participation of the people and their organizations in decisions at all levels of social, political and economic decision-making is termed

16 See also Doronila (ed.) (2001).

17 See for example Sheehan (2001); Mitchell (2001).

18 In the headlines of most newspapers in the Philippines throughout the summer period 2005.

19 During the Marcos era, the executive dominated the other two branches. After People's Power, the newly elected Congress had to learn legislative work, and it took time to strengthen this branch to the extent necessary in a parliamentary democracy (Source: As a trainee of the German Friedrich Naumann Foundation, I participated in the development of the legislative training of the congress members in 1988).

a right (Article XIII, Section 16), and the state is obliged by law to facilitate the establishment of adequate consultation mechanisms.

Former President Aquino sought to forge strategic alliances between the state and NGOs; she appointed a number of NGO leaders to the cabinet and presidential committees and established NGO Liaison Desks in 18 government departments and five specialized government agencies (Gonzales 2001, 278). NGO representatives were also invited to participate in central development planning. Her successor, Fidel V. Ramos, continued this line and strengthened state–NGO relations further. Under Joseph Estrada, these relations became more problematic. Some leaders from civil society were recruited as secretaries and other leading positions in the government, and some factions continued to have access to decision processes, whereas other factions seemed to lose this access.[20] Under President Arroyo, NGO influence weakened markedly.

In 1991, the Local Government Code was adopted, providing the regions with increased political power. The local government system consists of provinces, cities, municipalities and *barangays*. The province is the highest level of local government. It comprises cities (for urban areas) and municipalities (for rural areas). The lowest level of local governance in cities as well as municipalities is the *barangay*. All officials at all levels are elected. Local-level governance comprises policies pertaining to health care, social welfare, natural resources and the environment (Gonzales 2001, 263).[21]

Despite all these measures to ensure participation, the Philippine political system is still dominated by political dynasties with strong influence on central decision-making processes (Teehankee 2001). Some of these are old family clans that have maintained their political power ever since Philippine independence, whereas others – new political leaders with professional political skills – emerged in the years after People's Power I.

A Unique Case Country?

The Philippines is a unique country – as indeed is any country in the world. Nevertheless, it has features that are similar to many developing countries.

Environment The country is rich in biodiversity in the wild as well as in the fields, like most countries in Southeast Asia, South Asia, parts of West Asia, huge parts of Africa and most of Latin America. Just as in so many other developing countries, it has abundant diversity of PGRFA, as indicated above.

---

20 Based on interviews with NGO representatives in the Philippines in 2002.

21 A distinct administrative autonomy has been provided for the areas of Cordillera and Muslim Mindanao. According to Republic Act No. 6766 (Republic of the Philippines 1989), these areas have been given the status of autonomous regions where their indigenous peoples can exercise legislative, executive and judicial functions and maintain and develop their own economic, social and political institutions, based on their cultures (Gonzales 2001, 263).

Demography Demographic patterns are similar to those of most developing countries, with more or less rapid increases in population growth and increasing urbanization.

History The cultural heritage of the Philippines is unique, whereas the history of colonization is known throughout almost all the developing world. Nevertheless, different colonial powers created different structures and influenced cultures in various ways. The Philippines has much in common with the Latin American countries that were also colonized by Spain (for example in religion and societal structures), but 40 years under the United States provided the Philippines with other types of structures as well.

Poverty The Philippines is not among the poorest countries in the world in economic terms, but neither is it among the wealthier developing countries. Its poverty rates are comparable with a large number of developing countries, and it is quite typical also with regard to its ranking in the Human Development Index, although there are exceptions – for example, literacy rates are quite high.

Debt The country is deep in debt, as are most countries in the developing world, with structural adjustment programs more or less the same throughout the Third World. Trade liberalization is the direction of economic policy.

Civil society The history of the peaceful People's Power is unique to the Philippines, and provides one of several explanations for the vibrancy of its civil society. Nevertheless, many other countries have experienced various forms of liberation processes, particularly in connection with the struggle for independence, and there are similar features of civil society structures in many other developing countries.

Governance The political system is democratic, but does not always function according to democratic intentions – as in many other developing countries. Corruption is a problem – as in most developing countries. Corruption may follow different patterns in different countries, but the main features are probably similar.

Agriculture In the Philippines, as in most developing countries, agriculture is the largest of all sectors. The crop structure resembles that found throughout Southeast Asia, and the farming systems have many similarities. A relatively large proportion is basically subsistence agriculture with multiple cropping, with differences between rain-fed and irrigated agriculture, as in most developing countries. Diversity among and between crops is substantial in some areas, but genetic erosion has proceeded to a great degree in other areas.[22] The general problems of agriculture in the Philippines are similar to those found throughout the developing world: soil erosion, poor soil

---

22 There are few data available on changes in crop genetic diversity in the Philippines. According to the PCSD (2000), more than 3,000 traditional rice varieties were wiped out by the high-yielding varieties of rice introduced from 1966 on, and by related inadequate use of chemicals in agriculture.

quality, inadequate water supply, non-availability of seeds and propagating material, and marginal extension services (see Dar 1997, 7; Cramb et al. 2000).

The situation of farmers Rural poverty is a huge problem, as in most developing countries, but the situation in the Philippines is particularly extreme due to the distorted structure of land ownership.[23] The result – deep poverty – is comparable across countries, and it hampers improvements in agricultural production. Farmers are caught in a vicious circle, as in so many countries of the South.

In sum, the general framework conditions are – despite the uniqueness of the Philippines – not unlike those of many countries in the developing world. This provides a good basis for the discussion of the further relevance of the findings of this study.

Philippine Policies Pertaining to PGRFA Management

In this section we will look at the most relevant policies pertaining to PGRFA management, as expressed through official acts and departmental orders. This overview is necessary as a background for the embedded case analysis, since we will refer to these acts and regulations in various settings. The section concludes with an overview chart of the division of labour between the state entities established through all these policies.

In introducing the acts, we also embark on the first step of the analysis – to analyze their relevance with regard to PGRFA management in general, and the identified management issues in particular. There are many ways of categorizing the acts and regulations pertaining to PGRFA management. Based on the legislative landscape in the Philippines, Table 9.1 sets out one approach to presentation and analysis of the most relevant acts and orders.

Agricultural Policies

Philippine agricultural policy has one vital focus: *food security*. This is defined in the Agriculture and Fisheries Modernization Act (AFMA) as meeting the food requirements of the present and future generations of Filipinos in substantial quantity, and ensuring the availability and affordability of food to all, either through local production or importation.[24] Self-sufficiency is an important sub-goal, and particularly in rice and white corn.[25] The three documents of legislation presented here provide the foundations for Philippine policies regarding general agricultural development, import practices and the increase of rice production. However, no mention is made of PGRFA management – which indicates the low priority accorded by the Philippine agricultural authorities. These documents are included in this study because they

23 For an account of the situation of Filipino farmers, see Andersen 2007, 290–293.

24 Republic of the Philippines (1997b), Republic Act 8435, Section 4.

25 Republic of the Philippines (1997b), Republic Act 8435, Section 4; Department of Agriculture (1999); and <www.da.giv.ph/welcome.html> Programs: Blueprint for food security.

provide the basis for understanding Philippine agricultural policies, which in turn is a key to understanding the country's policy towards PGRFA management.

Table 9.1 Philippine policies pertaining to PGRFA management

Agricultural acts and orders providing the policy framework	Specific legislation on the conservation and use of PGRFA	Other major laws and regulations on conservation[a]	Regulation on access, benefits and ownership
Agricultural and Fisheries Modernization Act (AFMA) ***Republic Act 8435, 1997*** Agricultural Tariffication Act ***Republic Act 8178, 1995*** Hybrid Rice Program ***Administrative Order 25, 2001*** Implementation of the Program ***Executive Order 76, 2002***	Seed Industry Development Act ***Republic Act 7308, 1992*** High-Value Crops Development Act ***Republic Act 7900, 1995*** Coconut Preservation Act ***Republic Act 8048, 1995***	National Integrated Protected Areas System Act (NIPAS) ***Republic Act 7586, 1992*** National Biodiversity Strategy & Action Plan ***Memo. Order 289, 1995*** Philippine Agenda 21 (PA 21)[b] ***Memorandum Order No. 399, 1996*** Indigenous Peoples' Rights Act (IPRA)[c] ***Republic Act 8371, 1997***	Executive Order 247 on bio-prospecting (EO 247) ***Executive Order 247, 1995*** Wildlife Resources Conservation and Protection Act[d] ***Republic Act 9147, 2001*** Plant Variety Protection Act ***Republic Act 9168, 2002*** *Related:* Intellectual Property Code ***Republic Act 8293, 1997***

Notes

[a] This category covers diverse policies that can be characterized as being 'comprehensive'.
[b] PA 21 is relevant to most of the topics, but is placed here since it is a comprehensive policy.
[c] IPRA is also relevant for the category of access and benefit sharing (EO 247 and Wildlife Act).
[d] The Wildlife Act is also relevant with regard to nature conservation, but is placed here since it affects legislation pertaining to PGRFA on access and benefit sharing (EO 247).

The Agriculture and Fisheries Modernization Act[26] was approved on 22 December 1997 and came into effect 9 February 1998. The Implementing Rules and Regulations

26 Republic of the Philippines (1997b), Republic Act 8435.

for AFMA were adopted on 10 July 1998.[27] AFMA is the foundation for all Philippine policies pertaining to agriculture and, as such, is the most important law in this context. It is an all-encompassing law that covers the establishment of agricultural and fisheries development zones, modernization plans for agriculture and fisheries, credit facilities, irrigation, information and marketing support services, other infrastructure, production standardization and consumer safety as well as human resources development, research development and extension services. In addition it provides for rural non-farm employment, including a basic needs program, rural industrialization, and for trade and fiscal incentives. Despite this holistic approach to agricultural development, it does not address PGRFA management. The topic is not mentioned as such, and the closest AFMA comes to indicating anything in this direction, is in providing for an increase in the number of farms engaged in diversified farming (Section 15 g). 'Diversified farming' is, however, not defined.

Why is PGRFA management not addressed in AFMA? There are at least two reasons: (1) the main AFMA objective is to increase food production in order to meet a fast-increasing nutrition demand, and (2) PGRFA management is a matter of biodiversity management, which is the responsibility of the Department of Environment and Natural Resources (DENR).[28] The priority of food production is evident for a country that is struggling to meet basic food requirements – but the fact that its most important agricultural legislation does not address PGRFA management is a central finding for our analysis, and provides a clear indication of the low priority of this issue within Philippine policy. This is particularly important when we realize how crucially the poorest majority of farmers depend on PGRFA diversity.

In order to increase food production, AFMA aims at transforming agriculture and fisheries from a resource-based to a technology-based industry (Section 2). Central here is a more equitable distribution of opportunities, income and wealth, with the expansion of productivity the ultimate goal. One instrument for achieving this is to promote high-value crops, but these are often genetically homogeneous and therefore contribute to genetic erosion by supplanting farmers' varieties (Department of Agriculture 1995, 55). Much of AFMA focuses on how to promote such a development. However, there is also a provision providing for the reduced use of agro-chemicals that are harmful to health and the environment (Section 15 h) and another provision linking the AFMA to Philippine Agenda 21 (Implementing Rule 15.1). This is an important foundation for the Integrated Pest Management (IPM) that is emphasized as a policy in Philippine agriculture. One element of IPM is to improve farmers' seed selection, with a view to general quality and to resistance against pests and diseases. Depending on how this element is implemented, seed selection might contain a component of PGRFA management in terms of strategic *in situ* management and utilization for the maintenance and development of genetic

27 Republic of the Philippines (1998a), Administrative Order 6.

28 According to nearly all my interviewees at the Department of Agriculture in May 2000 and March 2002.

diversity.[29] As such, IPM is the agricultural policy that comes closest under AFMA to *in situ* management of PGRFA in the Philippines.

Most of the budget of the Department of Agriculture goes into implementation of AFMA. In 2002 nearly 26 billion pesos (approximately USD 520 million) were budgeted for AFMA implementation.[30] The general impression in March 2002 was nevertheless that Philippine farmers were still struggling to get by on incomes well below minimum wage requirements, due to falling rice prices, and that credits from the Philippine authorities were largely unavailable.[31]

The Agricultural Tariffication Act[32] was approved on 28 March 1996. 'Tariffication' refers to the lifting of all existing quantitative restrictions (such as import quotas or prohibitions) imposed on agricultural products, and replacing these with tariffs (Section 3 h). 'Tariffs' refer to a tax levied on a commodity imported from another country, enabling control of import and government revenues (Section 3 g).

The Agricultural Tariffication Act was a measure enacted to comply with the Philippine commitments to the World Trade Organization as negotiated under the Uruguay Round. Whereas import restrictions had been used to protect domestic agricultural production, they were now to be lifted to enable trade liberalization. As a compensating measure, tariffs were introduced to bridge the gap between low world market prices and high domestic prices. For rice, as the most vital staple crop in the Philippines, special provisions were introduced regarding two types of import: (1) a minimum import volume, as provided for under the WTO, was introduced to permit access to Philippine markets; and (2) imports were allowed to meet the demand resulting from shortages of rice or calamities. Prices were to be set by the government regardless of existing floor prices and subsidies.

It is difficult to determine the effects of this policy for PGRFA management. For crops other than rice and white corn, it has probably made little difference in terms of PGRFA management, since import prices are adapted to domestic prices through tariffication. For rice production, the situation is more complex. Farmers will most likely have to rely largely on seed saving and exchange, buying new seeds only when absolutely necessary to maintain yield-vigor, due to low prices on farm produce, *inter alia* resulting from the act.

The Hybrid Rice Program was launched by Administrative Order No. 25 of 12 December 2001, in an effort to boost rice production in the Philippines. Called the 'flagship of the Arroyo administration', it aimed at making self-sufficiency possible by 2004 (Boroña 2002a). Hybrid rice is seen as an important precondition for achieving substantial increases in rice production. Targets for hybrid rice production

---

29 This is not explicitly expressed in the available documentation, and 'seed selection' could also refer to selection of commercially offered high-yielding seeds with genetic homogeneity. However, genetic diversity is generally an important means of promoting resistance against pests and diseases, when chemical inputs are reduced. It is therefore assumed to be an implicit component of the IPM – an assumption underscored by several interviewees. Due to the lack of explicit policy formulations in this regard, PGRFA management can nevertheless be assumed to be of lower priority than other IPM components.

30 *Manila Bulletin*, 9 March 2002, 4.

31 *Manila Bulletin*, 12 March 2002, 10.

32 Republic of the Philippines (1996), Republic Act No. 8178.

in the major irrigated provinces were set at 135,000 hectares for 2002; 200,000 hectares for 2003; and 300,000 hectares in 2004.

For implementation, Executive Order No. 76 was adopted on 4 March 2002.[33] The Philippine Rice Research Institute, PhilRice, a core entity in the new program, was moved from being a branch of the Department of Agriculture and became a line entity directly under the President, to ensure effectiveness of its work. In addition, the Order provides for channeling a considerable amount of financial resources from AFMA implementation in order to implement the Hybrid Rice Program.

Under favorable conditions, and when the full input package of irrigation, fertilizers and pesticides is used, hybrid rice gives high yields – vital for meeting the nutrition demands of a fast-growing population. There have been success stories of production increase, but also bad experiences, due among other things to bacterial leaf blight and insufficient irrigation (Boroña 2002b).

Despite the production increase in many areas, there are also disadvantages (Boroña 2002a). The most important drawback in a PGRFA perspective is that hybrid vigor produces high yields only in the first generation of crops, and then farmers must buy new seeds each season in order to continue growing rice with high yields. This will not only reduce *in situ* management of PGRFA, but also make farmers dependent on a fragile system, because seed supply is insufficient in many areas of the Philippines (Department of Agriculture 1995, 57; Boroña 2002a).

Many farmers regard the hybrid rice as risky, since they have to carry the economic burdens in case of crop failure (Boroña 2002b). Critics of the Hybrid Rice Program maintain that it is not a technology for small-scale farmers, precisely because there is too much risk involved (Kuyek 2000, 16). According to IRRI data (ibid.), the maximum attainable yield in the Philippines between 1991 and 1993 was estimated at 6.3 metric tons per hectare, whereas the average yield was approximately 2.05 metric tons per hectare. The challenge for most farmers would not be how to increase their maximum attainable yield, but how to bridge the gap and increase their too-low average yields, according to the critics.

The number of hybrid rice varieties is very limited. Until 2002 there was only one variety – Mestizo or PSB Rc72, developed at IRRI – and it was susceptible to bacterial leaf blight during the wet season (Boroña 2002a). In 2002 four more varieties were introduced, two of which are supposed to be blight-resistant. Like other high-yielding rice varieties, hybrid rice has a complex composition of genetic traits, and can as such be regarded as genetically diverse. However, it replaces an often richer diversity of rice varieties with only one or a few genetically extremely homogeneous varieties. That is why the flagship of the Arroyo government, the Hybrid Rice Program, will most likely contribute to the erosion of PGRFA in rice, unless complementary measures are introduced.

Policies on Conservation and Utilization of PGRFA

In 1995 the Philippine Department of Agriculture reported to the FAO that existing laws did not provide incentives to farmers to conserve traditional seed varieties

33 Republic of the Philippines (2002a), Executive Order 76.

(Department of Agriculture 1995, 55). On the contrary, they wrote, the government provided incentives to farmers who used the modern high-yielding varieties. The hybrid rice flagship of the Arroyo administration, which has been central to agricultural policies since 2002, indicates that the incentive structures have not changed in this regard, as we have seen above. Philippine Agenda 21, which might have been expected to change this picture somewhat, has not made any significant difference, for reasons explained below. In general, there is little political awareness of the importance of proper management of PGRFA for agriculture. The focus has remained on using available material to develop high-yielding crops, as expressed in the Seed Industry Development Act and the High Value Crops Development Act analyzed here. An exception to this overall picture is the Coconut Preservation Act, as we shall see.

The Seed Industry Development Act[34] was approved on 27 March 1992. Its aim was to accelerate the development of the seed industry by conserving and developing domestic plant genetic resources, encouraging organizations in the sector, investing in the seed industry and encouraging the private sector to engage in seed research and development as well as mass production and distribution.

A National Seed Industry Council was established, headed by the Secretary of the Department of Agriculture, and consisting of the heads of major public institutions in the seed sector together with two representatives from farmers' associations and one representative from the seed industry. The Council was to serve as the lead council in all work to promote the seed industry. It was to be assisted by a Technical Secretariat, tasked, *inter alia*, with establishing seed standards and systems for evaluation, registration and commercial release. To operate the system and carry out these services, National Seed Quality Control Services was established within the Bureau of Plant Industry, which is a central body in PGRFA management. In addition, a network of seed centers – the National Seed Network – was established at the Bureau of Plant Industry and at major agricultural colleges, tasked with producing sufficient quantities of registered seeds of all varieties developed by the government sector.

Whereas the Seed Industry Development Act has conservation of PGRFA among its aims, there are no operative provisions for this purpose. All operative provisions focus on developing the highest-yielding varieties on the basis of available resources, in order to meet national food demands.

The High Value Crops Development Act[35] was approved on 23 February 1995. Its purpose is to promote the production, processing, marketing and distribution of 'high-value crops'. This term refers to crops other than rice, corn, coconut and sugar ('traditional crops') and includes coffee and cacao, various fruit, root and vegetable crops, legumes, spices and cut flowers. When developed as export crops in suitable areas, these are expected to augment the country's foreign exchange earnings significantly and contribute to farmers' incomes. This is the ultimate goal of the policy. To enable farmers to invest in high-value crops, a fund was established with

34 Republic of the Philippines Senate and Congress (1992), Republic Act 7308.

35 Republic of the Philippines (1995a), Republic Act No. 7900.

an initial amount of one billion pesos (approximately USD 20 million) to facilitate credits and funding of necessary infrastructure.

Seeds are to be distributed by the Department of Agriculture, in co-ordination with the state universities and colleges, as well as the Department of Trade and Industry and Farmers' Organizations (Section 9.i). Seeds may also be imported free of duty, although they are subject to quarantine laws. Additionally, the Act foresees the establishment of experimental stations and seed farms for the development of varieties suited to local agro-climatic conditions and markets that are assumed to provide greatest value-added to high-value crops (Section 11.h). This latter element points to an aspect of PGRFA management for high-value crops.

The High-Value Crops Development Act, confirmed also by the AFMA, has been followed up with two programs by the Department of Agriculture. The first was the *Gintong Ani* – the High Value Commercial Crops Program; more recently, under the Arroyo government, has come the *Ginintuang Masaganang Ani* (GMA) – the High Value Commercial Crops Program.[36]

The new program aims to be more market oriented, and also to produce crops that can replace imports, thereby saving foreign exchange. It includes no reference to the establishment or maintenance of experimental stations or seed farms, as addressed in the original Act, even though access to good-quality seeds is identified a main bottleneck.[37] Whereas the High-Value Crops Development Act contained a distinct PGRFA management component, this perspective has been lost in the implementation of the Act.

The Coconut Preservation Act[38] was approved on 7 June 1995. Its importance must be seen against the backdrop that one quarter of the Philippines' arable agricultural land is planted to coconuts, and the crop is among the country's top export earners.[39] Due to declining production of coconuts, and the massive cutting of coconut trees without replenishment, the government decided to take action with this particular Act. From the day the Act was approved, it became prohibited to cut any coconut trees except under special conditions, and then a permit to cut might be issued upon application. A precondition for any permit, in addition to the application fee, would be that at least the same number of coconut trees would be planted in advance, to counterbalance the loss. Only the Philippine Coconut Authority would have the authority to grant permits. Violations of the Act would result in imprisonment for at least one year, maximum six years. Alternatively or in addition, a fine of 50,000 to 500,000 pesos (between USD 1,000 and 10,000) could be imposed.

The Coconut Preservation Act is aimed at conserving coconut trees in general, and does not distinguish between monocultures and genetic diversity in coconuts (*cocos nucifera*). The huge monoculture plantations of genetically uniform coconuts, which dominate in many areas in the Philippines, do not as such contribute to PGRFA

36 See <www.da.gov.ph/>.

37 Department of Agriculture (1999): <www.da.gov.ph/> GMA-High Value Commercial Crops Program, Chapter II, Industry Overview.

38 Republic of the Philippines (1995b), Republic Act No. 8048.

39 Philippine Coconut Authority (1999): <www.pca.da.gov.ph/> Magnitude of the Industry.

diversity. However, the Philippine Coconut Authority maintains that coconut production becomes truly economically viable only if farmers develop systems of inter-cropping with fruits, such as banana, durian and papaya, various vegetables, and other suitable crops. Thus, coconut production can comprise diversity between crops, but the Coconut Preservation Act itself is not specifically goal-oriented with regard to PGRFA management.

Other Major Policies Pertaining to Conservation and Sustainable Use

Whereas there has been marginal attention to PGRFA management in mainstream Philippine politics, the emphasis on nature conservation, 'sustainable development' and the rights of indigenous peoples is more visible in policy formulations. These policies are relevant to PGRFA management for various reasons, to be further explained below.

During the presidency of Fidel Ramos, these issues enjoyed relatively high priority. This emphasis decreased, however, under President Joseph Estrada, although the environment was still on his 10-point priority agenda. Environmental issues do not feature on the 10-point agenda of President Gloria Macapagal-Arroyo,[40] but are nevertheless accorded some attention in other official documents on government priorities. All in all, the trend has been towards a decrease since the time of Ramos.

As early as in 1987, the year after the fall of dictator Ferdinand Marcos, the Philippines Department of Environment and Natural Resources was re-organized and the Protected Areas and Wildlife Bureau was created. When the report of the World Commission on Environment and Development, *Our Common Future* ('the Brundtland Report'), was issued in 1987, it was thoroughly debated in the Philippines.[41] This gave rise to the first draft of a comprehensive Philippine Strategy for Sustainable Development,[42] which was discussed among a broad range of stakeholders in May 1988. After a lengthy process, the Strategy was approved in 1989 and implementation began. The country was therefore well prepared for the United Nations Conference on Environment and Development (UNCED) in Rio de Janeiro in June 1992. Shortly before the CBD was opened for signature in Rio, the National Integrated Protected Areas System Act was approved by President Corazon Aquino in Manila. Subsequently the Philippines signed the CBD in Rio. Already, in September that same year, the multi-stakeholder Philippine Council for Sustainable Development (PCSD) had been created to develop the Philippines Agenda 21. In October 1993, the Philippines ratified the CBD.

40 From the official web-site of the President at <www.news.ops.gov.ph/pgma_10point-agenda.htm>.

41 I stayed in the Philippines for four months in 1988 and assisted *inter alia* in the preparation process and implementation of the seminar 'Joining Hands Towards Sustainable Development', held on 9 July 1988 at Manila Midtown Hotel. The seminar brought together government, business and civil society representatives to discuss the prospects for sustainable development in the Philippines, taking the Brundtland Report as the point of departure.

42 DENR 1988.

In April 1994, the Philippine Strategy for Biological Diversity Conservation (PSBDC) was formulated, and in 1995, President Ramos issued Executive Order 289, which directed the integration of the Strategy into the sectoral plans of the government (Meniado et al. 2002, 234). This was also the year when Executive Order 247 on bio-prospecting was signed (see below). In terms of policy development, then, the Philippines emerged as a pioneer in implementation of the CBD and the Rio Declaration. From 1995 to 1997 a Philippine Biodiversity Assessment Report was prepared, which led to the formulation of a comprehensive National Biodiversity Strategy and Action Plan (NBSAP), approved in June 1997. In the meantime the Philippine Agenda 21 had been developed in a broad-based participatory process, formally put into effect by President Ramos through Memorandum Order 399 in September 1996. From 1998, implementation took off with regard to Philippine Agenda 21 as well as the NBSAP, with the latter an integral part of the former. Projects based on the National Integrated Protected Areas System Act were the most important efforts at implementing the NBSAP, indicating that the Act represented a substantial foundation for the Strategy and Action Plan. (Meniado et al. 2002, 218, 225).

In the meantime, an Indigenous Peoples' Rights Act had been formulated and approved in 1997. This was another piece of landmark legislation: it provided for the protection of the rights of indigenous peoples to their ancestral domains, including the natural resources located there. Among many other rights, the Act covers the rights of indigenous peoples to manage and conserve natural resources within their ancestral domains, as well as to negotiate the terms and conditions for the exploration of natural resources, including biological and genetic resources. Also here, the Philippines was among the first countries to provide comprehensive legislation on the rights of indigenous peoples (Castro 2000, 35). Implementation, however, was to prove beset with barriers.

In the following, we shall see how these acts and policies directly and indirectly provided framework conditions for PGRFA management.

The National Integrated Protected Areas System Act[43] – or the NIPAS Act – was approved by the Senate and House of Representatives on 6 February 1992, three months before the CBD was adopted in Nairobi. President Corazon Aquino approved it on 1 June that same year, that is, four days before the CBD was opened for signature in Rio de Janeiro. The NIPAS Act provides for a comprehensive system for nature conservation in remarkably outstanding areas as well as biologically important public lands – indicating that the Philippines was engaged in questions of biological diversity and nature conservation also before the CBD. The Implementing Rules and Regulations for the NIPAS Act were signed on 29 June 1992 as Administrative Order No. 25 of the Department of Environment and Natural Resources (DENR).

The NIPAS Act and its implementation stand as a cornerstone of the Philippine National Biodiversity Strategy and Action Plan, adopted on 5 July 1995 through

43 Republic of the Philippines, Senate and House of Representatives (1992), Republic Act 7586.

Memorandum Order 289,[44] as a follow-up of the CBD. The Plan has six major components:[45]

1. expanding and improving knowledge on the characteristics, uses, and values of biological diversity;
2. enhancing and integrating existing and planned biodiversity conservation efforts, with emphasis on *in-situ* activities;
3. formulating an integrated policy and legislative framework for the conservation, sustainable use and equitable sharing of the benefits of biological diversity;
4. strengthening capacities for integrating and institutionalizing biodiversity conservation and management;
5. mobilizing an integrated information, education and communication (IEC) system for biodiversity conservation;
6. advocating stronger international co-operation on biodiversity conservation management.

These elements may be highly relevant and even necessary for protecting non-domesticated biological diversity, but none of them addresses the management of domesticated agricultural plant diversity. Since the NIPAS Act is the most central instrument for implementation of this strategy (PAWB 1998a, 10–13; PAWB 2002), we will in the following concentrate on that Act.

In 1998 there were 26 protected areas under the NIPAS Act, covering a total area of 1,442,740 hectares, including seascapes (PAWB 1998b, 3/26). By 2004 there were 290 sites all over the country that had been classified under various categories of protected areas. Among these were: national parks, national marine parks/reserves (67), game refuges and bird sanctuaries (8), wilderness areas (16), watershed areas (85), mangrove swamp forest reserves (27), and tourist zones/marine reserves (56).[46] With so many protected areas, there is the danger that they will remain protected only on paper ('paper parks') – and this is probably the situation for most of the sites. For selected priority sites, however, comprehensive efforts have been undertaken to make the protected areas a reality in establishing and maintaining the structures and processes necessary for protection. Substantial achievements have been reached under the NIPAS Act for the conservation of non-domesticated biological diversity, but PGRFA was not covered. As this is the major policy in the field of conservation, it indicates that non-domesticated resources have had some priority in the Philippines, whereas the *in situ* on-farm management of PGRFA has not received anywhere near the same attention.

Philippine Agenda 21: The Philippines was among the first countries to respond to the Agenda 21 process initiated at the Rio Conference, which was also promising with regard to PGRFA management. President Fidel Ramos invited 18 civil society

44 Republic of the Philippines (1995e), Memorandum Order 289.

45 Formulated at the official web-site on biological diversity by the Department of Environment and Natural Resources <www.psdn.org.ph/nbsap/page9.html>.

46 According to the official web-site of the Department of Environment and Natural Resources at <www.psdn.org.ph/nbsap/page7.html>.

leaders returning from the conference for a dialogue on sustainable development (Perlas 2000, 184). One result of this dialogue was the establishment of the Philippine Council for Sustainable Development (PCSD), as early as 1 September 1992 by Executive Order No. 15. The Council was created under the Philippine National Economic and Development Authority (NEDA), with its Director General as Chairperson and with a secretariat located at NEDA. It was to consist of environment-committed representatives from thirteen departments (ministries) and the office of Energy Affairs, as well as seven representatives from civil society. This composition reflected the broad mandate given to the Council in integrating the concerns of sustainable development into all areas of political life. The PCSD was given the power and functions to ensure implementation of the commitments entered into by Philippines with regard to the Rio Conference.

On 5 July 1995, after lengthy discussions and preparations in the Council, President Ramos initiated a broad consultative process involving various sectors and stakeholders all over the country.[47] The process led to a comprehensive 163-page strategy for action (PCSD 1997),[48] launched in Manila on 26 September 1996. In the foreword the president wrote:

> Rightfully, it can be expected, for many among us, the issue of sustainable development has become a matter of survival. I also see now how the Philippine Agenda 21 has engendered a process that empowers our people to be the architects of their own future. By doing so, they have also charted the destiny of our nation.

The president showed great engagement in the process, and was highly supportive of the work of the PCSD. The same day as the strategy was launched, he issued Memorandum Order 399 Directing the Operationalization of the Philippine Agenda 21. In this way, the strategy was formally adopted and devised to provide overall direction for sustainable development in the Philippines. The PCSD was to oversee and monitor implementation of the Philippine Agenda 21. Also that same day, President Ramos issued Executive Order No. 370: *Strengthening the Philippine Council for Sustainable Development,* aimed at strengthening the PCSD by including business and labor representatives as well as more representatives from civil society and government, and by broadening its mandate, power and functions. The PCSD now got a tri-sectoral approach, aimed at helping to mobilize business and civil society as well as the authorities in a partnership for sustainable development.[49] Its leadership was divided between the chair from NEDA and a co-chair from civil society, and decision-making was to be based on consensus.

Philippine Agenda 21 addresses two aspects of PGRFA management. First, it highlights the need to strengthen germplasm and seed banks for indigenous Philippine species. Second, it pinpoints the necessity of re-introducing the use of traditional pest- and disease-resistant varieties in order to reduce dependency on

47 Republic of the Philippines (1995d), Memorandum Order No. 288.

48 Philippine Council for Sustainable Development (1997).

49 See also PCSD (1999a).

inorganic chemical inputs in agriculture.[50] However, these requirements are not explicitly followed up in the priority action plan for 1996/97, or in the proposed regional agendas for the various parts of the country. The closest the regional agendas come to PGRFA management is in addressing the need to promote *sustainable* or *organic* farming – but these concepts are not defined, and it is uncertain if PGRFA management is understood as a component of such farming systems.

The main achievements of Philippine Agenda 21 have probably been the establishment of tri-sectoral institutional structures for the promotion of sustainable development at all levels of the political system[51] and a general heightening of awareness on to sustainable development.[52] The Agenda process has also contributed to framing policies in the Philippines, including implementation of the NIPAS Act and the IPRA in a larger context of sustainable development (see for example PCSD Coordinating Secretariat 1997, 42). The targets of Philippine Agenda 21 are, however, still far from being achieved. Nevertheless, there are several success-stories at the local level (PCSD 1998; 1999b; 2000). Among these are achievements with regard to PGRFA management, carried out by civil society organizations in various communities around the country (PCSD 2000).

In December 2001, developments took a new direction. Gloria Macapagal-Arroyo had assumed power after the People's Power II revolt, in which Joseph Estrada had been dismissed from the presidency. Arroyo soon engaged in the de-composition of the PCSD, as proposed by NEDA.[53] On 10 December 2001, she issued Executive Order No. 62 titled *Further Strengthening the Philippine Council for Sustainable Development*, according to which only two departments were to be represented in the Council: the National Economic and Development Authority (NEDA) and the Department of Environment and Natural Resources (DENR). Department of Agriculture was no longer represented. Civil society got five members; labor and business one each. This move was justified as necessary for strengthening the work of the PCSD, making the structure more responsive and focusing strategic interventions (as set out in the Preamble). However, it also narrowed the scope of action down to economic development policies as provided for by NEDA, and environment policies as provided for by the DENR. Obviously, the broad-based participation of stakeholders, which had provided for ownership of the process, as well as the mainstreaming of sustainable development in all sectors of society

50 In point 6.3 under Lowland/Agricultural Ecosystem (Philippine Council for Sustainable Development 1997, 55).

51 There are regional as well as local councils for sustainable development, and there are councils pertaining to different issue areas. These are the Committee on Social and Economic Dimensions; Committee on Conservation and Management of Resources for Development; Committee on Strengthening the Role of Major Groups; and Committee on Means of Implementation. In addition there is a separate civil society council and a business council for sustainable development.

52 This conclusion is based on various interviews with stakeholders in the Philippine Agenda 21 process and my assessment of the documents and internet-presentations resulting from that process.

53 Perlas, N. (2001), 'NEDA Blocks Sustainable Development, *Philippine Star* (9 November).

(including agriculture) which had been a key feature of the Agenda, were priorities no longer.

In protest against this move, Nicanor Perlas, director of the Center for Alternative Development Initiatives (CADI), Co-Chair of the PCSD from civil society organizations, and a driving force behind the PA 21 process, withdrew from the Council. Today, the Philippine Agenda enjoys only marginal attention in the country's politics.

The Indigenous Peoples' Rights Act[54] was approved on 29 October 1997. The Act, often referred to as IPRA, declares that the state shall recognize and promote all the rights of indigenous cultural communities and indigenous peoples. A central component of the Act is the rights of indigenous peoples to their ancestral domains – terrestrial lands, inland waters, coastal areas, and all the natural resources therein, including biological and genetic resources and sub-surface resources like minerals. Through the Act, indigenous communities are entitled to claim ownership over their ancestral domains, to develop lands and natural resources and to remain in their territories. The Act provides for the right of indigenous communities to manage and conserve the natural resources in their ancestral domains. Further, it gives them the right to negotiate the terms and conditions for the exploration of natural resources within their ancestral domains, for the purpose of ensuring ecological, environmental protection and the conservation measures pursuant to national and customary laws (Chapter III, section 7.b). On the other hand, access to biological and genetic resources in ancestral domains and to indigenous knowledge related to the conservation, utilization and enhancement of these resources is allowed, provided that the indigenous peoples give their free and prior informed consent to such access (Chapter 6, section 35). The Implementing Rules and Regulations go one step further, and declare that indigenous peoples are entitled to benefit from the utilization, extraction, use and development of lands and natural resources within their ancestral domains (Rule III, part 2, Section 2a). The Implementing Rules and Regulations also state that violations pertaining to the provisions on prior informed consent are strictly forbidden and subject to penalties (Rule VI, section 17). As such, the Act corresponds to all the obligations under the CBD Article 8j on indigenous peoples (see section on traditional knowledge on page 142).

A salient point for realization of these rights is the issuance of Certificates of Ancestral Domain Titles (CADT). These Certificates are provided upon the fulfillment of a range of requirements that include maps, pictures, historical accounts, anthropological data, and write-ups of names and places derived from the native dialect of the community in question.

For implementing the Act, a National Commission on Indigenous Peoples (NCIP) was established as an independent office directly under the President, consisting of indigenous commissioners. The NCIP formulated the Implementing Rules and Regulations for the IPRA, and approved them 9 June 1998. The Commission was to be the entity in charge of providing Certificates of Ancestral Domain Titles (CADT). However, the first years were beset with problems:

54 Republic of the Philippines (1997c), Republic Act No. 8371.

- In 1998, newly elected President Joseph Estrada appointed a new Chairman of the NCIP, although the first Chairman had been appointed for a period of three years (Castro 2000, 39). After heavy protests, the President withdrew this appointment.
- Later same year, a former Supreme Court Justice, together with an attorney, filed a petition to the Supreme Court, claiming the IPRA to be unconstitutional (ibid., 43). In particular, they held that sub-surface resources (minerals) belonged to the state, and that indigenous peoples could therefore not claim any rights to them. A full two years later, the Supreme Court finalized the process and dismissed the petition on a divided vote.[55]
- Still that same year, a task force was set up to investigate the work of the Commission after certain complaints, and all funds were withdrawn pending the results of the investigations (Castro 2000, 41). As of August 2001, there was still no clarity in this matter, and the funds were still withdrawn (see previous footnote).
- In addition, President Estrada appointed a new assistant on matters concerning indigenous peoples, and established two new bodies in addition to the one investigating NCIP.

There were obviously powerful forces against the new Act, partly fronted by the President himself. Due to all these constraints, NCIP could not become operational, and, as of 2000, had still not been able to issue one single CADT (Castro 2000, 39). It had 181 approved Certificates of Ancestral Domain Claims, issued by the Department of Environment and Natural Resources based on an Administrative Order No 2 from 1993 and which were eligible for CADTs, covering 8.5 per cent of the total land area of the Philippines. In 2001, at least one CADT was issued (Ferrari 2002), but lack of funds was still an obstacle. In 2002, another two CADTs were approved, and nine were approved in 2003.[56]

This is an important background for understanding the effects of the IPRA for PGRFA management. The Act and its rules and regulations have little effect, as long as indigenous peoples cannot claim their rights because they have not received Certificates of Ancestral Domain Titles. Only when such Certificates become a reality will the IPRA truly come into effect. So far, there have been few achievements related to the IPRA, except for the approval of the implementing rules and regulations of the Act, and a general increase in awareness of the situation and rights of indigenous peoples in the Philippines. In December 2002 a UN evaluation

55 Information provided by Legal Rights and Natural Resources Center Inc – Kasama sa Kalikasan and Friends of the Earth Philippines at <http://firstpeoples.org/land_rights/Philippines/what_new.htm>, dated 2001.

56 According to the National Commission on Indigenous Peoples, 15 July 2003, 'List of CADTs and CALTs Issued', at <www.ncip.gov.ph/news/newsdisplay.php?articleid=66>.

mission found that indigenous peoples had not yet experienced any benefits from the IPRA.[57] Transgressions of ancestral domains in the Philippines have continued.[58]

Despite the good intentions of safeguarding the interests of indigenous peoples in the Philippines, opinion differs as to whether the Act really does serve those interests. Debate has been heated, with representatives of the indigenous peoples also split. We cannot enter into the full discussion here, but will simply highlight the issues that concern PGRFA management (based on Castro 2000, 46–51):

- In some areas indigenous communities are living side by side with non-indigenous communities, and it is not regarded as legitimate that one group of residents should possess rights while the other would have none. For example, only the indigenous people in the area would then be allowed to decide over issues such as conservation and sustainable use of PGRFA, as well as access to these resources by the other residents and outsiders, and the sharing of benefits arising from their use.
- It is not always clear how prior informed consent can be legitimately achieved. Customary laws may not sufficiently provide for such processes. As a result, in several areas a culture of bribery is unfolding with regard to mining concessions, which can be regarded as an issue similar to access to PGRFA in terms of prior informed consent.
- Indigenous communities now 'belong' under a poorly functioning NCIP; this results in a situation where other authorities no longer feel responsibility. There have been cases where indigenous people have been denied access to buy seeds from Municipal Agricultural Offices, and were referred to the NCIP – where no seeds were available.

Although some representatives of indigenous peoples maintain that they would have been better off without the IPRA, most people seem to support the idea (Castro 2000). However, they demand political priority for its implementation as well as improvements in the implementing rules and regulations, to enable the intentions of the Act to be realized.

Regulation of Access, Benefit Sharing and Intellectual Property Rights to PGRFA

Our last category of policies comprises those pertaining to control over, and ownership claims to, PGRFA. In the first of these two sub-categories, we find the EO 247 on bio-prospecting and the Wildlife Act. Adopted in 1995, the EO

---

57 Press Release, Philippines (11 December 2002), 'Visit to the Philippines of Professor Rodolfo Stavenhagen, UN Special Rapporteur for the Human Rights of Indigenous Peoples', at <www.tebtebba.org/tebtebba_files/ipr/stavenhagenpress.html>.

58 Information provided by Legal Rights and Natural Resources Center Inc – Kasama sa Kalikasan and Friends of the Earth Philippines at <http://firstpeoples.org/land_rights/Philippines/what_new.htm>, dated 2001; and Press Release, Philippines (11 December 2002), 'Visit to the Philippines of Professor Rodolfo Stavenhagen, UN Special Rapporteur for the Human Rights of Indigenous Peoples', at <www.tebtebba.org /tebtebba_files/ipr/stavenhagenpress.html>.

247 applied to all kinds of genetic and biological resources, including PGRFA. However, implementation encountered many hurdles, and in 2001 the Wildlife Act was approved, replacing substantial parts of EO 247. It simplified the process for bio-prospecting applications, but only concerning non-domesticated plants and wild animals. Uncertainty arose as to what was left for EO 247 to regulate, especially how to deal with bio-prospecting for PGRFA. In my interviews with high-level officials in the Department of Agriculture, March 2002 (see interview list), it became clear that they were not aware that the new Wildlife Act would have any implications with regard to EO 247.

In the second sub-category we find the Intellectual Property Code of the Philippines and the PVP Act. The 1997 Intellectual Property Code of the Philippines contains a separate law on patents, which excludes plant varieties from patentability – but not parts and components of plants. The 2002 PVP Act provides for the protection of plant breeders' rights to plant varieties.

EO 247 on prospecting for genetic and biological resources: On 18 May 1995, Philippine President Fidel Ramos signed Executive Order 247 on the prospecting of biological and genetic resources.[59] The implementing rules and regulations of EO 247 were approved by the DENR Secretary on 21 June 1996 as Department Administrative Order No. 96–20. Thereby EO 247 entered into effect as the first comprehensive regulation worldwide to implement the CBD provisions on access and benefit sharing.

As the title indicates, EO 247 regulates prospecting for genetic and biological resources, and thus also PGRFA. The overall goal is the conservation of these resources, their sustainable use and their benefit for national interests. Additionally it aims to promote the development of local capability in science and technology in order to achieve technological self-reliance in selected areas.

Bio-prospecting is understood as 'research, collection and utilization of biological and genetic resources for purposes of applying the knowledge derived therefrom to scientific and/or commercial purposes'.[60] EO 247 covers all forms of bio-prospecting – whether in the public domain, on private properties or on ancestral lands, whether undertaken by local or foreign individuals, entities or organizations, and whether the results are to be used for the public good or for private commercial interests. EO 247 also addresses the traditional uses[61] of biological and genetic resources by indigenous and local communities: these are exempted from the requirements set in the Order. EO 247 and its implementing rules and regulations do not explicitly address the issue of access to genetic and biological resources maintained in gene banks. However, the national gene banks of the Philippines distribute seeds according

59 Republic of the Philippines (1995c), Executive Order 247 (EO 247).

60 Republic of the Philippines (1995c), Executive Order 247, Definition of Terms, Appendix A.

61 According to EO 247 (definitions), *traditional use* refers to the customary utilization of biological and genetic resources by local communities or indigenous people in accordance with written or unwritten rules, usages, customs and practices traditionally observed, accepted and recognized by them.

to their own established procedures, so it is reasonable to infer that access to gene bank collections cannot be made subject to EO 247.

In general the Order provides that bio-prospecting can be accepted only following the approval of an application for a research agreement between the collector(s) and the government. There are requirements for minimum terms concerning, *inter alia*, limits of samples to collect, information requirements, technology co-operation and benefit sharing. Also minimum requirements are specified to conform with environmental protection laws and regulations. There are two forms of research agreements: commercial and academic research agreements. Both types are to be considered by the Inter-Agency Committee for Biological and Genetic Resources (IACBGR), which was established through the Executive Order. The Inter-Agency Committee is also to monitor and enforce compliance with the Research Agreements, as well as to co-ordinate further institutional, policy and technological development. The Committee and its secretariat were located to the Protected Areas and Wildlife Bureau (PAWB) of the Department of Environment and Natural Resources (DENR). This location has become problematic after the approval of the Wildlife Act, which replaces the EO 247 with regard to non-domesticated resources (see below), as there will be no applications to the Inter-Agency Committee of relevance for PAWB after its entry into force. All applications relevant to PAWB will be dealt with according to the Wildlife Act, and considered by the Secretaries of the relevant departments, not the Inter-Agency Committee. Therefore, and particularly in view of the scarce resources of PAWB, there will probably be low motivation for providing secretarial services to the Inter-Agency Committee concerning PGRFA-related applications.

The Wildlife Act[62] was approved by President Gloria Macapagal-Arroyo on 30 July 2001. It took almost three years to work out its implementing rules and regulations in a joint process involving the Department of Environment and Natural Resources, the Department of Agriculture, and the Palawan Council for Sustainable Development. On 18 May 2004, the implementing rules and regulations were approved by the Secretaries of the two departments and the Chairman of the Council as Joint DENR-DA-PCSD Administrative Order No. 01.

The Act aims at conserving the country's wildlife resources and habitats for sustainability (Section 1). More specifically, it seeks to protect wildlife species and their habitats, promote ecological balance, enhance biological diversity, regulate the collection of and trade in wildlife, comply with international commitments in this area, and initiate or support scientific studies on the conservation on biological diversity.

According to the Act, by *wildlife* is meant wild forms and varieties of flora and fauna, in all development stages, including those in captivity or being bred or propagated (Section 5a). This formulation could leave some doubt as to whether the Act applies to agricultural biodiversity. In the scope of application of the Act (Section 3) it is stated that the provisions of this Act shall be enforceable for all wildlife species found in all areas of the country, including protected areas under NIPAS Act, and critical habitats. It is further emphasized that the Act also shall apply to exotic species which are subject to trade, are cultured, maintained and/or bred in

62 Republic of the Philippines (2001c), Republic Act No. 9147.

captivity or propagated in the country – which could again raise some doubt as to whether agricultural resources are included. The implementing rules and regulations bring some clarity into the matter. They state that domesticated or propagated species such as (but not limited to) livestock, poultry and common ornamental plants[63] are excluded from the coverage of this Order (Rule 3.3).

Another source for uncertainty as to whether the Act applies to agricultural plant genetic diversity is the fact that the Department of Agriculture participated in the formulation of the Act during the whole process, and has co-signed it. However, the entities involved from the Department of Agriculture were those dealing with fisheries, under the Undersecretary for Fisheries. Those in the Department of Agriculture who are in charge of agriculture were not involved in the process.[64] In implementation of the Wildlife Act, the Department of Agriculture is to have jurisdiction over all declared critical aquatic habitats; and all aquatic resources, including but not limited to all fishes, aquatic plants, invertebrates and all marine mammals, except dugong (Section 4).[65] Thus it seems reasonable to infer that agricultural biodiversity does not come under the jurisdiction of the Department of Agriculture for implementation of this Act – only fisheries are under the jurisdiction of the Department of Agriculture in this context – and that agricultural biodiversity is therefore not covered.

Even if the Wildlife Act does not directly affect the management of agricultural biodiversity as such,[66] it affects it indirectly, by altering the scope of EO 247. It may also affect implementation of EO 247, since the Inter-Agency Committee established for implementing the Order is institutionally located within the Department of Environment and Natural Resources (DENR). As EO 247 is no longer relevant to DENR, the latter may not be particularly motivated to continue implementing it with regard to agricultural resources. The Inter-Agency Committee is not vested with any responsibility pertaining to the Wildlife Act. Under the Wildlife Act, the respective Department is entitled to assess and decide on bio-prospecting proposals.

The Intellectual Property Code of the Philippines[67] was approved by President Fidel Ramos on 6 June 1997. Its overall goal is to protect and secure the exclusive rights of scientists, inventors, artists and other 'gifted citizens' to their intellectual property and creations, particularly when this is beneficial to the people (Section

63 These are to be determined by the DENR and DA upon the recommendation of the National Wildlife Management Committee, which was created through Rule 6.1 of the implementing rules and regulations.

64 According to my interviews with high-level officials in the Department of Agriculture.

65 The Department of Environment and Natural Resources (DENR) is vested with jurisdiction over all terrestrial plants and animal species, all turtles, tortoises and wetland species, including but not limited to crocodiles, water-birds, all amphibians and dugongs. In the province of Palawan, the jurisdiction is vested with the Palawan Council for Sustainable Development (Republic of the Philippines (2001c), Republic Act 9147, Article 4).

66 The Wildlife Act does apply to wild relatives of agricultural plants, and as such to these components of PGRFA. However, our focus is on agricultural biodiversity, and here the Wildlife Act is relevant only indirectly.

67 Republic of the Philippines (1997a), Republic Act No. 8293.

2).[68] For this purpose, an Intellectual Property Office has been established (Part I), and legislation provided on patents (Part II), trademarks, services, marks and trade names (Part III), as well as copyright (Part IV).[69]

The Law on Patents (Part II of the Intellectual Property Code) is relevant for our analysis. It provides that any technical solution of a problem in any field of human activity which is new, involves inventive steps and is industrially applicable shall be patentable (Section 21). However, excluded from patent protection are plant varieties or animal breeds or essentially biological processes for the production of plants or animals, although micro-organisms and non-biological and microbiological processes are defined as patentable.

It is further emphasized that these provisions are not to preclude the Congress from considering the enactment of a law to provide *sui generis* protection of plant varieties and animal breeds and a system of community intellectual rights protection. In this way, the Act was brought in harmony with the IPRA Act, which was close to approval, and opened the way for the formulation of the PVP Act. Provided that a PVP act could be put in place, the formulations of the patent law pertaining to PGRFA were compatible with Article 27.3.b of the TRIPS Agreement.

Although plant varieties are excluded from patentability, there is a trend in the Philippines towards accepting patents on specific processes of plant breeding and on parts and components of plant varieties. E. Rey Negre, Intellectual Property Specialist at the Intellectual Property Office of the Philippines, points out that interbreeding or crossbreeding processes are not patentable, but that processes for improving the yield properties of plants or suppressing or enhancing their growth are (Negre 2004, 27). He highlights the process for pruning plants as an example. Although biological by nature, the process can be regarded as technical, and thus patentable, he argues. Executive Director Patricio P. Faylon of the Philippine Council for Agriculture, Forestry and Natural Resources Research and Development (PCARRD) reported in 2004 that they had isolated six coconut genes through a gene-screening strategy, for further sequencing and patenting.[70] These examples show that patents concerning PGRFA are an issue in the Philippines, although there have been few examples so far. Patents pertaining to living organisms like plants are a highly controversial subject in the Philippines, for political as well as religious reasons. The same conflict lines are evident with regard to the PVP Act.

The Plant Variety Protection Act[71] was approved by President Gloria Macapagal-Arroyo on 7 June 2002. Its implementing rules and regulations were adopted on 24 February 2003 as Administrative Order 07 of the Department of Agriculture.

The overall goal of the PVP Act is to contribute to food security in the Philippines (Section 2). For this purpose, the Act provides for the protection of exclusive rights over plant varieties to those who have bred them, 'particularly when beneficial to the

68 A brief look in the registry of intellectual property rights in the Philippines shows that 'gifted citizens' *of the Philippines* are a minority among the holders of intellectual property rights: the picture is dominated by foreign firms and multinational corporations.

69 A separate part (Part V) details the final provisions.

70 *Manila Bulletin*, 14 November 2004.

71 Republic of the Philippines (2002c), Republic Act No. 9168.

people' (ibid.). In other words, the basic underlying assumption is that the protection of plant breeders' rights will contribute to higher levels of food security. This is a controversial assumption in the Philippines, with good arguments on both sides. We return to this discussion in the second case study, when we analyze the influence of various actors on the formulation of the Act.

With the PVP Act, plant breeders' rights are introduced as a means to encourage the participation of private enterprises and provide incentives to necessary investments in the development of new plant varieties, and to securing exclusive rights of scientists and other gifted citizens to their intellectual properties and creations. A breeder is defined as the person who bred, or discovered and developed a new plant variety, or their employer (Section 3, c). Further, the Act is to be implemented in a way supportive of the obligations of maintaining a healthful ecology in accord with the rhythms and harmony of nature.

The criteria for granting a Certificate of Plant Variety Protection follow the UPOV guidelines, as explained in Chapter 7: Novelty, distinctness, uniformity and stability.

Any breeder may apply for a Certificate of Plant Variety Protection (Section 17). If the country of nationality of the foreign citizen or company affords similar privileges to Filipino citizens, the Act obliges Philippine authorities to issue a Certificate of Plant Variety Protection, provided that the conditions in the Act are met (Section 23, national treatment).

Holders of Certificates of Plant Variety Protection have the exclusive right to authorize the production and reproduction, conditioning for the purpose of propagation, offering for sale, selling or marketing, exporting, importing and stocking for any of the mentioned purposes (Section 36). This right also extends to harvested material (Section 38). Following UPOV 1991, the right in addition extends to varieties that are not clearly distinct from the protected variety, and essentially derived varieties[72] (Section 39).

As provided for in UPOV 1991, there are exemptions to the rights of the plant breeders (Section 43). These are, however, wider than in UPOV 1991. As in UPOV 1991, acts done for non-commercial and/or experimental purposes are exempted. Also in line with UPOV 1991, acts done for breeding other varieties are exempted, but only if these new varieties are not essentially derived or not clearly distinguishable from the protected variety. The difference from UPOV 1991 lies in the wording as well as the exemptions regarding small-scale farmers. Farmers' traditions of saving, using, exchanging, sharing and selling their farm produce are termed a *right*, and are exempted from the rights of the plant breeders, provided that a sale is not for the purpose of reproduction under a commercial marketing agreement. The exemption also extends to the exchange and sale of seeds among and between farmers, provided that this is done for reproduction and replanting in their own land. In contrast, UPOV emphasizes that the legitimate right of the breeder must be protected, and makes it optional to governments to permit farmers to reuse their harvest from protected varieties for propagating purposes on their own land holdings. However, the formulations of this provision in the Philippine Act leave

72 These terms are explained in Chapter 7.

substantial scope for interpretation, as we will discuss in section 9.3.5 in analyzing Philippine legislation and policies on farmers' rights.

A registrar – *The National Plant Variety Protection Registrar* –is established to process applications, issue Certificates of Plant Variety Protection and carry out all other tasks necessary to implement the Act. As a transition arrangement, the Director of the Bureau of Plant Industry is to serve as acting Registrar.[73]

To oversee the implementation of the Act, a National Plant Variety Protection Board is to be established, headed by the Secretary of the Department of Agriculture. The board is to be composed of the directors of the central government institutions relevant to agriculture and the seed industry, the president of the Philippine Seed Industry Association (private business representative), and two representatives from civil society. One of the latter representatives is to be nominated by the Secretary of Agriculture from a federation of small farmers' organizations. The other representative is to be nominated by the National Academy of Science and Technology from the scientific community. The composition of the board and the procedures for nominating the representatives from civil society are controversial, and will be further discussed in connection with our second case study. For a transitional period for up to three years from the entry into effect of the Act, however, the National Seed Industry Council, which was created under the Seed Industry Development Act (see page 229), is to act as the board. Through the latter act, accredited farmers' organizations have two representatives, to be nominated by themselves and to be appointed by the Secretary of Agriculture.

An important question with regard to legislation on plant breeders' rights is how to establish an overview over *prior art*, that is already existing plant varieties, in order to ensure the novelty of a plant variety for which a Certificate of Plant Varity Protection is sought. There is no chapter or section in the Act that exclusively addresses the question of *prior art*, but there are several provisions under different sections, which, taken together, establish the policy on *prior art*:

- **Database:** One of the duties of the Registrar is to institutionalize, maintain and continuously update a database of existing plant varieties collected from foreign and local databases (Section 74, D). The Registrar shall also include community inventories in the database.[74]
- **Community inventories** are to be established by farming communities and *bona fide* farmers (Section 72). Among the tasks of the board is to encourage farming communities and *bona fide* farmers to develop such inventories.
- **Publication:** The Board shall publish a Plant Variety Gazette to all concerned parties. It is to be distributed free of charge, and in the major dialect understood in the locality, to small-scale farmers and indigenous communities (Section 73). All Certificates of Plant Variety Protection shall be published in this Gazette within thirty days of issuance (ibid., Article 57). Objections to the entitlement must be filed within a period to be prescribed by the Board from the date of its publication.

73 Republic of the Philippines (2002c), Republic Act 9168, Section 76.

74 Republic of the Philippines (2003), Administrative Order 07, Article 12.

In short, this means that farmers are made responsible for registering their own varieties; and that the burden of proof rests with farmers and farmers' organizations, if a breeder applies for protection of a plant variety which is essentially derived from, or identical to, a farmers' variety. It is the farmers who will have the work and expenses involved in establishing community inventories, informing the registrar, obtaining and reading the Gazette, and filing objections when necessary. If so, they will have to prove that the variety existed before the issuance of the Certificate. That may be difficult, since one criterion for novelty is that the variety has been discovered and developed, and it is uncertain how much or little a breeder must add in terms of 'novelty' to qualify for certification.

Implementation of the PVP Act was still in the transitional period by the end of 2004, with the transitional bodies at work. Institutional infrastructure was under development, and information workshops were conducted to prepare potential applicants with their options and procedures (see Golez 2004, 20–23).

Domestic Policy Decisions and Goal Achievements Pertaining to PGRFA Management

For each PGRFA management element defined in Chapter 3,[75] we will now analyze the state of policy decisions and goal achievements during the period from 1992 to 2004, and then, using the Six Cs model presented in Chapter 4, seek to identify the relative effects of international agreements on these developments. The methodology for these analyses is explained in detail in Chapter 4.[76] We will in each section start out by presenting a brief summary of findings to provide the reader with an overall orientation before delving into the details of the analysis. Finally, we will wrap up the sub-chapter with an overview over the findings and some preliminary conclusions.

In-situ Management of PGRFA in the Philippines

Whereas Philippine policies reflect an almost total lack of awareness of the *in situ* management of PGRFA, considerable attention has been paid to the *in situ* management of non-domesticated biological diversity through the NIPAS Act and the National Biodiversity Strategy and Action Plan. As a major CBD implementing measure in the Philippines, the NIPAS Act and its implementation correspond to the emphasis given to non-domesticated species in the CBD process. Domesticated PGRFA was left out of the Philippine implementation of the CBD with regard to *in situ* management.

In situ management of domesticated PGRFA has been carried out mainly by NGOs, and despite – not because of – government policies. Some initiatives have been initiated independently of international regimes, but most of them were inspired

75 The management element 'utilization of plant genetic resources for food and agriculture' will – due to limited space – not be dealt with separately, but as part of '*in-situ* management' and 'farmers' rights'.

76 A detailed interpretation of the values pertaining to each management component is found in Andersen (2007, 333–401).

by the IU, the Global Plan of Action, the Agenda 21, and/or the CBD. In some cases these regimes served as catalysts for funding, and there is also reason to believe that the international agreements have motivated the donor support granted to such activities.

However, the mainstream policies in the Philippines are still overwhelmingly in favor of replacing local varieties with genetically homogeneous high-yielding varieties, without any official support measures to ensure care of the disappearing crop varieties. The overall picture is therefore basically negative, although there are some positive nuances.

Coverage of policy decisions pertaining to in situ management The only policy formulation in the relevant acts and regulations that directly addresses the *in situ* management of domesticated PGRFA is found in the Seed Industry Development Act (Section 2a). It states, as a declaration of policy, that the government for the purpose of promoting and accelerating the development of the seed industry *inter alia* shall conserve, preserve and develop the plant genetic resources of the nation – but there are no operative provisions for conservation of these resources. This means that conservation of PGRFA remains a normative rule, without any operative implications. The Coconut Preservation Act is aimed at one crop, but one of important economic value for the country, since coconuts and coconut products are major export goods, and a high share of lands is planted to coconuts. The Act aims at maintaining the number of coconut trees (in order to uphold coconut production), but not at maintaining genetic diversity in coconuts. Although the Act does not address genetic diversity, it is likely to have a positive effect on the maintenance of such diversity in coconuts. The NIPAS Act is the most comprehensive document in terms of *in situ* management of biodiversity, but it does not address domesticated plants. Mention is made of PGRFA management in the Philippine Agenda 21 (Point 6.3), but only in terms of the re-introduction of traditional varieties, and is not targeted specifically at their conservation. On the other hand, the decisions to establish the Institute of Plant Breeding in 1976, PhilRice in 1985, the National Committee on Plant Genetic Resources (NCPGR) in 1986, and the National Network for the Conservation and Sustainable Use of Plant Genetic Resources for Food and Agriculture (PGRFA Network) in 2000 can be seen as contributions to *in situ* management.

The NCPGR was established under the Department of Science and Technology (DOST) in 1988 to provide direction and implementation of the guidelines for PGRFA conservation in the country. The NPGRL was to act as a secretariat for the NCPGR (Department of Agriculture 1995, 28), as well as serving as a repository for PGRFA and co-ordinating and monitoring conservation efforts in the Philippines. In 1989 several crop germplasm networks were initiated to increase conservation efforts. As of 2002, the NCPGR was formally still in existence, but had not met for several years due to financial constraints.[77]

The PGRFA Network was initiated by the Department of Agriculture through the Bureau of Agricultural Research in 2000 in order to fill the gap after the NCPGR, by

77 According to communication with Nestor C. Altoveros, Assistant Director for Plant Genetic Resources at the NPGRL in Los Baños, 15 March 2002.

seeking a new approach to organization. The new network was to establish a national system for collecting, conservation, regeneration, multiplication, characterization, evaluation, documentation and utilization of PGRFA of economic importance to the Philippines, thereby contributing to the conservation of biological diversity. To attain these objectives, the network was to review relevant programs and policies, develop a national integrated research and development program for PGRFA, strengthen the capabilities of member institutions, and conduct other activities. An advisory board was to recommend policies and directions for the national program, and a co-ordinator to monitor the overall activities of that program. The Institute of Plant Breeding provides administrative support for co-ordination and monitoring efforts, whereas the NPGRL and its Centers are the core entities in policy implementation, according to their mandates. A documentation unit is to serve as the centralized information management system of the network, and a policy unit develops policies and undertakes advocacy and information dissemination.[78] All these activities are, however, dependent on funding, which has been scarce.

Several acts affect the *in situ* management of domesticated PGRFA, but without actually addressing the issue. One of these is the High Value Crops Development Act, which addresses diversity between crops, but fails to touch on diversity within crops or the *in situ* management of these crops. The Act is implemented along with the AFMA. The AFMA itself is the most important Act in this regard, since it covers all crops and is backed by considerable economic power. As long as no support measures are introduced to manage the diversity of domesticated PGRFA *in situ*, this legislation represents a threat to genetic diversity in the fields, since it aims at the broad-based introduction of genetically homogenous varieties. The Integrated Pest Management program (IPM) under the AFMA may have provided some balance to this picture, but it is so small compared to the rest of the AFMA, and its *in situ* management component is so invisible, that any difference must be deemed insignificant in this context. The Hybrid Rice Program is also economically powerful, and is targeted at replacing old varieties with the most extremely genetically homogeneous of seeds – hybrids.[79] It targets only one crop, but that is the most important one in terms of nutrition and political stability, and covers the largest share of the agricultural lands, as well as involving the greatest number of farmers in the Philippines.

The coverage of policy decisions regarding the *in situ* management of domesticated PGRFA is, as a point of departure, total. All crops, all agricultural lands and all farmers are addressed in the totality of acts analyzed here.

Compatibility among policy decisions pertaining to in situ management As indicated, we find great discrepancies among and between policies. The norms of the Seed Industry Development Act together with the more operative Coconut

---

78 According to communication with Nestor C. Altoveros, Assistant Director for Plant Genetic Resources at the NPGRL in Los Baños, 15 March 2002.

79 Genetic homogeneity refers to the genetic composition of a crop variety and not to its genetic diversity *per se*. It means that all plants of that variety have the same genetic composition, without any variation among them – whether this composition comprises a broad or a narrow genetic base.

Preservation Act are the only policy decisions that pull in the direction of conservation and maintenance of domesticated PGRFA, aside from the establishment of the NCPGR and the PGRFA Network, which have elements of *in situ* management in their programs. All other acts mentioned above, including the rest of the Seed Industry Development Act, promote the replacement of the traditional diversity of PGRFA with high-yielding genetically homogeneous varieties of crops. The policies pull in different directions, and are deemed not compatible.

Conduciveness of policy decisions pertaining to in situ management The emphasis of policy decisions is on the introduction of genetically homogeneous plant varieties. This is detrimental to the *in situ* management of diversity in PGRFA, as long as there are no measures to ensure that these resources are maintained. The few provisions on conservation – including the Coconut Preservation Act, and elements in the programs of the NCPGR and the PGRFA Network mentioned above – are insignificant in the overall context. Therefore the general policy picture regarding *in situ* management is clearly negative.

Relative effects of international regimes on policy decisions In the various acts of legislation and related documents there are no references whatsoever to indicate any links to the international agreements in focus in this book. When the NCPGR was established it was mainly in response to domestic needs, but the 1996 FAO Global Plan of Action for the Conservation and Sustainable Utilization of Plant Genetic Resources for Food and Agriculture (GPA) provided some inspiration.[80] There is reason to believe that the PGRFA Network is also mainly a response to domestic needs, but that it will be regarded as an instrument in the follow-up to the ITPGRFA – as well as the TRIPS Agreement. The latter because it seeks to establish an inventory of crops grown in the Philippines, that is *prior art* – an important condition for assessing the novelty of plant varieties, a criterion for plant breeders' rights. All in all, and taking into consideration all policies regarding *in situ* management, we find that only marginal lines of influence can be traced back to the international regimes studied.

We find no correspondence between the mainstream policies affecting the management of domesticated PGRFA and the provisions regarding this management in the CBD and the IU/ITPGRFA. As for the formulation on conservation and preservation in the Seed Industry Development Act, it does correspond with the aims of conservation and sustainable use in the CBD and the IU. However, without operational provisions, the formulation is of little value. The Coconut Preservation Act does not correspond to the international agreements, as it is aimed not at preserving biological diversity, but at maintaining the country's coconut production. Only the NCPGR and the PGRFA Network are examples of correspondence – with the IU/ITPGRFA, the Global Plan of Action as well as the CBD. Since these are marginal in the larger context of policies pertaining to *in situ* management, the overall correspondence must be said to be low.

---

80 Based on an interview with Prof. Nestor Altoveros, Institute of Plant Breeding, Assistant Director of the NPGRL, at the institute, 15 March 2002.

As we have seen, the international regimes in question have had only very marginal influence on Philippine policy on *in situ* management of PGRFA, only to some extent with regard to the NCPGR and the PGRFA Network. However, this marginal effect must be considered positive for the *in situ* management of PGRFA.

Coverage of goal achievements regarding in situ management Just as the policy decisions diverge, so do the goal achievements. As we have seen, all the *mainstream policies* for increased agricultural production – and largely without approaches to *in situ* management of PGRFA – suffer from lack of funds and from institutional obstacles. This has resulted in limited outreach as compared to the targets, and thereby in limited goal achievements. The main outreach is in the lowlands, in areas with good conditions for agriculture. This is also where the best results have been achieved in terms of higher yields. A few efforts have been made with regard to *in situ* management (see for example Bao-Rong 1998). It has not been possible to get a total account of *in situ* efforts on the part of the Philippine Authorities, for example through the NCPGR and the PGRFA Network. However, my interviews in the Philippines in 2000 and 2002 revealed that the resources for such activities are infinitesimal, and there are no indications that this situation has changed since then.

More interesting are the initiatives for *in situ* conservation that, independently of government policies, have contributed to goal achievements of a normative rule. Three civil society organizations are central here: MASIPAG, SEARICE and CADI.[81]

Magsasaka at Siyentipiko Para sa Pag-unlad ng Agrikultura (MASIPAG)[82] is a farmer-led network of people's organizations, NGOs and scientists working for the sustainable use and management of biodiversity through farmers' control of genetic and biological resources, agricultural production, and associated knowledge. Farmer-led participatory plant breeding is a central component here, and the organization has done extensive work in the collection, identification, multiplication and maintenance of rice and corn. MASIPAG started its work to counter the genetic erosion in rice as early as 1986.[83] In a co-operation involving farmers, scientists and social scientists from civil society organizations, MASIPAG sought to train farmers to become rice breeders. According to MASIPAG, the intention was to enable farmers to regain control over their seeds and strengthen their capacity to select the varieties most suitable for their purposes and environments. Conservation of seeds and traditional

81 Also GRAIN should be mentioned. Genetic Resources Action International (GRAIN) is an international NGO based in Barcelona, Spain, with a field office in Los Baños, the Philippines. It promotes the sustainable management and use of agricultural biodiversity based on people's control over genetic resources and local knowledge. Central to its approach is the conviction that the conservation and use of genetic resources is too important to be left to scientists, governments and industry alone. Therefore it seeks to support farmers and community organizations in their efforts towards sustainable management of PGRFA. In the Philippines, GRAIN has been particularly engaged in policy issues.

82 The name translates as *Farmer-Scientist Partnership for Development.*

83 The information on MASIPAG is based on interviews with employees and board members of the organization in March 2000 (see attached list of interviewees), presentation leaflets and other documents, most notably: PCSD 2000.

technologies were vital elements in this work. Also maize, another staple crop, was brought into the program. The organization worked in co-operation with people's organizations in various parts of the Philippines, and thus rapidly spread its activities to various provinces in Luzon, the Visayas and Mindanao. Up to 2002 (when the organization split in two), there were several thousand members, and several trial farms in those provinces.[84] In Mindanao alone, some 500 rice varieties have been collected and tried, and the trials have resulted in nine locally adapted varieties considered pest-tolerant and pest-resistant that have been reproduced for farming (PCSD 2000, 13).

Southeast Asia Regional Initiatives for Community Empowerment (SEARICE) is focused on community-based conservation and development of plant genetic resources. It is based in the Philippines but operates throughout South and Southeast Asia. In the Philippines, the organization has been actively engaged in policy developments regarding access and benefit regulation, as well as plant breeders' rights – in addition to its community-based activities for conservation and development of plant genetic resources in various areas of the Philippines. SEARICE has also been involved in networks on the conservation of plant genetic resources. In 1992, SEARICE established the Community-based Native Seeds Research Center Inc. (CONSERVE) as a non-stock, non-profit organization.[85] Its flagship program was on Community Plant Genetic Resources conservation and development, aimed at the collection, conservation, research, development and utilization of PGRFA in partnership with farmers.[86] In 1993 a collection expedition was conducted, which yielded 298 traditional rice and 14 maize varieties. Ten rice varieties were given to selected farmer-curators for conservation and adaptability testing. As a result 10 traditional varieties were added to those already maintained by the farmers. By 1995, CONSERVE had a total of 140 farmer-curators in Cotabato, Mindanao (Department of Agriculture 1995, 28–29). In 2000, SEARICE started a similar project in the far south of Mindanao, in Sultan Kudarat; by 2002 the project was established in five communities.[87] SEARICE was deeply involved in the negotiations leading to the ITPGRFA at the FAO during the 1990s, with its former director René Salazar as a driving force in the Philippine delegation. The CONSERVE project must be understood in this light.

SEARICE is also one of the co-ordinating organizations of the Community Biodiversity Development and Conservation Program (CBDC), an international initiative aimed at understanding and strengthening farmers' systems of PGR conservation and development in particular, and biodiversity management in general. The CBDC came into being as a result of a meeting on on-farm management and

84 It has been difficult to obtain exact figures. Information source: MASIPAG (2002) at <www.masipag.org/> and interviews with representatives from several civil society organizations, including MASIPAG (see interview list).

85 The information on SEARICE is based on interviews with present and former directors, observers, several presentation brochures and material, internet presentations at the web-sites of SEARICE and the Development Fund in Norway, as well as other documents, most notably the Department of Agriculture (1995, 28–29).

86 SEARICE (2002) at <www.searice.org.ph/conserve.htm>.

87 SEARICE (2002) at <www.searice.org.ph/sk.htm>.

development of PGRFA convened in connection with the Keystone Dialogues on plant genetic resources in Oslo in 1991.[88] The program recognizes the intimate links between biodiversity conservation and local sustainable development, and works for biodiversity conservation through decentralized approaches to promote local control of resources, and strengthen the autonomy of farmers. After the adoption of the CBD (1992) and later the Global Plan of Action (1996), the CBDC saw its efforts as contributions to the implementation of these agreements – in addition to the IU. The program is being implemented by organizations from Africa, Latin America and Southeast Asia and is supported by institutions from several industrialized countries.

The CBDC project in the Philippines is located on the island of Bohol and is co-ordinated by SEARICE. It is aimed at broadening the genetic base in farmers' fields for rice, maize, yam, sweet potato, and cassava, through assisting farmers in breeding crop varieties suitable to their environments. *In situ* conservation of PGRFA is a central component of the project, which seeks to enhance farmers' knowledge and skills on participatory plant breeding and provide them with better access to diverse PGRFA.[89]

The Center for Alternative Development Initiatives (CADI) has been the lead agency in the Philippine Agenda 21 from the side of civil society. Ever since the 1990s it has also, via the Don Bosco Center, been providing technical assistance to an increasing number of farmers in Mindanao who grow biodynamic[90] rice, conserving and utilizing several local varieties, including varieties of red rice.[91] In 2000 the number of farmers involved in biodynamic agriculture had reached 1,300.

Bottom–up initiatives for *in situ* management of PGRFA serve as important examples for farmers and scientists piloting approaches for *in situ* management of PGRFA, and for conserving crop varieties that would otherwise be in danger of disappearing. Altogether, their coverage in terms of saved crop varieties and awareness-raising is evident. Also the outreach in geographic terms is comprehensive, as the projects have reached several provinces in Luzon, the Visayas and in Mindanao. On the other hand, coverage in terms of number of farmers and size of agricultural land for targeted *in situ* management of PGRFA is relatively low in the larger context.

Against this background, the overall value with regard to the coverage of goal achievements regarding policy decisions on *in situ* management of PGRFA must be said to be limited.

---

88 According to personal communication with Cary Fowler (in June 2005) and Jaap Hardon (in August 2005), who together convened that meeting at the concluding day of the Keystone Dialogue in Oslo. For information on the Keystone Dialogue, see Chapter 5 and Keystone Center (1991). For further information on the CBDC see <www.cbdcprogram.org/final/frames/phases.htm>.

89 SEARICE (2002) at <www.searice.org.ph/cbdc-bohol.htm>.

90 Biodynamic farming is agriculture based on a holistic world-view, combining principles of ecology and cosmology with a deep sense of spirituality.

91 CADI 2002 at <www.cadi.ph/Selected_Accomplishments.htm>. The information on red rice is self-experienced during my field trip to the Philippines in 2000.

Compatibility between goal achievements in in situ management As we have seen above, there are great discrepancies between the two categories of achievements. Goal achievements regarding the mainstream policies affecting PGRFA management concern the increased use of high-yielding varieties, replacing local varieties, and without compensating measures for PGRFA management. Bottom–up approaches to the *in situ* management of PGRFA aim at the opposite: Using local varieties to maintain the diversity of PGRFA and enable farmers to control their farming systems and select varieties suitable to local conditions. The introduction of genetically uniform high-yielding varieties without compensatory measures of PGRFA diversity is seen as a threat to such a kind of farming. Therefore, the two approaches are deemed to be diverging.

Conduciveness of goal achievements in in situ management It follows that the goal achievements from the mainstream policies are negative for the *in situ* management of PGRFA as long as no measures are undertaken to halt *in situ* genetic erosion, whereas the bottom–up initiatives emerge as positive. Since, however, the former goal achievements have so much broader coverage, the totality must be seen as negative. Nevertheless, we must also account for the importance of the diverging, conducive goal achievements contributed by bottom–up initiatives elaborated above.

Relative effects of international regimes on goal achievements Implementation of the mainstream policy cannot be said to have any links to the international regimes in question. It has not corresponded with them, neither are they conducive in this regard. As the NCPGR was hardly functioning, due to lack of funds, and the PGRFA Network had not really started its work by 2004, there is little to say about these initiatives in this context. For the NGOs working in the field of *in situ* management of PGRFA, however, the CBD, the IU/ITPGRFA and the Agenda 21 were important sources of inspiration and in some cases even catalysts for funding. These are the main activities in the Philippines regarding *in situ* management and have been in line with the respective provisions of the international agreements, as well as being conducive to *in situ* management as defined in Chapter 3. Thus, we can conclude that it is possible to trace some lines of causation back to the international agreements in question, that there was some correspondence between the goal achievements and the international regimes in this regard, and that the effects of the international regimes, although limited, were positive.

Ex Situ Conservation of PGRFA in the Philippines

Thus far, *ex situ* conservation has not been a politicized issue in the Philippines. It does not seem to attract political attention at all, and policy formulation in this area has been left to implementing agencies. Specific programs for *ex situ* conservation activities during the period studied could not be identified, but a new effort to organize the institutions in a network under the Bureau of Agricultural Research was to result in a program for these activities. Another effect of the policy vacuum in this field is the lack of financial resources devoted to *ex situ* conservation. Whereas considerable efforts have been made to collect and store huge amounts of PGRFA in Philippine gene banks, malfunctioning storage facilities may have destroyed or reduced the value of some, perhaps many, of these accessions. Lack of documentation has added to the problem. Goal achievement has therefore proven negative for some parts of the collections. However, there is also a brighter side. The major staple crop, rice, has been taken better care of than the rest, and new efforts have also been started for other crops, in terms of reorganization as well as renovation of storage facilities.

All in all, there have been considerable efforts for *ex situ* conservation of domesticated PGRFA in the Philippines despite the lack of designated policies, and despite the government's focus on one commodity, rice. The overall picture is therefore basically positive, despite severe constraints in implementing plans and programs. As such, it is in line with the provisions regarding *ex situ* conservation in the CBD as well as the IU and ITPGRFA, although there are no signs of attribution to these agreements except for some inspiration from the Global Plan of Action.

Coverage of policy decisions pertaining to ex situ conservation There is no single act on *ex situ* conservation of PGRFA in the Philippines. A few acts and regulations do address *ex situ* conservation, most notably the Seed Industry Development Act. It declares, as a policy of the State, that the plant genetic resources of the nation shall be conserved, preserved and developed (Section 2a). As pointed out above, there are, however, no operational measures in the act for that purpose. The Act covers all species and varieties of PGRFA and has thereby full coverage in terms of this norm. Also the Philippine Agenda 21 (p. 53) addresses the need to strengthen germplasm conservation, particularly of indigenous Philippine varieties – but without further operationalizing this norm. In addition, the decisions to establish the Institute of Plant Breeding in 1976, PhilRice in 1985, the National Committee on Plant Genetic Resources (NCPGR) in 1986, and the PGRFA Network in 2000 can be regarded as contributions to the *ex situ* conservation of PGRFA (Department of Agriculture 1995, 28). All these institutions have *ex situ* conservation of PGRFA in their mandates. Aside from PhilRice, with its own gene-bank activities, the NPGRL was to serve as the national base collection center and duplicate documentation center. The crop-specific centers were intended to maintain active collections and duplicate gene banks for mandated crops, as well as contributing to the national documentation system. In addition NPGRL was to co-ordinate and monitor conservation efforts in the Philippines. With the creation of the Bureau of Plant Breeding in 1976 the government sought to spearhead the efforts of the Department of Agriculture for plant genetic resources conservation.[92]

Philippine policy on *ex situ* conservation is mainly normative and organizational. More detailed policies and operationalization are left to the institutions in charge. As a point of departure, the policies cover all PGRFA. However, the importance paid to the various crops differs significantly. During the period studied, highest priority was given to rice, whereas other crops were only marginally addressed in terms of

92 According to a presentation brochure on the National Network for the Conservation and Sustainable Use of Plant Genetic Resources for Food and Agriculture, published by the Institute of Plant Breeding. The date is not stated, but the brochure was received at the Institute of Plant Breeding in March 2002.

policies. The combination of a lack of designated policies for the direction of *ex situ* conservation and the lack of priority for conservation of a huge number of crops indicates a medium level of coverage.

Compatibility between policy decisions pertaining to ex situ conservation Whereas the policies differ in terms of priorities given to the various crops, there are no conflicts among or between them. They are all compatible and pull more or less in one direction.

Conduciveness of policy decisions pertaining to ex situ conservation We have seen that the various policies, despite differing priorities, point in a direction conducive to *ex situ* conservation, so the general policy picture regarding *ex situ* conservation must be said to be positive.

Effects of international regimes on policy decisions *Ex situ* conservation started in the Philippines before the international agreements in question (IU/ITPGRFA and the CBD) had been established, and there seems to have been an awareness of the importance of this issue even since the 1970s. After the 'People's Power I' movement in 1986, efforts were intensified. There are no signs of attribution to international agreements in any documents concerning policies on *ex situ* conservation. According to those of my interviewees who were involved in these processes, the international agreements did not have much effect on the internal work, although some themes from the Global Plan of Action were picked up in the work plan of the NPGRL.[93] The policies correspond with the provisions of the IU, the ITPGRFA as well as the CBD on *ex situ* conservation, although they are not strong in the sense of regulatory or operational rules. They provide for national efforts (as per the CBD), as well as international co-operation (as per the IU/ITPGRFA) on *ex situ* conservation, the latter particularly in rice. This is conducive to PGRFA management, but can generally not be seen as an effect of the regimes in question. Thus, we can conclude that only very marginal lines of influence relate to the international regimes in question, even though the policies do correspond substantially with these regimes – and that the very few traceable effects have been conducive to the management of PGRFA.

Coverage of goal achievements pertaining to ex situ conservation It is difficult to assess the coverage of goal achievements when the policy decisions are as vague as with *ex situ* conservation. Basically, all we can do is to seek an account of the state of PGRFA conserved *ex situ* and then discuss the findings in light of our guidelines outlined above.

The most comprehensive account of *ex situ* conservation in the Philippines was prepared as an input to the 1996 FAO Technical Conference on Plant Genetic Resources in Leipzig (Department of Agriculture 1995, 29). According to that report, a total of 45,898 accessions were maintained and conserved in Philippine cold gene banks, field gene banks and in *in vitro* facilities. The NPGRL maintained and

93 According to Prof. Nestor Altoveros, Institute of Plant Breeding, Assistant Director of the NPGRL in an interview at the institute 15 March 2002.

conserved a collection of 32,446 accessions of varieties of 396 plant species. Most varieties were considered indigenous in that they are grown only in the Philippines. PhilRice conserved 3,123 accessions of traditional, modern and wild rice varieties. The remaining accessions were stored by regional centers (often agricultural colleges or universities in the regions) as well as crop-specific centers. NPGRL holdings cover cereals, fiber crops, food legumes, forage/pasture, fruit trees, nut trees, oil crops, ornamentals, plantation crops, root crops, small fruits and vegetables. Food legumes and vegetables make up the bulk of these holdings (two thirds of the accessions).

Sources document differing numbers of accessions under the auspices of the various institutions. According to its own 1999 presentation, PhilRice stores approximately 6,000 samples of rice, almost twice the number reported to the FAO in 1995 (PhilRice 1999, 23). An account presented to me at the Institute of Plant Breeding in 2000, which dated back to 1997, showed that the germplasm holdings of the NPGRL contained 41,333 accessions from all parts of the Philippines, almost 10,000 accessions more than reported to the FAO in 1995.[94] Another account identifies 29,134 accessions of varieties of traditional vegetables, covering 80 species and sub-species, held by the Bureau of Plant Industry, the Philippine Root Crop Research and Training Center and the NPGRL.[95]

There might be several explanations for these discrepancies. For example, the ways in which accessions are counted according to species or institutions might differ, and/or there might have been changes over time. For our purposes, these figures offer an indication of *ex situ* activities in the Philippines, but cannot sufficiently show coverage, since we cannot know what proportion of the entire crop genetic diversity has been conserved. The new organization of *ex situ* conservation through the PGRFA Network was an attempt to meet the problems inherent in the old structure, including the lack of funds.[96] Whereas the lack of funds is a clear indication of the low priority given to this matter, the reorganization indicates that *ex situ* conservation was not perceived as satisfactory among the stakeholders during the period studied. The fact that measures are being undertaken to improve this situation, through the establishment of new organizational structures, could indicate a positive development.

To complete our picture of goal achievements regarding *ex situ* conservation of PGRFA, as far as possible based on our data, we should also look at the situation with regard to storage facilities, regeneration, documentation and availability in terms of active collections. The storage facilities for rice are designed for medium-term storage, with room temperature of 2–4°C and stable humidity: this should, according to PhilRice (1999, 23) preserve seeds for 25 to 40 years. It has not been possible to obtain information on the state of documentation regarding the PhilRice

94 List of germplasm holdings of the NPGRL (as of 1997) given to me by National Scientist Dr Dolores Ramirez at the Institute of Plant Breeding in May 2000.

95 Altoveros, Nestor (2000), The Conservation and Utilization of Traditional Vegetables in the Philippines. Unpublished document from the Institute of Plant Breeding, given to me by National Scientist Dr Dolores Ramirez at the Institute of Plant Breeding in May 2000.

96 According to Prof. Nestor Altoveros, Institute of Plant Breeding, Assistant Director of the NPGRL in an interview at the institute 15 March 2002.

collections. IRRI is taking care of long-term storage, where the seeds are frozen to –18°C. PhilRice as well as IRRI have active collections for distribution upon request. PhilRice is known to have the best facilities for gene-bank conservation in the Philippines after IRRI.

For other crops there are mainly two categories of cold storage: base collection for long-term storage, and medium-term storage of active collections for distribution (Department of Agriculture 1995, 33). Base collections contain varieties of peanut, winged bean, soybean, maize, sorghum, mungbean, snap bean and tomato, stored as seeds in units where temperature is maintained at –18 to 20°C. Medium-term units contain a greater variety of crops, stored at 0 to 5°C. All seeds are dried and kept under controlled moisture conditions (ibid.). The Department of Agriculture (1995, 33) reports that leakages in the refrigeration system of the units for cold storage are often difficult to detect, however. Repair and maintenance of cold storage facilities require considerable financial support. In 2001, the Bureau of Agricultural Research reported[97] that three of five chambers at the NPGRL for cold storage were non-functional, and that funds had been granted for an urgent upgrading. Such information indicates that there have been severe deficiencies with regard to cold storage facilities at least from 1995 to 2001, and perhaps longer. The consequences for the germination potentials of the stored seeds may have been serious, depending on whether samples could be moved into the two remaining units, but no data are available on that point.

Field gene banks have been established for banana, root crops, fiber crops, sugar cane, plantation crops, tropical fruits and nut species (ibid.). It was not possible to find documentation on the conditions of these field gene banks. Four institutions maintain *in vitro* collections, covering some horticultural species, *abaca*, banana, potato, sweet potato, taro, yam, garlic and shallots (ibid., 34). Also the conditions of these facilities are not reported.

It is difficult to get exact figures on the state of characterization, evaluation and documentation – which can be taken as an indication of the documentation problem. It seems that the Department of Agriculture had the same problem. The information available in the 1995 report indicates that there are considerable backlogs with regard to documentation (Department of Agriculture 1995, 34–35, 102–104). This is a major problem with regard to access to and utilization of the resources, to which we return in the next sections.

According to the Department of Agriculture (ibid., 36), regeneration of germplasm takes place when the amount stored falls below the required level, or germination becomes lower than 85 per cent. Given our information on non-functional cold storage units, however, it seems reasonable to assume that it is difficult to keep a sufficient account of the accessions which have lower germination than the required 85 per cent, and that regeneration is therefore not keeping pace with these objectives.

Our information on goal achievement is not precise, but nevertheless indicates that there is a discrepancy between rice and other crops, with rice receiving considerably more attention with regard to *ex situ* conservation than other crops. On the other hand, great efforts have been made regarding the collection of germplasm for other

97 *BAR Chronicle* 2:17, 1–15 September 2001, 6.

crop species. The malfunctioning of gene-bank facilities and lack of documentation should be accounted for under the rubric of conduciveness to *ex situ* conservation. Thus we may conclude that there has been high coverage of goal achievements, as compared to the policies.

Compatibility between goal achievements pertaining to ex situ conservation As seen, there are great discrepancies in goal achievement between the various institutions engaged in *ex situ* conservation. All efforts seek to pull in the same direction, but the fact that some storage facilities have been malfunctioning for years and that there are backlogs regarding passport data is negative for *ex situ* management. Therefore, the goal achievements must be said to be diverging in practice.

Conduciveness of goal achievements pertaining to ex situ conservation As indicated above, it is not evident that our findings are conducive to *ex situ* conservation. A substantial collection of PGRFA is conserved in Philippine gene banks. Nevertheless, in cases where genetic material is stored in facilities unable to provide the requisite temperatures and moistures, the situation could also be understood as negative to *ex situ* conservation. However, we do not have sufficient information on how these problems are dealt with, or how much damage they have caused to existing collections during the period studied. Another factor pointing in a basically negative direction is the lack of documentation on the characteristics and evaluation of a number of accessions – a serious constraint to access and utilization of these resources. Again, we do not know how extensive this problem is. On the other hand, the rice collections are comparatively well cared for.

All factors taken together, the situation can be described as basically positive, since efforts all point in the direction of *ex situ* conservation. However, we also know that the quality of some of these efforts, and perhaps a substantial part of them, has not been sufficient, so *ex situ* conservation in these cases must have deteriorated.

Effects of international regimes on goal achievements It has not been possible to trace any lines of influence from international regimes on the goal achievements presented here, except for the inspiration from the Global Plan of Action on the work of the NPGRL – and even that could not help in avoiding the problems caused by severe financial constraints. We have found limited correspondence to the international regimes in question, due to the malfunctioning of storage facilities and the documentation problem. As the very limited effects of international regimes were not able to bring about any positive change, we may conclude that they were neutral.

Access to PGRFA in the Philippines

In seeking to comply with the provisions in the CBD on access and benefit sharing, EO 247 appears to emphasize benefits, to the detriment of access.[98] Thereby it comes in conflict with the IU and the ITPGRFA. As we will see in Chapter 10, scarce

98 We return to the benefit perspective in the next section.

resources were allocated for implementation of EO 247, and it was impossible to keep the processing time within the estimated timeframe of five months. Although the regulation was meant to cover all bio-prospecting in the Philippines, information did not reach out to all stakeholders, and adequate enforcement mechanisms were not in place. Therefore the intended full coverage could not be achieved, and several (perhaps many) breeders continued accessing PGRFA without following the requirements of EO 247. Hence, implementation of this regulation was only marginally in line with the provisions of the CBD on access to PGRFA.

The provisions of EO 247 posed barriers to research in the Philippines and to foreign collaboration. They have proven especially problematic for Philippine bio-prospectors without foreign counterparts, for non-profit research, and for students. The problems involving collection for conservation purposes have been lessened somewhat, but have still represented barriers to important activities for the conservation and maintenance of PGRFA, which have remained hampered by scarce resources.

On the other hand, it is important to bear in mind that EO 247 was a pioneering work – the very first attempt to implement the provisions of the CBD on access and benefit sharing. Since then, lessons have been drawn and assessments carried out. The resulting Wildlife Act represents a simplification of the rules – however, without covering PGRFA, and thus creating a policy vacuum for the facilitation of access to PGRFA after 2004.

Coverage of policy decisions on access to PGRFA EO 247 on the prospecting of genetic and biological resources is the core act with regard to access to PGRFA in the Philippines. The IPRA (Indigenous Peoples Rights Act) specifies the particular conditions for gaining access to PGRFA in ancestral domains, in line with EO 247, and stresses that prior informed consent must be obtained in accordance with customary law. Since the IPRA overlaps with EO 247, the latter is more comprehensive and the IPRA has hardly been operationalized, we will concentrate on EO 247 here. The Wildlife Act does not address access to domesticated PGRFA, although that might affect the implementation of EO 247, for reasons discussed earlier. However, the potential effects of this new legal situation would be evident only towards the end of the period under study here, since the implementing rules and regulations for the Wildlife Act were not adopted until 2004. The PVP Act regulates access from another angle, the exclusive right of breeders over the new plant varieties for which they have Plant Variety Protection Certificates. This means that Certificate holders are in a position to set conditions for access to and use of new varieties of plants, and thereby restrict their accessibility. However, there are exemptions from these rules, enabling breeders and farmers – under certain conditions – to use their harvests of protected varieties for reuse, exchange and sales or for breeding purposes.

In terms of PGRFA, the EO 247 provides full coverage but does not affect the PGRFA in gene banks, since they have already been collected. However, it does apply to efforts towards collecting PGRFA for gene bank storage and use. All in all, the coverage of the regulation with regard to PGRFA must be deemed to be high.

When it comes to users of PGRFA, EO 247 is targeted at bio-prospecting for scientific as well as commercial purposes. Thus it covers public as well as commercial breeders and plant genetic scientists, whether local or foreign. Farmers and indigenous peoples are not affected by the legislation in terms of obtaining access to PGRFA, as long as they are not involved in scientific or commercial activities regarding bio-prospecting of these resources. Since access legislation is normally targeted at bio-prospecting for scientific and/or commercial purposes, the coverage of the policy decisions with regard to users must be defined as high.

Adopted in 2002, the PVP Act introduced a new situation with regard to access to PGRFA. This Act is dealt with in greater detail below, revealing that the coverage of this legislation is also high.

In sum, we must conclude that the coverage of the Philippine legislation pertaining to access to PGRFA is high.

Compatibility between policy decisions pertaining to access to PGRFA Until 2001, the relevant acts were EO 247 and the IPRA Act. These were fully compatible, pulling in the same direction in their way of regulating bio-prospecting.

With the adoption of the Wildlife Act in 2001, a new situation arose, and there is reason to ask whether the new Act is compatible with the remaining parts of EO 247. Both acts have the same targets. The main differences are that the Wildlife Act does not address domesticated PGRFA and that it provides a simplification of the rules of EO 247. The reason why their compatibility could be questioned is that the Wildlife Act seems to remove the foundations for implementation of EO 247 for PGRFA, as explained earlier. We return to that point below. Also the PVP Act of June 2002 pointed in a different direction with regard to access, as it contributed to limiting access to PGRFA.

Conduciveness of policy decisions for access to PGRFA Under EO 247, all forms of bio-prospecting require a research agreement with the Philippine government, represented through a relevant ministry, such as the Department of Environment and Natural Resources (DENR), the Department of Health (DOH), the Department of Agriculture (DA) or the Department of Science and Technology (DOST). There are two forms of research agreements: Commercial Research Agreements are designed for bio-prospecting that is intended, directly or indirectly, for commercial purposes by private persons and corporations, as well as all foreign or international entities, and are valid for three years; Academic Research Agreements, valid for five years, are designed for bio-prospecting intended for primarily academic purposes, and are reserved for recognized Philippine universities and academic institutions, domestic governmental entities, and intergovernmental entities. All private persons and commercial firms are legally assumed to have commercial motives, and must therefore obtain a Commercial Research Agreement (La Viña et al. 1997, 3).

The procedures for obtaining a commercial or academic research agreement, as outlined in EO 247 and its implementing rules and regulations, are comprehensive.[99] First the applicant submits a letter of intent to the designated Inter-Agency Committee on Biological and Genetic Resources. An attached research proposal must state the

99 Additional information on the interpretation of these documents is taken from La Viña et al. 1997, 4–9.

purposes, source of funds, duration, and a list of biological and genetic materials and amounts to be taken. The Technical Secretariat of the Inter-Agency Committee conducts a screening to determine whether EO 247 applies to the research proposal. If the answer is yes, the applicant completes an application form and sends it to the Inter-Agency Committee. Attached to the application form should be a letter of acceptance from Filipino counterpart(s); a letter of endorsement from the head of the institution where the applicant is affiliated or another relevant reputable institution; company or institution profile; an environmental impact assessment, as determined by the Technical Secretariat; a code of conduct; and such other documents as may be required from the government agency with jurisdiction over the project. The applicant must also pay an application/processing fee.

As a next step, prior informed consent must be ensured: the applicant submits a copy of the research proposal to the recognized head of the local or indigenous cultural community or communities that may be affected. These are given 60 days to react to the proposal, and no other action may be undertaken with regard to the application until this period has lapsed. As a result the applicant may obtain the required Prior Informed Consent Certificate from concerned communities or entities, such as from the recognized head of indigenous people, Municipal or City Mayor of the local government unit, Protected Areas Management Boards, or private land owners. The certificate is then submitted to the Technical Secretariat.

Against this background, the Technical Secretariat conducts an initial review and evaluation of the application and submits its evaluation and a draft research agreement to the Inter-Agency Committee within 30 days of receipt of all the documents from the applicant.[100] The Inter-Agency Committee submits its recommendation to the government agency that has jurisdiction over the project. The secretary/head of that government agency approves or rejects the Research Agreement. Upon approval, the applicant pays a bio-prospecting fee determined by the Inter-Agency Committee. The government agency that approved the application is then to transmit the signed research agreement to the Technical Secretariat, which in turn shall furnish a copy to the applicant as well as to other individuals or entities concerned. When the applicant has received this document, the process is finalized and accessing the resources can start.

This is a demanding set of procedures. The requirements are perhaps feasible for professional applicants, but may represent a hurdle for others, particularly Philippine institutions with low institutional capacity. They certainly do not provide any incentives for collaboration between Philippine and foreign researchers. Also for gene banks with limited resources, this procedure may pose an additional hurdle to organize collecting efforts, even when prior informed consent is not required. Moreover, the procedures are time-consuming. According to a note from the Technical Secretariat of the IACBGR,[101] speedy handling of the evaluation and decision process would at best take five months. This is a highly optimistic

100 The Inter-Agency Committee is designed to convene at least once every third month.

101 Technical Secretariat of the IACBGR (2001): Status of Executive Order 247 Implementation. Internal brief received during a visit at the Secretariat/PAWB in 2002, p. 7.

assumption, given the institutional capacity in Philippine state institutions, and the estimate does not include the work of the applicant. Even five months is a long time in a breeding context, and verges on the unacceptable in connection with the need to respond to crop pests and diseases from season to season.

In addition to the work and time related to the application process, there are further requirements that restrict the scope of access and add additional obligations. There must be a limit to the number of samples to be collected, and the approved list of plants and samples is to be followed strictly. A complete set of all specimens collected is to be deposited with the National Museum, and access to these specimens and all relevant data is to be allowed to all Filipino citizens. Information must be provided to all concerned parties with regard to discoveries and commercial use of the material; and the terms and conditions must be fixed for payment of royalties to concerned parties if commercial use is derived from the resources. Moreover, the Philippine government can unilaterally terminate the Agreement, and is thus not bound to it. Regular status reports must be submitted to the Inter-Agency Committee.

There are different provisions concerning technical co-operation between foreign and Philippine counterparts.

Whereas these terms may again be understandable in light of the benefit-sharing goals, some of them pose a heavy workload to breeders seeking access to PGRFA. It may be difficult to know in advance which plant varieties to look for, and not very practical to have to adhere to that list if unexpected and promising plants should be discovered. Not least for gene-bank collectors, this rule is not very conducive. The additional reporting requirements add to the bureaucratic burden.

All in all, the policy does provide for access to PGRFA – in theory, but facilitation is not organized in an efficient manner. Much depends on how the legislation is implemented. All in all, we must conclude that the policy is negative with regard to accessibility of PGRFA. Just how negative it is will depend on implementation.

As to the situation after 2001, the negative value is strengthened. With the removal of the basis for implementing still-existing EO 247 for PGRFA, it is reasonable to expect that any applications for access to PGRFA will be hampered through lack of institutional capacity for processing them. Plant breeders' rights as provided for under the PVP Act will further limit access to PGRFA, as will be further elaborated below.

Effects of international regimes on policy decisions As we will examine further in Chapter 10, EO 247 is largely derived from the CBD, partly as a reaction to the TRIPS Agreement, and most lines of influence can be traced back to the CBD. As for the PVP Act, most lines can be traced back to the TRIPS Agreement in combination with the UPOV Convention, as we will see below and in Chapter 10.

As analyzed in Chapter 6, access to genetic resources is a secondary norm under the CBD, subordinate to the core norm of equitable benefit sharing. Independent of its implementation, EO 247 can be said to be in line with the CBD in this regard, as a policy. It could be discussed whether the policy decisions are conducive to access to genetic resources as set out in the CBD, which stipulates that the countries shall endeavor to create conditions to facilitate access to genetic resources and not impose restrictions that run counter to the CBD objectives (Article 15.2). The requirements for access under EO 247 are comprehensive, but the intention behind the rules is to provide for facilitation of access within what was felt to be a reasonable time. The CBD is not explicit with regard to pace of access procedures. There is therefore no reason to conclude that the provisions of EO 247 would not be in line with the CBD. On this basis, we conclude that there is a substantial match between EO 247 and the CBD. The same goes for the match between the PVP Act and the TRIPS Agreement together with the UPOV Convention, as we will see below.

Under the IU and the ITPGRFA, however, expeditious access to PGRFA is a core issue. The EO 247 is not in line with that. Since 2001, it has also not been in line with the Multilateral System under the ITPGRFA: this is understandable due to its earlier adoption, but will require amendment when implementing the ITPGRFA in the country. We must conclude that there is a low match between EO 247 and the IU/ITPGRFA.

With regard to the conduciveness of regime effects, our focus is on the CBD, since that is the international regime from which EO 247 is largely derived, and on the TRIPS Agreement, as that is the international regime that gave rise to the PVP Act. When compared to our criteria on conduciveness of access policies for PGRFA management, as set out in Chapter 3, we must conclude that EO 247 and the PVP Act are not conducive to the facilitation of access to PGRFA.

Coverage of goal achievements pertaining to access As the implementation of the PVP Act did not really get underway until the end of 2004, we will concentrate on the EO 247 here. Though it was celebrated as the first regulation of its kind in the world, implementation of EO 247 has been beset with problems. As a result, only five proposals for bio-prospecting in the Philippines had been approved and signed as of April 2002, whereas 15 commercial research proposals and 25 academic research proposals had been filed (Liebig et al. 2002, 36). Five of the filed applications were withdrawn due to the bureaucratic procedures, and two of these applicants, both in collaboration with the US National Cancer Institute, decided to move to another Asian country instead of the Philippines because of the new regulation (ibid.). For the bulk of applications it was not sure as of 2002 whether they were still pending or had in practice been withdrawn, since the applicants had stopped responding to requests for further documentation from the Technical Secretariat of the Inter-Agency Committee (ibid.). Only a few proposals pertained to plant genetic resources, most were related to medicines or for other purposes.

We do not know how many samples of agricultural plant genetic resources were actually collected for bio-prospecting in the Philippines between 1995 and 2002. According to my interviews with many stakeholders in March 2002, EO 247 was not known to all relevant parties, and some breeders collected and used plant genetic resources without being aware of the regulation.[102] Some breeders had heard about it but did not realize that it pertained to agricultural genetic resources as well. Other parties knew about the regulation but made no effort to apply it, simply ignoring it

102 Some breeders I spoke with during my stay in the Philippines in 2002 had not yet heard about EO 247, and I spent part of my interview time informing them about the regulation.

instead. My impressions are confirmed by other studies on the topic (Liebig et al. 2002; Swiderska et al. 2001). Information and enforcement was a problem from the beginning, due to the scarce resources available to the Inter-Agency Committee (see case study at the start of Chapter 10).

All in all, goal achievement has been limited to the establishment of the Inter-Agency Committee for the implementation of EO 247, the reception of 40 applications, approval of six of these applications and signing of five of them. That some firms have reacted by ending research collaboration and shifting to other countries is also a response to the regulation, and should be considered in the context of the coverage of goal achievements – in negative terms. It is difficult to determine whether the achievements qualify for low/no or medium coverage of goal achievement, since we do not know the total amount of the accessions made. Taking the processed applications as points of departure, we would probably have to label the coverage of goal achievement as low to zero. If we took as our point of departure the number of applications, it could be argued that this qualifies for medium coverage of goal achievement, since the number of applications is substantial. However, this measure does not say anything about the quality of goal achievement, which is dealt with under conduciveness below, but indicate solely the extent to which target groups comply with the regulation. All in all, it seems reasonable to label the coverage of goal achievements as medium due to the relatively high number of applications filed.

Compatibility between the goal achievements pertaining to access EO 247 has been the core regulation pertaining to access to PGRFA since 1995. Although the IPRA is hardly relevant, we can note that there are no conflicts between the two, as the EO 247 provisions on prior informed consent are in line with the IPRA, and have been followed up in the few cases which have been approved, according to my interviewees from the IACBGR and its secretariat. The two can thus be said to pull in the same direction.

As for the Wildlife Act, there was no conflict with regard to goal achievements during the period studied, since the implementing rules and regulations were not adopted until 2004. The IACBGR approved at least four applications for bio-prospecting, and initiated two reviews of other applications after the adoption of the Wildlife Act in March 2001.[103] However, all these were filed before the Act had been adopted. It is reasonable to assume that, since then, new applications are filed directly with the respective departments, as long as they pertain to wildlife or the other species specified in the Act. For applications concerning PGRFA, it is reasonable to believe that they are to be directed to the IACBGR, which has not been explicitly closed down by the new act or its implementing rules and regulations. Probably the system is currently not working, because the IACBGR has been out of function after the implementing rules and regulations of the new Wildlife Act were adopted in 2004.

103 According to a table of research applications under EO 247 in Liebig et al. (2002), 79.

As our period does not extend past 2004, we must conclude that the Wildlife Act did not hamper the implementation of EO 247 with regard to PGRFA during the short time from its enactment and to the end of the period under study here.[104] Therefore, we can conclude that there was compatibility between all goal achievements regarding access to PGRFA in the Philippines between 1992 and 2002.

Conduciveness of goal achievements pertaining to access Though rather marginal, the goal achievements from the implementation of EO 247 were not conducive to efficient facilitation of access, in view of the length of the application process and the bureaucratic procedures. The fact that only six out of 40 proposals were approved from 1995 to 2002 is a clear indication of that. In addition, the application process for the approved proposals took from one to five years, which is much too long, not least compared to the 5-month estimates from the Technical Secretariat of the IACGBR mentioned above (Liebig et al. 2002, 36).

There are several factors that can explain the length of the application process. Considerable time was necessary to obtain prior informed consent, due to the trust-building measures necessary, the distribution of information and the 60-day waiting time to enable potential responses from those consulted. The IACBGR has in periods not been operational due to meetings not taking place, or low attendance without the requisite quorum. This has slowed down some of the application processes appreciably. There is, however, no evidence that the IACBGR slowed down the processes by any decisions or requirements exceeding its mandate (Liebig et al. 2002, 41). The problem was the inability to achieve the necessary quorum. Secretaries responsible for signing the agreements after their approval have in most cases taken an unacceptably long time to accomplish this task (Liebig et al. 2002, 36). For example, the University of the Philippines system had to wait for one and a half years for signature after the agreement had been approved, and could not start their activities in this period, due to the delay. In other cases, the waiting time for the Secretaries' signatures was from four months and up to a year.

As for the bureaucratic procedures, most applicants as well as potential applicants considered them too burdensome (see for example Swiderska et al. 2001; Liebig et al. 2002; Benavidez 2004; and the internal note from the Technical Secretariat of the IACBGR referred to earlier). The result was that applicants withdrew their applications or stopped responding to the inquiries of the IACBGR, and some potential applicants did not even try. Those who did not try may have chosen to continue their practice without seeking approval, or they may have terminated their activities in the Philippines.

In one case, the application of the Philippines National Museum, together with the Coral Reef Research Foundation and the National Cancer Institute of the United States remained stuck in the Bureau of Fisheries and Aquatic Resources (BFAR) for more than a year (Liebig et al. 2002, 39). A similar situation arose when Siliman University applied for a Commercial Research Agreement together with

104 Also after the adoption of the Wildlife Act, the IACGBR was in function, seeking to process the applications still under consideration – as a kind of transition period to the enforcement of the Wildlife Act.

the University of California at San Diego (ibid.). In two cases, the applications were withdrawn, and the US counterpart decided to end its collaboration with the Philippine partner and move its activities to another Asian country (ibid., 36).[105] From all sides, the Technical Secretariat of the IACBGR at PAWB has been held to respond too slowly to applicants. In Chapter 10 we will analyze why this was so, with a particular view to the lack of resources and political will as major factors.

For the period in focus, we must conclude that the goal achievements were not conducive to the facilitation of access to PGRFA.

Effects of international regimes on the goal achievements As we will see in Chapter 10, some lines of influence can be traced back from these goal achievements to the CBD, particularly due to development co-operation for the purpose of supporting the realization of EO 247 as a CBD implementing measure. Also, there was considerable international attention in the initial years, as EO 247 was seen as a pioneering effort for CBD implementation with regard to access and benefit sharing. However, the core reasons for the lack of accomplishments with regard to goal achievements are to be found in the institutional capacity for implementation, as we will see in Chapter 10.

The goal achievements were not in line with the provisions of the IU or the ITPGRFA on access to PGRFA, and they were hardly in line with the CBD, since they did not facilitate access – except for in a few cases. Thus there was low to zero match between the goal achievements and the international regimes.

Development co-operation support offered for the implementation of EO 247 as a CBD implementing measure was aimed at making implementation more efficient. For several reasons, to be elaborated in Chapter 10, this did not work out. Therefore the effects of the CBD must be said to be neutral in this regard.

Sharing of Benefits from the use of PGRFA in the Philippines[106]

The EO 247 provided a highly sophisticated system for benefit sharing, covering a wide variety of benefits, beneficiaries and a broad time span. The picture was somewhat distorted, due to the Wildlife Act in 2001 and the PVP Act in 2002, but these new developments did not really impact on the implementation of EO 247 during the period under study here. As such, EO 247 corresponds with the CBD – from which it is derived – in terms of policy intentions, and with regard to the benefit-sharing arrangements reached.

---

105 The US National Cancer Institute ended its collaboration with the Philippine partners and collected the required specimens in other Asian countries.

106 Access to genetic resources alone is not sufficient to generate benefits to be shared. A condition is that the accessed resources are used in such a way that they produce value. The utilization of genetic resources should thus ideally have been dealt with before coming to benefit sharing in this chapter. However, due to limited space, this PGRFA management component is dealt with as part of the sections on *in situ* management and farmers' rights. For a separate analysis of the utilization of PGRFA in the Philippines, see Andersen (2007, 365–73).

Table 9.2 Minimum terms for benefit sharing under Executive Order 247 and its implementing rules and regulations

Access to collected and/ or new varieties of plants	Transfer of technology or /and relevant equipment	Monetary benefits and/or development co-operation
• A complete set of the specimens collected shall be deposited with the National Museum of the Philippines or a designated governmental entity, such as the NPGRL, for the purposes of agricultural species. • Access to collected specimens and relevant data shall be allowed to all Filipino citizens and the Philippine governmental entities whenever these specimens are deposited abroad. • All discoveries of commercial products derived from Philippine genetic or biological resources shall be made available to the Philippine government and the local communities concerned. • In cases of endemic species, a statement must be provided ensuring that the technology is made available to a designated Philippine institution and can be used commercial and locally without paying royalty to the collector. Where appropriate, other agreements may be negotiated on this issue.	• In cases where a commercial collector is a foreign citizen or entity, Filipino scientists shall be actively involved in the collection and research activities at the cost of the commercial collector. • Where applicable and appropriate as determined by the IACBGR, Filipino scientists shall – at the cost of the commercial collector – be actively involved in the technological development of a product derived from biological or genetic resources collected from an area in the Philippines. • The commercial collector shall be encouraged to avail of the services of Philippine universities and academic institutions, and can be required to transfer equipment to a Philippine institution or entity. • For commercial research agreements, the collector shall donate some of the equipment used for the project to the Philippine government agency, institutions or universities concerned.	• All benefits resulting directly or indirectly from the bio-prospecting activities shall be shared equitably upon mutual consent among the Philippine government, communities concerned and collector. • A separate agreement shall be made for the transfer of royalties and technology, ensuring benefits for the respective local or indigenous communities, and for conservation measures. A detailed description of anticipated immediate and long-term compensation is foreseen in the Research Proposal. • For cases of commercial use of the obtained genetic resources, the agreement shall include a provision for the payment of royalties to the Philippine government, individuals or local communities. • In cases of commercial research agreements, the collector shall submit an ecological rehabilitation bond to be deposited and the amount to be determined by the IACBGR.

Despite the sophisticated rules and regulations, no benefits were achieved with regard to PGRFA. Why was this so? From interviews with all kinds of stakeholders, my impression is that EO 247 was either not known, and therefore not followed, or it was deemed too demanding for plant breeding, and was therefore ignored. However, it might be that project ideas were not pursued due to the barriers posed by EO 247, as perceived by breeders. When it comes to PGRFA, EO 247 has, in sum, been much ado about nothing. It has hampered access to PGRFA without providing anything in return.

Coverage of policy decisions on benefit sharing pertaining to PGRFA The legal and administrative regulation of benefit sharing regarding PGRFA utilization is dealt with in EO 247 on bio-prospecting, and in conjunction with the IPRA as regards indigenous peoples' rights to resources on ancestral lands. Also the Wildlife Act and the PVP Act became relevant, but they did not become fully operational until the end of the period and will therefore be dealt with only briefly. These are the same acts and orders as for the regulation of access to PGRFA. Normally, access and benefit sharing are seen as two sides of the same issue, and are definitely closely interrelated. Nevertheless, potentials for conflict between the two were identified in Chapter 8, and that is why we analyze legislation on access and on benefit sharing separately in this part of the study. EO 247 and its Implementing Rules and Regulations provide for minimum terms for benefit sharing between collectors, the Philippine government and local/indigenous communities, as shown in Table 9.2.

As the table shows, all forms of benefit sharing are covered with these minimum terms. Whereas the provisions on access and sharing of technology are quite specific, those on the sharing of equipment and of monetary benefits are more open. There is no specification of the share of equipment to be left with Philippine counterparts, or of minimum shares of royalties to be channeled to the Philippines. Nor is the equitable sharing for the immediate and long-term specified. These are all issues for negotiation and agreement upon mutually-agreed terms. Nevertheless, the benefit-sharing provisions under EO 247 must be said to be quite all-encompassing.

Also when it comes to target groups for benefit sharing under the provisions of EO 247, the picture is one of substantial coverage, as Table 9.3 shows.

The EO 247 clearly provides for benefit sharing in all phases of a project, and most explicitly when reference is made to immediate, medium and long-term benefit sharing in Section 8.1.11 of its Implementing Rules and Regulations. The condition is that access is provided – and that is a problem in many cases. It could therefore be argued that EO 247 does not establish a framework that sufficiently facilitates access, and thus does not sufficiently address benefit sharing – since benefit sharing depends on access in the first place. Clearly, there is a discrepancy between the normative rules of EO 247 to provide for benefit sharing, and the regulatory rules, which makes this difficult due to the heavy bureaucratic workload combined with weak institutional capacity. When discussing EO 247 as a policy, however, we should give weight to its intention, which is to facilitate access as well as benefit sharing. Therefore, we can infer that EO 247 does as such provide for full coverage of this policy with regard to benefit-sharing arrangements.

Table 9.3 Target groups for benefit sharing as provided for under EO 247[107]

BENEFITS	BENEFICIARIES				
	National authorities	Local authorities	Scientific institutions	Local communities	Individual citizens
Deposit specimens in museums/gene banks	X				
Access to collected specimens	X				X
Making available any discoveries	X			X	
Sharing of technology for endemic species			X		
Scientific collaboration			X		
Availing services from academic institutions			X		
Donation of equipment			X		
Conservation measures		X			
Sharing of monetary benefits	X			X	
Royalties				X	X
Performance compensation	X				

The IPRA Implementing Rules and Regulations state that indigenous peoples have the right to benefit from the extraction and utilization of *inter alia* PGRFA, and to negotiate the terms and conditions for such activities (IPRA Implementing Rules and Regulations, Rule III, Part II, Section 2a; IPRA, Section 7b). At least 30 per cent of all funds received for such activities shall be channeled directly to development projects within the ancestral lands to benefit the indigenous peoples (ibid.). Since the Act was hardly operational during the period under study, we will not delve into its details.

The Wildlife Act does not address domesticated PGRFA, but might indirectly affect the implementation of EO 247 with regard to these resources. However, these potential effects would appear only towards the end of the period in focus or later,

107 Inspired by a table in Liebig et al. (2002, 49).

since the implementing rules and regulations for the Wildlife Act were not adopted until 2004.

Also the Plant Variety Protection Act did not become operational within the period studied, and will not be dealt with in detail. However, it is interesting to note that the Act does not oblige applicants for plant breeders' rights to disclose the sources of the genetic resources they used in the breeding process, or to document any benefit-sharing arrangements. In theory, this is not detrimental to benefit-sharing arrangements – but neither is it conducive. In practice it opens for the possibility of providing plant breeders' rights to breeders who have not followed the legislation for bio-prospecting, since there is no obligation to document such compliance – as long as EO 247 does not have any effective enforcement mechanisms.

Maintaining our focus on EO 247, we can nevertheless hold that there was full coverage of the legislation with regard to the regulation of benefit sharing regarding PGRFA utilization in the period in focus.

Compatibility between policy decisions on benefit sharing pertaining to PGRFA As shown above, there is full compatibility between EO 247 and the IPRA and thus for all relevant legislation up to 2001. After that, the picture became more complicated. The Wildlife Act removed the foundation for implementation of EO 247 and may therefore threaten its further implementation, whereas the PVP Act did not directly conflict with EO 247, but could and can be used in a conflicting manner, as discussed above.

Conduciveness of policy decisions for benefit sharing pertaining to PGRFA The rules of EO 247 on benefit sharing all aim at providing for benefit-sharing arrangements, and are as such conducive towards this end. On the other hand, it could be discussed whether EO 247 is so demanding that potential benefit sharers are frightened off by the regulation. This discrepancy between the normative and regulatory rules makes it difficult to conclude with regard to conduciveness. At this stage we will – with regard to policy decisions – give weight to the intention with the EO, which is to regulate for benefit sharing. Then we will deal with difficulties of following the regulatory rules in the sections on goal achievements below. At this stage, we thus hold that the policy decision regarding the adoption of EO 247 is conducive to benefit sharing.

After 2001, the situation changed. Since the Wildlife Act removed the foundation for implementation of EO 247, without providing an alternative, this must be understood as clearly negative for benefit sharing. In addition, the PVP Act is potentially negative with regard to benefit sharing, as explained above. Nevertheless, EO 247 still existed as a policy. Altogether, the situation after 2001 is best described as neutral with regard to benefit sharing.

Effects of international regimes on the policy decisions EO 247 is the central regulation with regard to benefit sharing. As shown above, not all lines of influence can be traced back to the CBD, but most of them can. The way in which benefit sharing is provided for at the policy level must also be deemed fully in line with the CBD. As a point of departure the influence was also positive with regard to benefit sharing based on PGRFA utilization, provided that access to these resources was made possible. Since this was the intention behind the policy, we conclude that the effects were positive. We investigate these findings further in depth in Chapter 10.

Coverage of goal achievements pertaining to benefit sharing All five research agreements on bio-prospecting concluded by the end of 2002 contain provisions on benefit sharing. None of these agreements included bio-prospecting on PGRFA, so we must conclude that no benefits were achieved with regard to PGRFA during that period. Since, however, these are the only examples of the implementation of EO 247 so far, we will briefly outline some of the goal achievements with regard to pharmaceutical products (based on Liebig et al. 2002). The research agreements contained various combinations of monetary and non-monetary benefit-sharing provisions, with broad coverage as to content, target groups as well as timing. Nevertheless, Liebig et al. (2002, 46) found that scientific institutions were the main beneficiaries of the agreements, whereas any benefits to the local communities were rather marginal or unspecified. They furthermore found that negotiations on prior informed consent and on benefit sharing had largely been kept separate, and suggested this as an explanation for the negotiation output for local communities.

Another aspect that should be taken into consideration is that some Philippine institutions lost their foreign co-operation partners as a result of EO 247 (Liebig et al. 2002, 50), as shown above. Philippine institutions complain that they lost benefits as a result.

As we can see, EO 247 proves to have had potential for regulating benefit sharing for research co-operation on pharmaceutical products, but at a price that was just too high when compared to the lost opportunities for collaboration. The benefits gained would probably have been gained without EO 247. All this pertains to pharmaceuticals. We do not find a single example of benefit sharing from bio-prospecting on PGRFA during the period in focus.

Compatibility of goal achievements with regard to benefit sharing As no goals have been achieved with regard to the sharing of benefits derived from the use of PGRFA, the question of compatibility is irrelevant.

Conduciveness of goal achievements with regard to benefit sharing The lack of benefit sharing is clearly not conducive to 'benefit sharing' as defined in Chapter 3.

Effects of international regimes on the goal achievements As we will see in Chapter 10, some lines of influence can be traced back from these goal achievements to the CBD, particularly due to development co-operation for supporting the realization of EO 247 as a CBD-implementing measure. Also, there was considerable international attention in the initial years, as EO 247 was seen as a pioneering effort for implementing the CBD with regard to access and benefit sharing.

The EO 247 is partly in line with the CBD when it comes to benefit sharing from genetic resources for pharmaceutical purposes. It is in line with the CBD in ensuring benefit-sharing arrangements as parts of research agreements. It is only partly in line with the CBD because it fails to meet the CBD goal of promoting access, and thereby limits the chances of generating benefits to be shared. When it comes to PGRFA, however, EO 247 has in practice not yet proven at all relevant with regard to benefit-sharing arrangements.

Farmers' Rights in the Philippines

In their report to the FAO International Technical Conference on Plant Genetic Resources in Leipzig, 1996, the Philippine Department of Agriculture wrote: 'The concept of "farmers' rights" needs to be interpreted and explained based on the realities and perception of Filipino farmers' (Department of Agriculture 1995, 55). So far, however, no authoritative interpretation or explanation has been put forward. In the Philippines, farmers' rights are usually associated with land rights and the massive problem of landlessness among farmers. Farmers' rights pertaining to PGRFA were nevertheless relatively strong up to 2001, as compared to the years since then.

Philippine policies on farmers' participation in decision-making were conducive, particularly so up to 2002, and farmers were relatively well represented in relevant forums up to 2001. The customary practices of saving, using, exchanging, sharing or selling farm-produced propagating material were in general legal up to 2002,[108] but became restricted through the PVP Act and for material protected with plant breeders' rights after that. Although the EO 247 provided for benefit sharing, which could have rewarded farmers for their contribution to the global genetic pool, this did not materialize, as we have seen above. Moreover, even it had materialized, it would only have benefited the very few communities that entered into benefit-sharing agreements –not all the millions of farmers who contribute to maintaining the diversity of PGRFA. The best examples of rewards are the projects carried out by NGOs in the Philippines, often with donor support, to strengthen farmers' seed systems through measures like community gene banks, participatory plant breeding and marketing of diversity based produce. However, the outreach of these efforts is still highly limited.

Whereas the CBD affected policies marginally, and goal achievements even less, the TRIPS Agreement with the UPOV Convention had considerable influence at the policy level, as we will see in Chapter 10. The IU/ITPGRFA had no traceable influence, except for some inspiration to the NGOs working to strengthen farmers' seed systems.

Coverage of policy decisions pertaining to farmers' rights No act in the Philippines is exclusively devoted to farmers' rights as such.[109] The only act that explicitly addresses farmers' rights is the 2002 PVP Act, which has provisions pertaining to

---

108 Some respondents claim that breeding companies in some cases have demanded that farmers sign contracts that prohibit the use of farm saved seeds from the propagating material they buy. I have not been able to validate this information. The possibility that such practice may have occurred should, however, be kept in mind.

109 Republic of the Philippines (1992a), Republic Act No. 7607, The Magna Carta for Small Farmers is not dealt with here, since it is seen to have been replaced by the AFMA.

the rights of farmers to their self-harvested propagating material. The Indigenous Peoples' Rights Act is also relevant, since many indigenous people are farmers, but it does not explicitly address farmers' rights. And EO 247 is relevant in a farmers' rights context, since it requires prior informed consent of local as well as indigenous communities to bio-prospecting, and provides for information on research findings to go to such communities, as well as for benefit-sharing arrangements. In cases where bio-prospecting would be directed at PGRFA, it is likely that farmers would be involved, and thus their rights are addressed indirectly, but not explicitly. In addition, all acts pertaining to PGRFA are relevant as to the question of farmers' participation in decision-making within the areas regulated by these acts.

The regulation that comes closest to providing for *rewards* to farmers for their conservation and maintenance of PGRFA is EO 247, with its benefit-sharing provisions. These do not explicitly address farmers or their rights, but since EO 247 covers PGRFA, they are relevant. Farmers who share their PGRFA with recipients capable of developing these resources into benefits could be among the indigenous and local people with whom the benefits are to be shared at the local level. However, EO 247 does not provide for any reward to all the other millions of farmers who contribute to maintaining the diversity of PGRFA in the fields without any benefit-sharing recipient on the other end: there are no policies aimed at that issue. Therefore EO 247 would reward only an infinitesimal proportion of the many farmers who contribute to maintaining plant genetic diversity in agriculture, so the coverage offered must be said to be marginal.

Until June 2002 there were no acts in the Philippines that directly addressed or limited the customary practices of farmers to save, use, exchange and sell PGRFA.[110] There was therefore no policy coverage in this sense up to that time, and farmers were free to continue their customary practices.

In the PVP Act, the traditional rights of small farmers to save, use, exchange, share or sell their farm produces of varieties protected under the act are recognized in the form of an exemption from the breeders' rights (Section 43 d). For an act based on the UPOV Convention, the term 'rights' in this context is quite radical. In UPOV the term used for such practices is 'privilege', which has quite different connotations. In addition, UPOV 1991, which provided a framework for the development of the act, does not allow exchange and sales of self-harvested PGRFA from protected varieties. Therefore the Philippine PVP Act would appear to be quite far-reaching in terms of farmers' rights, as compared to the UPOV Act, which formed a point of departure for its formulation.

However, several formulations limit the provision in the PVP Act. First of all, the National Plant Variety Protection Board is to determine the conditions under which the exemption shall apply, taking into account the nature of the plant cultivated, grown or sown. This means that the Board will also have to define the maximum 'size' of a small farmer. Furthermore, small farmers may exchange and sell seeds only for reproduction and replanting on their own land holdings. Whereas the term 'own

110 There were policies providing for incentives to farmers to use high-yielding crops and buy certified propagating material, but these were not aimed at restricting farmers' practices of reusing farm-saved seeds.

land holding' is taken from UPOV'91, it is difficult to apply in a Philippine setting, since most farmers are tenants and do not own the lands on which they grow their crops. The Board will therefore also have to determine how 'own landholdings' is to be defined. The Implementing Rules and Regulations[111] do not shed any further light on these questions. They only limit the exemptions further by stating that farmers are not allowed to sell any seeds from protected varieties under the trademark or trade name of the holder of the breeders' right. The crucial question is therefore how the Board will interpret this provision. There is only one farmers' representative on that Board, to be nominated by the Secretary of Agriculture from a small farmers' organization, and this procedure has proven to be controversial.

So far, we can conclude that since June 2002 there has been an act that deals with farmers' customary practices of saving, using, exchanging, sharing and selling PGRFA; further, it has full coverage in that it regulates all such activities. For large-scale farmers it will be prohibited to maintain customary practices regarding PGRFA which is protected with plant breeders' rights, under specific – not yet defined – conditions, whereas for small-scale farmers – including indigenous farmers – this will be allowed.

The question of policy decisions pertaining to farmers' participation in decision-making processes is a more complex one. Since People's Power in 1986, leading to the overthrow of President Ferdinand Marcos and the reintroduction of democratic rule, it has officially been the policy of the state to include NGOs in the development process. This is reflected in the Philippine Constitution of 1987, which terms the participation of the people and their organizations in decisions at all levels of social, political and economic decision-making a 'right' (Article XIII, Section 16). It further obliges the state to facilitate the establishment of adequate consultation mechanisms. The DENR issued general rules and regulations on NGO participation in DENR programs as early as 1989.[112] As for agriculture, AFMA specifies that the state shall provide all citizens, through direct participation or through their duly elected, chosen or designated representatives, with the opportunity to participate in policy formulation and decision-making by establishing and improving appropriate mechanisms and by giving them access to information.[113]

There are, nevertheless, no further operational rules on how participation is to be implemented for agricultural policies in general, with regard to farmers' rights concerning PGRFA in particular. In general, all the acts and policies discussed here provide for the participation of one or a few representatives of farmers' organizations or other relevant civil society organizations on the boards or councils that have been established to oversee implementation of these policies. Normally, representatives are nominated by the organizations in question and appointed by the respective secretary, but, as in the case of the PVP Act, there is also the possibility of the secretary nominating the representative from farmers' organizations. In such cases, there is reason to question how representative the nominated person actually is.

111 Republic of the Philippines (2003), Administrative Order No. 07.

112 DENR (1989), Administrative Order 120.

113 Republic of the Philippines (1998a), Administrative Order No. 6, Sections 2f and 3e.

Among the most important decisions for farmers and their representatives are those regarding control over PGRFA. Both EO 247 and the IPRA provide for prior informed consent by local communities and/or indigenous peoples, which would be farming communities for the case of bio-prospecting of PGRFA. However, the PVP Act allows plant breeders to apply for plant breeders' rights without having to disclose any information on the sources of the PGRFA they used for the variety, and leaves the burden of proof to the farmers.

Summing up our discussion, it is important to distinguish between the time before and after June 2002 (adoption of the PVP Act). Prior to June 2002 there were no policies concerning rewards to farmers for their conservation and management of PGRFA, or on the practice of farmers to freely use self-harvested propagating material. There were policies on peoples' participation in decision-making, including farmers' participation. These policies can be said to have had full coverage, there was full coverage with regard to one of three items: all groups of farmers directly or indirectly addressed through these policies were to be involved in decision-making at all levels pertaining to agriculture. They should also, according to the policies, have decisive power with regard to prior informed consent for bio-prospecting. Since these policy formulations are relatively strong and are even reflected in the Constitution, we will consider the level of overall policy coverage with regard to farmers' rights as medium.

The situation after June 2002 differs in that the PVP Act had been adopted, thereby regulating the utilization of self-harvested propagating material in various ways for all categories of farmers. Thereby two of the items listed above – pertaining to all groups of farmers in the country – were covered by the Philippine policy decisions after June 2002. As indicated above, this situation could qualify as high or medium coverage, depending on the strength of the policy decisions. Since the PVP Act covers all farmers, and any violations of a plant breeder's rights can be punished by imprisonment of not less than three years and not more than six years (Title VIII, Section 56), we may infer that this is a strong piece of legislation. As also the legislation on participation is considered relatively strong, we will conclude that there was high coverage of policies regarding farmers' rights after June 2002.

Compatibility between policy decisions pertaining to farmers' rights Policy decisions on farmers' rights are generally compatible up to 2002. After 2002 there are two interrelated sources of incompatibility: (1) the relation between the PVP Act and other acts concerning representation, and (2) EO 247 in relation to the PVP Act with regard to farmers' possibilities for participating in decision-making regarding control over PGRFA.

Since the PVP Act does not allow farmers' organizations to nominate their own candidate for its Board, we can conclude that it does actually not provide for farmers' representation in the real sense of the word. This is not in line with the overall policies of the country. However, since the overall policies of the country and all other policies are strict on this point, the general picture is largely that of compatibility on this issue, with the exception of the PVP Act.

When it comes to the practices of farmers to save, use, exchange, share and sell farm-produced seeds, the PVP Act is not in line with the policies on participation in decision-making. The PVP Act deprives farmers of their possibilities to decide over farm-saved seeds and propagating material from protected varieties. The exemptions for small-scale farmers may ease the situation, depending on how these provisions are interpreted and implemented. Since this act is so important with regard to farmers' rights, and farmers are deprived of any possibility to influence it at all levels, right up to the National Plant Variety Protection Board, this situation is best described as one of divergence between the PVP Act and the various acts pertaining to participation in decision-making.

On this basis we can say that policies until June 2002 were largely compatible, whereas the situation after 2002 reveals divergence.

Conduciveness of policy decisions pertaining to farmers' rights The policies on reward for farmers' representation in decision-making processes were generally conducive to farmers' rights up to June 2002, though they were not particularly strong. The absence of PVP regulation enabled farmers freely to use self-harvested seeds – and that is also positive in terms of farmers' rights. Thus, the situation can be said to be generally conducive. From June 2002 the picture changed: the PVP Act represents a move in the wrong direction with regard to farmers' rights, due to its limitations on farmers' customary use of seeds and propagating material and lack of farmers' representation.

Effects of international regimes on policy decisions As noted, the CBD was decisive for EO 247 and the TRIPS Agreement for the PVP Act. (This will also be explored in greater depth in Chapter 10.) There are no other documentable traces of influence from international agreements on the policies dealt with here.

The EO 247 affects farmers' rights only to a limited extent: it provides (implicitly) for rewards to a limited number of farmers or farming communities and for their control over PGRFA use in their fields, that is a form of participation in decision-making. In the broader context, the latter is one of several potential arenas for decision-making. Thus, EO 247 affected farmers' rights only to a limited extent, also as compared to the other policies discussed above. We may say that some, but not all, of the lines of influence can be traced back to the CBD.

The PVP Act limited the rights of farmers considerably with regard to their customary practices in agriculture and to some extent with regard to their participation in relevant decision-making processes. It was enacted because of the TRIPS Agreement, so the effect of the TRIPS Agreement on these policies must be said to be substantial after 2002. Also the effect of the UPOV Convention of 1991 was decisive, since the PVP Act is largely derived from that Act, for reasons to be explored in Chapter 10. Since, however, the PVP Act substantially affects one of the three components of farmers' rights (customary practices), the other to a limited extent (participation in decision-making), and does not affect the third type of right directly (benefit sharing), we can conclude that some lines of influence on the policies pertaining to farmers' rights in the Philippines can be traced back to the TRIPS Agreement and the UPOV Convention.

The EO 247 corresponds with CBD with regard to those provisions that are relevant to farmers' rights. As noted, the TRIPS Agreement does not specify what an effective *sui generis* system for plant varieties is, but economically strong forces have maintained that the UPOV Act of 1991 would provide the required framework for complying with Article 27.3(b) of the TRIPS Agreement. The PVP Act is largely in line with the UPOV Act of 1991, and parts of the text are copied from this Act. There is some divergence with regard to what the PVP Act calls 'farmers' rights'; however, exactly how great this divergence will be depends on interpretation. All in all, we may conclude that there is substantial match between the two.

Although there are no signs of attribution concerning the effects of the IU/ ITPGRFA on policies pertaining to farmers' rights in the Philippines, it would be interesting to discuss correspondence. However, this is difficult to assess, both because the agreed interpretations of the IU (see Chapter 5) are rather vague, and because ITPGRFA implementation is up to the national governments as far as farmers' rights are concerned. However the policy situation before June 2002 was clearly more in line with the two agreements than the situation after this date.

Despite their limited coverage, the effects of the CBD on policy decisions pertaining to farmers' rights as defined in Chapter 3 must be regarded as positive. The effects of the TRIPS Agreement and the UPOV Convention on policy decisions regarding farmers' customary practices with regard to PGRFA are deemed negative, for reasons to be further explained in Chapter 10.

Coverage of goal achievements pertaining to farmers' rights What of rewards to farmers for maintaining PGRFA in the fields – the first of the three types of rights identified above? As seen above, there was no benefit sharing based on EO 247 regarding PGRFA, and thus no rewards to farmers.

Here we should recall the bottom–up initiatives undertaken by several NGOs for *in situ* management of PGRFA and participatory plant breeding. Although these do not respond to any policies concerning rewards to farmers, they can be regarded as a kind of reward, since they help farmers to develop improved varieties based on the PGRFA they have saved. Geographic coverage is comprehensive, but the number of farmers involved is low. These projects are mainly important as examples of how it is possible to value farmers' efforts in the conservation of PGRFA in a way that supports and promotes these activities while also improving their livelihoods. This is a kind of reward that is not dependent on agreements between users and providers of specified PGRFA, but can as a principle accrue to farmers at large who take part in maintaining PGRFA, as far as financial and institutional resources allow. However, these achievements cannot be related to any of the policies dealt with in this section on farmers' rights, and are addressed above.

As to the second category of farmers' rights, there are no indications that farmers' practices with regard to the saving, exchanging, sharing and selling of farm-produced seeds and propagating material have changed during the period in focus in this study – also not in the short time that has elapsed after the entry into force of the PVP Act. The PVP Act did not become fully operational until the end of 2004, so it was probably too early to expect any goal achievement within our period of study.

As mentioned above, a problem with regard to farmers' or participatory breeding is that it is not permitted, according to the Seed Industry Development Act, to sell farm-developed seeds on a commercial basis, unless they are certified – and farm-developed seeds will normally not satisfy the requirements for certification. On this point, NGOs maintain that the seeds in question are not of an inferior quality, but that the government criteria differ from the standards set by farmers (Department of Agriculture 1995, 57). There are two representatives of accredited farmers' organizations in the National Seed Industry Council, which oversees activities under the Seed Industry Development Act, but NGOs claim that farmers were not involved in the government process of setting criteria for seed certification up to 1995; according to my interviewees, the situation did not improve thereafter.

This brings us to the complex issue of farmers' participation in making policy decisions. First of all, what does 'representation' mean? Farmers are not a homogeneous group, so there is a challenge involved in selecting only one or two institutions or persons to represent all farmers. There is also the question of what weight is accorded to the opinions of these farmers' representatives: Are they actually listened to, or are they merely tokens for the participatory approach to policy-making, mandatory under the Constitution? It was not possible to find all the answers to these questions during my fieldwork, as they differ considerably from forum to forum. The participation of farmers' representatives in the bodies most relevant from a farmers' rights perspective has been as follows:

- Farmers' organizations were represented in the National Committee on Plant Genetic Resources (NCPGR) until it faded out in 1999/2000 (see above), and had an active role in the Committee. Its successor, the National Network for the Conservation and Sustainable Use of PGRFA (PGRFA Network), had no farmers' representation, which is not in line with the provisions on participation in policy processes provided in the Constitution, the AFMA, and the Seed Industry Development Act, as outlined above. Since the two institutions were/are central in PGRFA management in the Philippines, farmers' participation is particularly important here.
- In the execution of EO 247, civil society has been represented with *inter alia* a representative from SEARICE, an organization focused on farmers. From my interviews with various stakeholders, it seems that SEARICE has had a constructive and central role in the process of implementing EO 247.
- Two farmers' organizations are represented in the National Seed Industry Council, which oversees the implementation of the Seed Industry Development Act. However, they were not involved in the development of criteria for certification, nor are they directly involved in certification matters.

In other words, farmers were quite well represented before 2000–2002, but not after. In addition, farmers were represented in various other forums before 2001, some of which changed their composition during this time, the most important being the Philippine Commission on Sustainable Development. As explained above, this Commission was radically reduced in terms of members, and important farmers' representatives went out. After 2001 the IACBGR became less relevant due to the new situation with the Wildlife Act, although some further decisions were made after 2001 to conclude ongoing processes. All in all, the coverage with regard to farmers' participation in policy processes regarding PGRFA can be described as moderate before 2000, and low since then. The tendency is clearly in the direction of less participation. On the other hand, these 'achievements' correspond with the policies in question, and as such represent substantial goal achievements with regard to participation.

Summing up these findings, we find no goal achievements with regard to rewards to farmers as indirectly set out in EO 247, and also no goal achievements concerning the limitations of farmers' practice to freely use farm-saved seeds protected by intellectual property rights. However, up to 2000 there was moderate to high goal achievement with regard to participation. On the whole, this qualifies as partial coverage with regard to goal achievements pertaining to farmers' rights.

Compatibility between goal achievements pertaining to farmers' rights Since the only goal achievements were made with regard to farmers' participation in decision-making processes, the question of compatibility is applicable only to this category of rights. It can be argued that farmers' participation in several central forums in the period up to about 2000/2001 was one of synergy, in the direction of farmers' general participation in decisions regarding PGRFA. However, farmers were not sufficiently included in some processes, among them seed certification. Since that is an important issue for farmers engaged in participatory breeding, it needs to be taken into account. This is, however, not in conflict with the policies of inclusion in the other forums, and is not incompatible. Altogether, we find a lack of compatibility up to 2001, which was nevertheless also not a conflict as such.

After 2001 and up to the end of 2004, there was a clearer discrepancy between representation of farmers in some forums, and no or low representation in other forums. However, as above, the goal achievements in this regard were not directly conflicting, only pulling in different directions.

Conduciveness of goal achievements pertaining to farmers' rights Altogether the conduciveness of the goal achievements was positive up to 2000, although there were some elements pointing in a negative direction, like lack of farmers' participation in practical certification policies. Then the situation deteriorated; from 2001, it became negative in terms of participation.

Effects of international regimes on goal achievements It is not possible to trace any line of influence from international agreements on these few goal achievements, except for some inspiration for NGOs, as explained above. As it would be to early expect goal achievements with regard to the PVP Act, we will not assess correspondence between the PVP Act and the TRIPS Agreement in this regard. Correspondence between the goal achievements pertaining to EO 247 and the CBD was marginal, limited to the representation of one farmer-related organization in the IACBGR. Thus, the effects must be said to be rather neutral as to conduciveness so far.

Summary of the State of Policy Decisions and Goal Achievements

PGRFA management is generally in a poor state in the Philippines, and there are no clear indications as to the future.

There is no political awareness of the importance of, and means for, *in situ* management of PGRFA. Indeed, all major policies point in the opposite direction, without any compensating measures. The only goal achievements with regard to *in situ* management have been initiated and carried out by NGOs. These projects may function as examples of how to manage PGRFA *in situ*, but they do not have enough outreach to make a difference for the *in situ* management of PGRFA in the country at large, nor can they counterbalance the lack of compensating measures from the Philippine authorities.

There is also not very high political awareness on the importance of *ex situ* management of PGRFA. However, since 1976, institutional structures have been shaped which to some extent take care of these needs, and as such the policies are positive. Nevertheless there has been a notable lack of financial resources for the *ex situ* conservation of all crops, except for rice, and this in turn has led to non-function of some cold gene bank storage facilities and perhaps a substantial loss of PGRFA. Also, there is a considerable backlog as to documentation of plant varieties collected in gene banks, as well as scope for improvement with regard to the information system on stored plant varieties. The system is generally conducive to the *ex situ* management of PGRFA, but malfunctioning has threatened parts of it. There is an expectation that the entry into force of the ITPGRFA may improve the financial situation, if donor agencies decide to support its implementation.

The situation with regard to access to PGRFA has deteriorated since the introduction of EO 247 in 1995. Plant breeders – public, private, domestic and foreign alike – have been obliged to apply for research agreements to get access to PGRFA grown *in situ*. Several breeders do not comply with these new rules, but those who do are faced with long and bureaucratic procedures or, alternatively, seeking access to PGRFA elsewhere. Also for conservation purposes, access requires a research agreement, albeit on simplified terms: nevertheless, this poses a serious constraint to these activities, due to scarce institutional and financial capacity in the Philippines. Farmers are not directly affected by the act in terms of access, but indirectly to the extent that they depend on plant breeders for their propagating material. Since the adoption of the Wildlife Act in 2001, there has been uncertainty as to the further implementation of EO 247, since the Wildlife Act covers access to genetic resources in wild plants and animals, but not to domesticated ones, whereas responsibility for the implementation of EO 247 used to rest with the Protected Areas and Wildlife Bureau (PAWB) of the Department of Environment and Natural Resources. There is no motivation for further implementation of EO 247 in PAWB. The Wildlife Act Implementing Rules and Regulations from 2004 did not solve this problem.[114] Since June 2002 there has also been the Plant Variety Protection Act, which as a policy regulates access to PGRFA from another angle. However, within the timeframe of this study it was too early to observe any particular effects of the new act.

The utilization of the diversity of PGRFA is likewise in a problematic state. Whereas there is some political awareness on the importance of diversification of crops, there is no awareness on the importance of diversity *within* crops when it

114 Neither did the new Joint DENR-DA-PCSD-NCIP (2005), Administrative Order No. 1, 'Guidelines for Bio-prospecting Activities in the Philippines'.

comes to PGRFA utilization. When a few national agricultural research centers and some NGOs contribute to the utilization of a greater diversity of plant varieties, this is not because of the policies but despite them. On the other hand, these projects can serve as examples of how it is possible to maintain genetic diversity within crops in PGRFA utilization – while also improving farmers' livelihoods.

The policies on the sharing of benefits arising from the use of PGRFA are comprehensive in all aspects, and should as such ensure that a considerable sharing of benefits takes place in the Philippines. However, no benefits were achieved following from EO 247 with regard to PGRFA. Quite the opposite: the regulation made access to PGRFA more difficult – and access is considered the most important benefit of all from the use of PGRFA. EO 247 was designed with a view to genetic resources in wild plants and animals, and was therefore not adapted to the particular conditions in plant breeding. The new situation with the Wildlife Act covering the bio-prospecting of genetic resources in wild plants and animals leaves PGRFA with a regulation that was not designed for agricultural resources, and which is no longer in practice – although it is still in force for PGRFA. This is a highly confusing legal situation. On the other hand, it opens a window of opportunity for reformulating EO 247 to fit better with the management needs for PGRFA.

Farmers' rights were taken care of indirectly up to about 2001. Until then, farmers were represented in some of the most important committees and other decision-making bodies, according to the Constitution and various other pieces of legislation. Also there were no acts limiting the practices of farmers to reuse their seeds and propagating material, or to exchange it, share it or sell it. Since about 2000, however, a process has developed whereby farmers and their representatives are less represented in the relevant decision-making bodies. They are not represented in the new national network on plant genetic resources for food and agriculture and also not – in the real sense of 'representation' – in the body to oversee the new PVP Act. In other forums as well, representation is decreasing, also in the Philippine Council for Sustainable Development, which oversees the implementation of the Philippine Agenda 21. The PVP Act limits farmers' use of protected varieties. On the other hand, it makes exemptions for small-scale farmers, referring to the practice of saving, reusing, exchanging, sharing and selling PGRFA as farmers' rights. Whereas this is positive as a confirmation of what farmers' rights are, the very fact that the new rules limit these rights for all those farmers who are not covered by the exemptions shows that the act is a step in the opposite direction.

Table 9.4 summarizes the state of policy decisions and goal achievements pertaining to PGRFA management discussed in this section.

Table 9.4 Overview of the state of policy decisions and goal achievements pertaining to PGRFA management in the Philippines[115]

Core elements in the management of PGRFA:	State of policy decisions pertaining to PGRFA management 1992–2004:			State of goal achievement pertaining to PGRFA management 1992–2004:		
	Coverage	Compatibility	Conduciveness	Coverage	Compatibility	Conduciveness
***In situ* management**	**A**	⇔	÷	**B**	⇔	÷ (+)
***Ex situ* conservation**	**B**	⇒	+	**A**	⇔	+ (÷)
Access to PGRFA	**A**	⇒ Since 2001: ⇔	÷	**B**	⇒	÷
PGRFA utilization	**A**	=	÷	**B**	⇔	÷ (+)
Benefit sharing	**A**	⇒ Since 2001: ⇔	+ Since 2001: ≈	**0**		÷
Farmers' rights	**B** Since 2002: A	⇒ Since 2002: ⇔	+ Since 2002: ÷	**B**	=	+ Since 2001: ÷

Preliminary Conclusions on Relative Effects of International Regimes on PGRFA Management in the Philippines

We have in this chapter made a brief assessment of all effects of the international regime constellation, as far as documentation and other data have allowed, but without thorough tracing of the lines of causation. This has been for the purpose of overview. In the next chapter, we explore the effects in depth with regard to the two most central policies as to PGRFA management. We have been able to identify substantial traces of attribution, and that allows us to compare our findings from the Philippines with the assumptions set out in Chapter 8 as to aggregate effects of the international regimes for developing countries.

In Chapter 8, we assumed that the aggregate effect of the regime constellations would be scant attention to *in situ* management of PGRFA. That is also what we found in the Philippines – but was this due to the aggregate effects of the international regimes in focus? We found marginal lines of influence from international regimes on the *in situ* management of PGRFA, and it is certainly difficult to find evidence for causal relations between international regimes and the lack of attention. In my interviews at the Department of Agriculture in May 2000 and March 2002, however, the general answer was that one had to await the new international agreement on

115 See Table 4.6 on p. 63 for an explanation of the symbols used here.

PGRFA before new policies could be developed in this area. On the other hand, it was made clear that *in situ* management of PGRFA would never become a high priority issue as long as increased food production was necessary to feed the rapidly growing population. In other words, the negotiations leading to the ITPGRFA provided a good excuse for scant attention, but the real reason was another set of political priorities. The results of these priorities were policies that did largely not correspond with the provisions of the IU/ITPGRFA and the CBD on *in situ* management of PGRFA. Nevertheless the marginal influence from the FAO Global Plan of Action provided positive input to the planning of *in situ* management under the NCPGR and the PGRFA Network, and we found that the FAO process as well as the CBD had substantially influenced bottom–up approaches for *in-situ* management. Therefore, we may conclude that the overall effects of the international regime constellations have been limited, but still positive.

In Chapter 8, we assumed that the aggregate effects of the regime constellations would be greater with an *ex situ* conservation focus on national gene banks since 1992. However, we found only marginal lines of influence from the international regimes in question on *ex situ* policy in the Philippines. Nevertheless, the Philippines has followed two tracks with regard to *ex situ* conservation: national gene banking and international co-operation (particularly in rice) which correspond with the CBD (emphasizing national gene banks) as well as the IU/ITPGRFA (emphasizing international co-operation on *ex situ* conservation). Probably an important reason for the international co-operation is that IRRI is located in the Philippines. A critical aspect for the non-rice gene banks was the lack of financial resources to maintain storage facilities and develop the collections. It seems that international support was either not sought or not available. International regimes had marginal influence on the goal achievements regarding *ex situ* conservation, basically in terms of inspiration from the FAO Global Plan of Action. As such, this must be deemed positive. Therefore, the overall effects of the international regime constellations can be said to have been very limited, but still positive.

In Chapter 8 we assumed that access to PGRFA would increasingly be limited from 1993 by legislation on access to genetic resources, and from 2000 by legislation on intellectual property rights. After 2004, access to specified crops would be ensured through the Multilateral System, we assumed, but that is after the period studied here. In general, our assumptions were shown to be correct for the Philippines. As we will see in Chapter 10, EO 247, which limited access to PGRFA decisively as a policy, is largely derived from the CBD, although not all lines of influence can be traced back to the Convention – and fury against the TRIPS Agreement provided an important impetus for the regulation. The IU/ITPGRFA did not affect this policy development in any traceable way, although availability and accessibility were/are central topics in these regimes. As a policy, EO 247 was in line with the CBD but not with the IU/ITPGRFA, and the policy was not conducive to access to PGRFA. Poor performance made the situation worse; moreover, due to the lack of facilitation for access, it proved to be not even in line with the CBD after few years (see Chapter 10). The PVP Act, which further delimits access from another angle, was largely derived from the UPOV Convention as the Philippine response to the TRIPS Agreement. The limitation of access is largely in line with the UPOV Convention (though not as strict as the UPOV 1991 Act), so there is reason to assume that it is also in line with the minimum requirements posed in Article 27.3(b) of the TRIPS Agreement. Thus, the influence of the international regime constellations on the facilitation of access to PGRFA has been considerable – and negative.

We did not develop separate assumptions with regard to the aggregate effects of the regime constellations on the utilization of PGRFA, partly because this is a topic that received international attention only in the course of the negotiations towards the ITPGRFA, and partly because it has parallels to *in situ* management, which is covered. We have nevertheless sought to identify the aggregate effects of the international regimes on utilization of the PGRFA in the Philippines, and have found a multifaceted structure of influence. There were marginal traces of influence leading to the IU/ ITPGRFA, but several to the CBD and the TRIPS Agreement – via EO 247 and the PVP Act. However, these regulations restrict access to PGRFA, so the influence was negative. Despite this overall trend, the influence of the FAO regimes and the CBD with regard to bottom–up initiatives for *in situ* management can also be assumed to have influenced PGRFA utilization positively. The bottom line can be said to have been a balance between negative (EO 247) and positive (inspiration to and support for NGO initiatives via the IU/ITPGRFA and the CBD) effects of international regimes up to 2002, but this turned negative with the PVP Act, which was adopted in order to comply with the TRIPS Agreement.

In Chapter 8 we assumed that the major benefit from the use of PGRFA – access to these vital resources – would be reduced from 1993, and that there would be few if any direct benefits to be shared from the use of PGRFA as a result of the international regime constellations. After 2004, it could be assumed that access would be ensured to the PGRFA from the Annex 1 crops, and that benefits from specified intellectual property rights would be warranted. The latter would, however, be beyond the scope of this book. Our assumptions as to the aggregate effects of the international regime constellation regarding the sharing of benefits from the use of PGRFA up to 2004 were principally correct. Largely derived from the CBD – in response to the TRIPS Agreement – EO 247 was targeted at all resources, and as such, the effects of the CBD on the policy were conducive to PGRFA-related benefit sharing. The praxis was, however, negative in terms of such benefit sharing, which also indicates that the overall effects of the CBD in this regard were negative, and not in line with the international agreements.

In Chapter 8, we assumed that little would have been done to realize farmers' rights in developing countries so far, but that more would have been achieved with regard to the enhancement and protection of traditional knowledge as set out in the CBD. Nevertheless, access legislation derived from the CBD could limit farmers' rights; and the introduction of intellectual property rights to PGRFA could substantially limit these rights further, particularly if introduced without legislation on the recognition of farmers' rights. In the Philippines these assumptions were found to be largely correct, and can generally be traced back to the regime constellations. The effect of the CBD is two-edged: On one hand EO 247 provides for prior informed consent for access to PGRFA, which indicates that farmers might have a say, and potentially also with regard to benefit sharing. On the other hand, EO 247 restricts access, which is an important precondition for the availability of adequate seeds also for farmers – a benefit. Furthermore, the approach of sharing benefits with the providers of PGRFA excludes from the scope for benefit sharing the large majority of farmers who do participate in maintaining PGRFA for future generations. Even if implemented properly, EO 247 would never have reached more than some farming communities – due to the limited demand for PGRFA. Thus it affects the realization of farmers' rights only to a limited extent. The TRIPS Agreement in conjunction with the UPOV Convention has substantially contributed to limiting the rights of farmers through the PVP Act. This may be largely in line with these agreements – but it is not conducive to farmers' rights. Interestingly, it seems that the IU/ITPGRFA had little effect on farmers' rights in the Philippines, except for its catalyst effects for bottom–up initiatives to support farmers in *in situ* management of PGRFA, including participatory plant breeding. In this context the CBD is also relevant.

All in all, the international regime constellations were found to be important for the situation of farmers' rights, but there were also several other domestic traces of causation. On the whole, we can say that the various effects from international regimes balanced each other up to 2002, but that they turned negative after the adoption of the PVP Act. Nevertheless, we should recall that the concept of 'farmers' rights' was included in the PVP Act, and that the exemptions were more radical than provided for in UPOV'91. This can partly be traced back to the negotiations on farmers' rights in the FAO, as Chapter 10 will show.

In sum, we have seen that international regimes have had considerable effects, particularly on policies but also to a certain extent on goal achievements regarding PGRFA management. This is summarized in Table 9.5.

Table 9.5 Relative effects of international regimes on PGRFA management in the Philippines[116]

Relative effects → ... **pertaining to ↓**	**Relative effects of the IU/ITPGRFA on the state of:**		**Relative effects of the CBD on the state of:**		**Relative effects of TRIPS (27.3.b) on the state of:**		***Aggregate effects of the international regimes***
	policy decisions	**goal achievements**	**policy decisions**	**goal achievements**	**policy decisions**	**goal achievements**	
***In situ* PGRFA Management**	**I/0** ≠ (+)	**II** − +	n/a	**II** − +	n/a	n/a	Limited but positive
***Ex situ* PGRFA Conservation**	**I/0** ≡ (+)	**I/0** − ≈	n/a	n/a	n/a	n/a	Very limited but positive
Access to PGRFA	n/a	n/a	**III** ≡ ÷	**II** ≠ ≈	**II** ≠ ÷	n/a	Substantial, negative
Utilization of PGRFA	**I/0** ≠ (+)	**I/0** ≠ +	**III** ≡ ÷	**II** ≠ ≈	**III** ≡ ÷	n/a	From neutral to negative
Benefit sharing	n/a	n/a	**III** ≡ +	**II** ≠ ≈	**II** ≠ +	n/a	Limited and negative
Farmers' rights	**I/0** ≠ (+)	n/a	**II** ≡ +	**I/0** ≠ ≈	**II** ≡ ÷	n/a	From neutral to negative

116 See Table 4.9 on p. 68 for an explanation of the symbols used here (n/a = not applicable).

Chapter 10

Mechanisms of Influence of International Regimes: Two Cases

As we have seen, the constellations of international regimes concerning PGRFA management have substantially affected Philippine policies. How then is it that international regimes can influence domestic policies in a country? Understanding these mechanisms of influence will put us in a better position to see what action is required at the international level. For our embedded case studies we have selected the two policies most directly influential for PGRFA management in the country – Executive Order 247 and the Plant Variety Protection (PVP) Act. Our analysis will be guided by the theory framework developed in Chapter 3 and the method set out in Chapter 4.

The First Case: Philippine Policy on Access and Benefit Sharing

As was shown in Chapter 9, the Philippine policy on access and benefit sharing, EO 247, was not conducive to PGRFA management, because of its overly bureaucratic procedures concerning access. The core intention of the policy – to provide for benefit sharing – was not achieved. Quite the converse: legal access to genetic resources – the most important of all benefits in a PGRFA perspective – was reduced. How could this happen? And how did the international regimes in question affect developments?

The Story of Executive Order 247

The story of Executive Order 247[1] is the tale of a bottom–up initiative originally taken by a small group of scientists who built strategic alliances with civil society organizations, particularly environmental NGOs and key people in relevant government agencies, and worked strategically to achieve a fully fledged regulation on access and benefit sharing regarding biological and genetic resources in the Philippines. This strategy proved successful in getting the policy adopted, but, as time would show, it was not conducive to a sustained policy on access and benefit sharing.

---

1 This presentation is based largely on the analysis of the participation in the formulation of the EO 247 in Swiderska et al. 2001, which provides a detailed and highly valuable account, and partly on information in La Viña et al. 1997. The subject was also taken up in my interviews in the Philippines in 2000 and 2002.

The prehistory to the EO 247 process dates back to the overthrow of Ferdinand Marcos in 1986, when a new era for the country emerged. Civil society initiatives flourished, environmental issues came on the agenda and there evolved a sense of nationalism, including the demand to regain control over the country's resources.[2] This demand was spurred not only by the experience of having a dictator widely known for his exploitation of the country, but also by the long colonial history. In 1987, the Protected Areas and Wildlife Bureau (PAWB) was established under the Department of the Environment and Natural Resources (DENR), vested *inter alia* with responsibility for regulating the collection of non-domesticated biological and genetic resources (La Viña et al. 1997, vi). Here it is important to note that the collection of PGRFA remained within the mandate of the Bureau of Plant Industry (Benavidez 2004, 154).

There were no rules or guidelines for PAWB's work with regard to the regulation of the collection of non-domesticated resources.[3] In 1990 government agencies together with academic institutions adopted a Memorandum of Agreement, the *Guidelines for the Collection of Biological Specimens in the Philippines*. These Guidelines addressed bio-prospecting of non-domesticated biodiversity, but they were basically an administrative co-ordination and permit system, and could not serve as a regulatory framework (La Viña et al. 1997, vi).

Parallel to these efforts, groups within the Philippine academic community played a seminal role in developing international guidelines for the ethical and sustainable utilization of Asian biological resources (ibid.). In February 1992, the Seventh Asian Symposium on Medicinal Plants, Spices and Other Natural Products (ASOMPS VII) was convened in Manila. This meeting marked a milestone in awareness-raising among Asian scientists with regard to bio-prospecting. The symposium resulted in the *Manila Declaration Concerning the Ethical Utilization of Asian Biological Resources*, which had an appended *Code of Ethics for Foreign Collectors of Biological Samples* as well as *Contract Guidelines*.[4] The Manila Declaration recognized national sovereignty over biological resources, the need for local scientists to be involved in research on biological resources, the need for countries to receive a fair share of the commercial benefits arising from such research, and to develop legislation on these issues. Whereas biological resources were not defined in the Declaration, the focus was clearly on medicinal plants. No mention was made of PGRFA.

Three months later, the Convention on Biological Diversity (CBD) was adopted in Nairobi, and the Philippines signed it at the later Rio summit. The CBD received considerable attention in the Philippines, where it was ratified in the following year after (October 1993), and various efforts were undertaken to pioneer its implementation, particularly with regard to protected areas. The year after the CBD had entered into force, in February 1994, a small group of Manila-based university

2 Based on Swiderska et al. 2001, 9 and my own experience during a four-month stay in the Philippines in 1988, working *inter alia* with development support for environmental initiatives.

3 Prior to 1987, the National Museum of the Philippines was the responsible entity regarding issues concerning the collection of biological samples (Barber and La Viña 1997).

4 Available at <http://dikti.org/p3m/mandec.htm>.

chemists,[5] acting on behalf of the Philippine Network for the Chemistry of Natural Products, took the initiative to draft a regulatory framework for bio-prospecting by non-nationals as a preparation for the next ASOMPS symposium (Swiderska et al. 2001, 9). They used the Manila Declaration as basis for their work. The initiative was motivated by the demand to regulate foreign bio-prospecting in the country for the purpose of chemistry, with a particular view to pharmaceuticals. It did not address PGRFA as such.

Against this backdrop, the group of chemists decided to proceed with the initiative by involving a legal expert. They commissioned Attorney-at-Law Antonio G. M. La Viña, who was then Professor of the University of the Philippines College of Law and had also worked with an NGO involved in ancestral land issues. In April, he revised the draft to include elements of the CBD, including the overall coverage with regard to biological diversity implicitly also covering PGRFA, and introduced two further significant changes (ibid.): (1) The scope was expanded to cover Filipino scientists and academic research in addition to foreign scientists; (2) rules were introduced concerning prior informed consent of local and indigenous communities. In addition, Dr La Viña advised the Philippine Network to propose the regulation as an Executive Order, and not as an Act of legislation: the latter would take several years, and it would also be easier to get the former through the required procedures (ibid: 10). An Executive Order needs only to be signed by the President – and President Ramos was known for his commitment to the Rio and CBD processes.These are the circumstances in which the first draft of the Executive Order was presented on 1 May 1994. The work had been funded by the UNESCO Regional Network for the Chemistry of Natural Products in Southeast Asia (La Viña et al. 1997, vii).

With the new draft Executive Order, the group of chemists now turned to consultation work. The first consultative meetings were held with members of the Philippine Network based in Manila, with other scientists from various universities in Manila, including some institutions involved in bio-prospecting, such as the Marine Science Institute (MSI) and the National Museum. These consultations were funded by the UNESCO Regional Network for the Chemistry of Natural Products in Southeast Asia. A committee on the Executive Order was then established at the University of the Philippines in Diliman (Quezon City, Metro Manila), and it included chemists and biologists.

The group of chemists now took the initiative to convene consultative meetings with officials from the Department of Science and Technology (DOST) and with government scientists from the National Academy of Science and Technology (NAST), an agency affiliated with DOST. In light of comments received from DOST and NAST, the draft EO was revised. As a result, DOST and NAST joined the core group of chemists in July 1994. Now the scientists had ensured that the views of academic stakeholders were taken care of, whereas the government officials could ensure consistency with the views of some central government scientific research bodies. Most notably the Philippine Council for Agriculture, Forestry and Natural

5 The chemists were from the University of the Philippines at Diliman, the Ateneo de Manila University and the University of Santo Tomas – all located in Metro Manila.

Resources Research and Development Board (PCARRD), also affiliated with DOST, was consulted for each draft (Swiderska et al. 2001, 11). In addition, one NGO was involved at this stage, the Earth Savers, acting on behalf of a network of individuals and organizations involved in environmental issues (ibid.).

In August 1994, around 100 participants attended a multi-stakeholder conference organized and funded by NAST at the University of the Philippines in Diliman. The main target group was still the academic and scientific community, and this time some scientists from other regions in Luzon than the capital area took part. The meeting was also attended by representatives from a range of government agencies, including DOST, the DENR and PAWB, as well as a few NGOs. After the conference, the DENR and PAWB were actively included in the drafting of the Executive Order. PAWB was later to become a lead agency in implementing EO 247.

Following the NAST conference, further consultative meetings were held with selected government institutions, NGOs and Peoples Organizations (POs).[6] Among the government institutions were the Departments (ministries) of Agriculture (DA), Health, Education and Foreign Affairs (ibid.). It is remarkable that the DA (which also covers fisheries) participated with units responsible for marine biodiversity. It seems that they were not aware of the implications of the proposed regulation for the management of PGRFA. The PCSD Council, which includes several NGOs and was involved in the meetings, had a decisive say in the finalization of the text, and its approval was made dependent on endorsement of the PCSD.

NAST presented a draft Executive Order at a meeting of a Cabinet Cluster of Agriculture and Agro-Industry composed of cabinet secretaries and top-level officials from agriculture-related line agencies (ibid., 12). A final multi-stakeholder seminar was held for consultations with the business community. This seminar, organized and funded by the Asian Institute of Management, was attended by between 40 and 50 participants, including representatives of the Pharmaceutical and Healthcare Association of the Philippines (which includes foreign members), some representatives from pharmaceutical and agrochemical companies, including international companies (among them, Monsanto), in addition to government officials, academics, scientists and NGO representatives (ibid). Against this background, the draft was again slightly revised before it was adopted by President Fidel Ramos in May 1995 as *Executive Order No. 247 Prescribing Guidelines and Establishing a Regulatory Framework for the Prospecting of Biological and Genetic Resources, Their By-Products and Derivatives, for Scientific and Commercial Purposes, and for Other Purposes*.

EO 247 did not become operational until its Implementing Rules and Regulations (IRR) were in place in 1996. In September 1995, the Inter-Agency Committee on Biological and Genetic Resources (IACBGR) was set up, as envisaged in Section 6 of EO 247. It was co-chaired by the DENR and DOST, and included representatives of the Departments of Agriculture (from the Bureau of Fisheries and Aquatic Resources), Health and Foreign Affairs, the National Museum, the academic/scientific

6 Philippine term for local level organizations in civil society, often organized with a bottom-up approach, whereas NGOs in the Philippine understanding tend to work more top-down from the national to the local level.

community, NGOs and indigenous peoples' organizations. The IACBGR appointed a small group consisting of scientists and legal staff from PAWB, the DENR and DOST to draft the IRR, with PAWB as lead agency. The Department of Health was also closely involved (ibid., 13). In January 1996, Dr La Viña was appointed DENR Under-Secretary for legal and legislative liaison, and took responsibility for drafting the IRR. Soon after that, a first draft was ready for circulation.

A new consultation process started, involving various government departments, the PCSD with its Sub-Committee on Biodiversity and the PCSD Council (which includes NGOs and POs), applied scientists (*inter alia* from PCARRD), industry groups and national pharmaceutical companies (ibid.). This time, funding was even more limited (all provided by the DENR), so the consultation process could not be so extensive. But discussions were heated, since the final text of EO 247 had now been presented to a range of groups who had not known about it – or shown much interest – until they discovered the implications of the regulation, after it had already been adopted (ibid., 17). Some groups of scientists and pharmaceutical groups were highly critical of the new regulation, and fought to limit the consequences through the IRR. Some indigenous groups, on the other hand, felt that the regulation did not go far enough and demanded that bio-prospecting be prohibited on ancestral lands (ibid.). Several environmental NGOs used the opportunity to provide input to the drafting of the IRR.

In June 1996, the IRR were approved as Department Administrative Order 96-20,[7] and implementation could start. Soon EO 247 became known in many parts of the world – particularly among people involved in CBD implementation – as a pioneer effort aimed at the realization of the CBD provisions on access and benefit sharing. Developments were followed, with great expectations. However, the road to implementation proved to be full of pitfalls, resulting in major difficulties and lack of accomplishments. In analyzing the mechanisms of influence in the next sections, we will seek to analyze the reasons for these problems.

Advocacy Coalitions and the Relative Influence of Actors Involved in PGRFA Management

In this section we will discuss the relevance of *advocacy coalitions* as a concept for describing the actor constellations pertaining to the EO 247 process and analyzing the relative influence of the various actors – as a mechanism of influence with regard to the relative effects of international regimes. 'Influence' here refers to the substantive contributions of several categories of actors (from civil society, state entities and business) to the formulation of the final text. We also ask how and why the various actors achieved the influence they did – and why PGRFA interests received so little attention. In this way, we will see how the actors engaged in the field of PGRFA influenced the process leading up to the adoption of EO 247.

A basic requirement for identifying an advocacy coalition, as described in the Advocacy Coalition Framework (ACF: see Chapter 3), is that it must be based on

7 Republic of the Philippines (1996a), Department Administrative Order (DAO) No. 96-20.

beliefs and not on common interests. In our case, it may seem difficult to determine whether the core motivation behind the coalition that emerged during the EO 247 process had to do with beliefs or common interests. Various actors were involved, some of whom were particularly active and contributed substantially to the development. These were the actors from civil society and state authorities, including from science. They all shared the core idea that Philippine resources should be controlled by the people of the Philippines. This idea became gradually more pronounced from consultation to consultation, and is clearly a common interest on behalf of the country. As such, it is, however, also based on a belief about what is perceived as best for the Philippines – a belief based on the shared experiences of centuries of colonialism and exploitation, as well as the newly overthrown dictatorship that had been headed by one of the world's most corrupt presidents.

Particularly in civil society and among some government entities, there was in the years after People's Power I a sense of shared commitment to rebuilding the country into a true democracy, one in which the people would have extensive control over the development and the resources of the nation. This was clearly the glue between the participants in the coalition that took shape during the EO 247 process.[8] It was much more than a common interest: it was a core belief, based on shared experiences of exploitation, on how the Philippines could best be developed into a true democracy and how the resources of the country could be utilized to the benefit of its people. As such, the coalition can be described as an advocacy coalition in the sense of the ACF.

And what of the little group of chemists who had started the development? In fact, their initial concern had been to ensure their participation in bio-prospecting activities. In its first stage, the process did not have such strong matching patterns with the ACF. And then, within only a few months, the process had taken a different direction. Philippine scientists were also subordinated to the proposed regulations, and prior informed consent by indigenous and local people was introduced. The initiators supported these changes – which indicates that they also shared the core beliefs of the emerging advocacy coalition.

The set of core beliefs included fundamental value priorities: There was scant concern as to the effects of the legislation for bio-prospecting activities, or access to biological and genetic resources in general, and little – if anything – was done to assess these effects. Democratic control over resources was clearly felt to be more important than the interests of foreign companies in exploiting the biological and genetic resources of the country. There was also a general belief that the new biotechnologies would rapidly spur the demand for access to genetic resources, and that a steep increase in benefits from the utilization of such resources could be expected. The main thing was to ensure that the Philippines would have a share in these resources.

8 Based on Swiderska et al. (2001) and my interviews with stakeholders in the Philippines in 2000 and 2002 – as well as impressions from my four-month stay in the Philippines in 1988, where I worked together with various academic institutions as well as NGOs as an intern of a German donor organization.

As presented in Chapter 3, the AFC is based on three basic assumptions concerning policy-oriented learning, and these match with our findings. The whole process leading to the adoption of EO 247 can be described as a *learning process* towards consensus building (Swiderska et al. 2001, 19). New elements introduced by various parts of the coalition were received and considered, resulting in input to or revisions of the text – partly or fully in line with the 'lessons' learned. Information that could challenge the basic beliefs was formally sought only towards the very end, when business actors were invited for consultations. During the process leading up to the adoption of DAO 20-96, critics brought such information and perspectives into the consultations with much emphasis. Nevertheless, as the DAO clearly indicates, the coalition resisted it,[9] in terms of the bureaucratic burden attached to access procedures. Thus, the coalition *resisted information that could lead to conflicting inferences*, in that for example the core beliefs might be found invalid or unattainable.

Finally, the authors of the ACF suggest that the *only way that policy core attributes of a coalition are changed is through shocks originating from outside* the sub-system of coalitions, utilized by a minority coalition in order to induce change. Recognition that EO 247 came to serve a barrier to bio-prospecting, due to heavy bureaucratic burdens and long processing time, can be described as a kind of slowly evolving shock. Particularly government agencies in the DENR took this experience seriously, and in response included new and simpler rules on bio-prospecting in the new Wildlife Act – thereby overruling EO 247 with regard to wild biological and genetic resources, with unforeseeable consequences for access to PGRFA. By that time, however, the original advocacy coalition no longer existed, because central actors in the old coalition were not involved in the process, as will be explained below. On this basis, we can infer that the ACF is applicable for the analysis of the EO 247 process. We will now see how it can help explain what happened.

Coalition building and the relative influence of actors from civil society The origin of EO 247 was quite extraordinary: the initiative to drafting a regulation was taken by a small group of scientists was a bottom–up process. It did not follow the more common procedure of proposing the development of a regulation to the authorities, who would then take the lead in formulating the regulation – involving relevant actors. In fact, in the Philippines, several legislative initiatives developed this way in the years after People's Power I.[10]

Actors from civil society were clearly the most influential groups with regard to EO 247. The group of chemists was the initiator: thus, it was an academic group and not environmental NGOs who started the process. Also the first consultation phase indicates that the focus of the initiators was on scientists. On the other hand, NGOs did have a say from early on. Later on several NGOs participated in the work, some of them with substantial inputs. The process was characterized by a high level of

---

9 This is confirmed by Swiderska et al. (2001, 20).

10 This is my general impression after several interviews and having read several draft acts of various kinds that were circulating among academics and NGOs.

consensus, as the participants shared basic norms and ideas about the purpose of the regulation.

It was Dr La Viña who brought the ideas from the CBD into the work, referring to the Philippine ratification of the Convention. He drew on the CBD to legitimate the process and ensure compliance, and made significant changes to the original version. One of these derived directly from the CBD: the introduction of detailed rules on prior informed consent of local and indigenous communities. As the first attempt worldwide to develop a CBD-based regulation in this area, it was indeed pioneering work.

The second change he made, which caused heated debates after the adoption of EO 247, was the inclusion of Filipino scientists in the target group for the regulation – the requirement that also Filipino scientists would have to apply for research agreements in order to carry out bio-prospecting in the country. The practical reason for this move was that foreign actors could easily enter into collaboration with Filipino scientists to get hold of biological and genetic resources.[11] Without specific regulation, such procedures would be legal – loopholes in the regulation of bio-prospecting. However, the result was a regulation that established high bureaucratic hurdles for Filipino scientists for the collection of biological and genetic resources for various purposes – whether they were working together with foreign scientists or institutions or not.

In retrospect we can see that the outcome was not conducive to the interests of Filipino scientists, since it introduced additional obstacles to their work without improving the prospects for participation in bio-prospecting together with foreign entities, due to bureaucratic barriers. We may say that the initiators of EO 247 got what they aimed for – a legal framework that regulated bio-prospecting of foreign individuals and institutions in the Philippines – but that the outcome proved to have a negative effect on their implicit overall objective, which was to improve the possibilities for taking part in the exploration of the biological and genetic resources of the Philippines.

The output in terms of provisions accounting for the anticipated interests[12] of indigenous peoples in EO 247 was substantial, even though indigenous communities had been only marginally involved in the process. Individuals like Dr La Viña, with his background from an NGO engaged in the situation of indigenous peoples, and the few NGOs with similar backgrounds, contributed significantly towards this end. Decisive factors were probably the reference to the CBD, combined with the general commitment of political leaders to the CBD and the Rio process, since the Philippines had no tradition of legislation on the rights of indigenous peoples.

What then did the various actors do with regard to PGRFA management – particularly the organizations and institutions that were involved in PGRFA management? SEARICE and CADI were the most central organizations in this regard. The text of EO 247 shows no indications of differentiating between various types of biological and genetic resources, such as PGRFA, or the requirements for obtaining

11 Based on interviews at PAWB in March 2002 (see interview list).

12 From Swiderska et al. (2001), it appears that the organizations represented the indigenous peoples on the basis of what they assumed to be the interests of these peoples.

access to such types of resources. From Swiderska et al. (2001) and my interviews with stakeholders in the process in May 2000 and March 2002, it seems that the organizations in question did not address issues of particular relevance for PGRFA management, and that they were not aware of the specific conditions concerning PGRFA management, as compared to other biological and genetic resources in this context. At that time, the idea that biodiversity should comprise wild, as well as domesticated, resources was quite new. Its implications for the management of and – particularly – access to PGRFA were generally not yet understood, and there were still high and undifferentiated expectations as to the benefits that bio-prospecting could bring to a country. It seems that these expectations were central to all the parties involved in drafting EO 247, and that there was no basis for differentiating the regulation with regard to PGRFA.

In addition, the TRIPS Agreement had just been finalized, and among most environmentalists there was deep opposition to the provisions on intellectual property rights to plants and plant genetic material – seen as closely interlinked with bio-prospecting activities. This perception among many coalition members, including those involved in agriculture, represented a further quality of the glue that kept them together, and spurred the commitment to emphasizing benefit sharing in the new regulation.

As noted in Chapter 3, Walzer (1992, 102–3) points out that a democratic civil society seems to require a democratic state, and that a strong civil society seems to require a strong and responsive state ('the paradox of the civil society argument'). Whereas the strength of the Philippine state can be questioned – depending on the criteria for measurement – the responsiveness was most evident in the period under which EO 247 took shape and was adopted. The responsiveness of the respective government entities can be explained by two factors:

- The new administration after the Marcos dictatorship and People's Power I sought to recruit individuals from civil society organizations and groups to join the government as 'progressive bureaucrats' (Swiderska et al. 2001, 7). These individuals contributed decisively to promoting awareness of the civil society agenda within the government ranks, and, in some cases, which were not politically controversial, they directly promoted the civil society agenda. Biodiversity conservation was one of these issues, and bio-prospecting was considered a part of that agenda. Therefore, members of the government agencies readily joined advocacy coalitions and were particularly responsive to initiatives based on this agenda.[13]
- The other factor was the special commitment of President Fidel Ramos to the follow-up of the Rio process, including implementation of the CBD. There was a high level of activity with regard to CBD implementation under his

13 My impressions from interviews confirm these perspectives. I spoke with many deeply committed individuals in several government agencies and learned how they had sought to include stakeholders from civil society in their processes, and respond to various initiatives and contributions from civil society.

> term in office. Government officials were probably most responsive to the initiatives and perspectives from civil society under the Ramos presidency.[14]

In other words, those progressive officials who took the lead on the government side shared the 'glue' of norms and ideas of the initiators, and became part of the advocacy coalition. The combination of such 'progressive bureaucrats' and a deeply committed president was particularly conducive to responsiveness to the initiative and input from civil society, and was probably a decisive factor in explaining their achievements. In this sense, state entities exercised considerable influence over the process.

It is, however, necessary to take a closer look at this influence. What happened was that the participants from the side of the government agencies acted largely as parts of the advocacy coalition, in resisting information that might have challenged the core values and norms of the coalition. They did not recognizably act as government agencies are normally expected to, balancing the various interests regarding the issue under consideration. As a result, the perspectives of the target groups for the legislation were not taken sufficiently into account, as was later confirmed.[15] This was an important reason why the regulation did not fulfill the promises embodied in its rationale and objectives, why it did not prove feasible.

Relative influence of actors from state authorities And what of the influence of state entities on the EO process in the context of PGRFA management? Awareness among state entities with regard to the particular requirements of PGRFA management seems to have been low at that time, as it was among the other parties involved. This may seem surprising, since the regulation was meant to cover all biological and genetic resources, including agricultural resources. For a regulatory framework with such consequences for crop development, it would have been reasonable to expect close and targeted consultations with representatives from the Bureau of Plant Breeding, the Institute of Plant Breeding together with the National Plant Genetic Resources Laboratory, the Bureau of Agricultural Research at the Department of Agriculture, PhilRice and other networks for crop improvement. Some of these institutions were slightly involved, and there was one meeting at the Department of Agriculture and one meeting of the Cabinet Cluster of Agriculture and Agro-Industries. However, the result was that the Bureau of Fisheries and Aquatic Resources was finally selected to represent the Department of Agriculture in the IACBGR – not any agencies vested with the management of plant and animal genetic resources for food and agriculture. This virtually defined the agricultural entities in the DA out of the issue.[16]

---

14 This impression is also confirmed by Swiderska et al. (2001, 14), who argue that the succeeding Estrada administration placed less emphasis on biodiversity conservation and strongly promoted business interests. They doubt that EO 247 would have been adopted under Estrada.

15 Internal note from PAWB: Status of Executive Order 247 Implementation. Undated, received from Ms. Altea Lota at PAWB 6 March 2002. The note was probably written before September 2001.

16 According to my interviewees at the Department of Agriculture, there is scant communication between the Bureau of Fisheries and Aquatic Resources and the agricultural

Why did the DA take this stand? One reason is that access and benefit sharing was not deemed relevant with regard to PGRFA management because the issue was primarily associated with wild resources and pharmaceuticals. Similarly, the Department did not prioritize the issue, because PGRFA management itself was not a priority. Finally, the DA did not feel responsible for PGRFA, which was seen as part of the biodiversity issues emerging from the CBD – and that was the responsibility of the DENR. In my interviews in the Department of Agriculture in 2000 and 2002, these explanations were confirmed. Probably a fourth factor should be added, as above: the generally low awareness among those involved in non-domesticated biodiversity and PGRFA issues that time – at the national as well as international levels – of the distinctions between the two when it came to management requirements.

Relative influence of actors from business Actors from the business sector were involved only towards the very end of the whole process, shortly before the text was to be submitted to the President. This timing indicates that the drafters did not attach much importance to their inclusion in the work. Moreover, the business representatives were not consulted in a targeted manner, but as a part of a multi-stakeholder workshop with various participants and perspectives. And so, substantial parts of the EO remained, with only minor changes in the draft that was to be presented to the President shortly after. As a result, business actors were largely critical of EO 247 and sought to minimize its effect during the consultations leading up to DAO 20–96. They were never part of the advocacy coalition.

Some stakeholders have later maintained that foreign companies involved in bio-prospecting in the Philippines should have been more actively involved, since they were targeted by the Executive Order, and would have the most relevant experience for assessing the draft text (Swiderska et al. 2001, 18). In particular, the private plant breeding companies organized in the Philippine Seed Industry Association (PSIA)[17] might have provided valuable input. We will never know what this could have meant, or whether it would have helped to avoid the bureaucratic burden that was to become the curse of EO 247. What this possible 'lost opportunity' confirms is that the advocacy coalition closed ranks against arguments from actors with differing norms, ideas and perceptions, and with conflicting interests. Such arguments were taken into account only marginally – if at all – as compared to the arguments of other consulted parties.

---

entities in the DA. Thus, the Bureau could not be expected to represent all interests of the Department, but was selected because of its responsibility with regard to fisheries and aquatic resources.

17 The Philippine Seed Industry Association (PSIA) is the most important business actor in our context from within the Philippines. It was formed in the late 1970s to respond to the need for seed that was spurred by the launching of the Green Revolution. Seed imports had increased rapidly and PSIA set up a protectionist lobby for the domestic seed industry. As of 2000, PSIA had 18 members, three of them affiliated with multinational corporations. Most member companies offer diversified seed baskets in terms of crops. Fourteen of them produce vegetable seeds of various species and varieties, while four specialize in corn, two in rice and one in sorghum.

Consequences for EO 247 implementation It seems that the EO 247 advocacy coalition was most effective with regard to agenda setting and advocacy under the favorable political conditions in the Philippines during the formulation process, but that their approach was not so fortunate when it came to finalizing practical policies in terms of the final text of EO 247 and DAO 20–96. The neglect of potential negative implications for other societal interests (access to biological and genetic resources for conservation purposes and for plant breeding) during the formulation process became a curse for implementation of the act, and thereby for the realization of their objectives, particularly as the advocacy coalition slowly dispersed. The participation of civil society in implementing EO 247 was limited to the organizations and institutions appointed to the IACBGR. These were the representatives from science, of SEARICE and People's Organizations. There was some engagement for implementing the policy in the initial years,[18] but after 1998, the civil society members of the coalition were no longer visible *qua* coalition in this context. The policy was in place, and most civil society actors probably felt that their job was done. New battles had to be fought as new issues came to the fore, such as the introduction of genetically modified organisms to the Philippines, illegal logging and other conservation issues, and the introduction of plant breeders' rights. Moreover, several of the environmental organizations relied on Northern donor organizations, which increasingly wanted more field activities in terms of conservation projects, and funding for political work dried out.[19] This development paralleled the general tendency in the donor community to involve NGOs in problem solving together with government authorities, increasingly with shared responsibility.

The participation of government agencies from the coalition in implementing EO 247 was limited to their participation in the IACBGR as well. However, only two government agencies participated actively in the IACBGR in terms of attendance, whereas the others were absent from a considerable number of meetings. The two were the DENR (PAWB) and the Department of Health (represented by its Philippine Institute of Traditional and Alternative Health Care) (Liebig et al. 2002, 40). After the government shift in 1998, the top levels of various government agencies were changed and core government actors from the coalition were no longer in position – as was the case with La Vinã, among others. After this point, there was not much left of the coalition. The IACBGR almost ceased to convene for nearly three years (only two meetings were held in this period), before a new series of meetings started in 2001 – this time to round off work in connection with the new Wildlife Act.

Whereas the advocacy coalition efficiently achieved its objectives in terms of EO 247 adoption, the strategy was not effective with regard to sustained goal

18 For example, through the 1997 manual on implementation of EO 247 developed by the Foundation for the Philippine Environment in collaboration with the World Resources Institute, with support from the Federal Ministry for Economic Cooperation and Development of the Federal Republic of Germany and reproduced by PAWB (La Viña et al. 1997).

19 I interviewed several representatives from environmental organizations in March 2002, and all interviews confirm this picture. SEARICE and CADI do not count among these organizations, and were generally concerned with policy issues along with their work in the field throughout the period studied.

achievements. This is one of the key lessons from the experience. If the objective of the coalition had been not only to get a policy adopted, but also to get it implemented in a long-term perspective, then the counter-arguments and suggestions should have been taken into account. That was supposed to be the role of state entities. However, as they were part of the advocacy coalition, and therefore more or less resisted arguments that seemed to go against core beliefs of the advocacy coalition, they failed to fulfill their role. If more consideration had been taken of these arguments and suggestions, the policy might have been framed in a more feasible way.

The implementation of EO 247 cannot be explained solely on the basis of the Advocacy Coalition Framework, not least because after a while there was no advocacy coalition left. Therefore we will now turn to the institutional capacity approach.

Institutional Factors Explaining the Adoption and Implementation of EO 247

The institutional capacity of the relevant government agencies is an important factor in explaining how EO 247 was adopted and, particularly, how it was later implemented. In this section, we analyze how institutional capacity affected the process before and after the adoption of the regulation, following the analytical framework set out in Chapter 3.

It is not easy to get an overview over the scope and depth of the formal competence vested in the government agencies (as recommended by Hanf and Underdal 1998, 165), since such information is normally not readily accessible in the Philippines. The overall impression from my interviews is that the level of expertise in government agencies at the national level was generally high and that the key individuals were knowledgeable, well-informed and highly competent. However, there is greater variation among the staff of the relevant agencies, particularly the DENR, at the regional level, and PAWB recognized early that capacity building was a prerequisite for effective implementation of EO 247.[20]

Lack of resources proved a serious constraint.[21] According to PAWB,[22] most of the funding utilized for implementing EO 247 was shouldered by PAWB out of its annual budget. Nothing was provided specifically for implementation over the DENR budget. Funding for implementation was intended to come from the member agencies of the IACBGR and from fees imposed on research agreements, according to the EO. In practice, there were almost no fees, since there were so few research agreements, and PAWB's repeated requests to the member agencies for funding were turned down on grounds of unavailability of funds, according to PAWB (ibid.).

20 As early as in 1998 a project was planned to accommodate that. The project is referred to further below (PAWB 2000).

21 Based on interviews in DENR, including PAWB, at PCARRD and the Institute of Plant Breeding, in addition to various observers from international donors.

22 PAWB note received from Ms. Altea Lota, PAWB, 6 March 2002, titled 'Status of Executive Order 247 Implementation'. The note is undated, but the content indicates that it was written in 2001.

In addition, the IACBGR member agencies had difficulties in the processing of applications. No additional financial or personnel resources were allocated to the agencies for this – a factor that central actors in the IACBGR see as a major reason for the slow progress in this regard (Liebig et al. 2002, 39). There was also no clear and effective mechanism to deal with monitoring and enforcement of Philippine biological resources brought out of the country (Benavidez 2004, 4). Financial and personnel resources are often two sides of the same coin, as financial resources make it possible to employ personnel. As the Secretariat for the IACBGR, PAWB simply did not have the capacity to compensate for the lack of resources for implementation.

Although financial and personnel constraints were substantial, help was not far away. In 1998 the German official development agency, Gesellschaft für Technische Zusammenarbeit (GTZ), started planning a project together with PAWB to increase institutional capacity for EO 247 implementation. The project included assistance for developing various guidelines, a management information system for bio-prospecting, training of regional core staff of the DENR in enforcing the regulation, and information activities to foster general awareness of bio-prospecting and its regulation. However, project planning was seriously delayed, for several reasons.[23] Two shifts of government, in 1998 and 2001, changed the political landscape and thereby the commitment to the project. With the new leadership under Estrada (1998 shift), project planning became drawn out, and implementation could not start until in 2000, by which time the stakeholders seemed to have given up the whole IACBGR. When Gloria Macapagal-Arroyo resumed power in 2001, shortly after project implementation had finally started, a new problem arose. Her signing of the Wildlife Act decisively reduced the importance of EO 247, and many actors felt that it had been replaced. Due to internal conflict in the IACBGR on the future course of work as well as the continued existence of the body itself, project planning had to be substantially reviewed and the implementation phase was prolonged from 2001 until 2003. After lengthy and conflict-laden negotiations, the focus of the project slowly turned to support for implementation of the new Wildlife Act instead.

The fact that financial resources, a total of Euro 99,000 – a considerable amount in the Philippine context – to increase institutional capacity were accessible but not accessed before it was too late, leads us to another factor with greater explanatory power: *the political 'clout' of the leadership.*

The role of key leaders from central government agencies in the formulation process was central to the achievements. The way in which leaders (particularly of NAST and later the DENR and PAWB) gradually took over responsibility for the processes that had started in civil society and realized these ideas through comprehensive consultation processes, shows commitment, entrepreneurship and strength. The combination of this engagement and President Ramos' commitment to the Rio processes was an important factor in explaining the adoption of EO 247.

Until 1998, the IACBGR worked according to its intentions, as outlined in EO 247 and DAO 96–20. But then, with the presidential shift from Ramos to Estrada,

---

23 Based on PAWB (2000) and Convention on Biological Diversity (undated, probably 2002): 'Access to Genetic Resources and Benefit-sharing – ABS Capacity Building Project', at <www.biodiv.org/programmes/socio-eco/benefit/project.aspx?id= 5678>.

the political leadership structures changed dramatically, including top levels of government agencies. Several leaders who had been promoters of the process, among them La Viña, were replaced with new people without the same commitment to EO 247, and environmental issues no longer received as much political attention. This post-1998 lack of commitment to the process was expressed primarily through non-attendance at IACBGR meetings and meetings of its Technical Secretariat, but also through slow progress of the processing of applications, including the signing of applications that had been approved (Liebig et al. 2002, 40).

As DENR Undersecretary, La Viña was the first Chair of the IACBGR. He was considered a motivating leader, committed to the task and following up on his duties (ibid.). His successor after the 1998 elections attached less importance to the IACBGR. For example, several times he called meetings that he failed to attend himself, and also called meetings that he cancelled too late. Liebig (ibid.) notes that this was felt by participants to be highly non-motivating, and may have contributed to the low meeting frequency and difficulties in achieving a quorum.

According to EO 247, the IACBGR was to convene at least quarterly. Until early 1998 members met almost twice as often. After the election of Estrada in 1998 there were no meetings until 1999, and only one meeting each in 1999 and 2000. In 2001 came greater activity, now under the new president after the overthrow of Joseph Estrada, Gloria Macapagal-Arroyo. Despite the increased meeting frequency, almost half of the meetings after 2001 failed to achieve a quorum, and only one meeting of the Technical Secretariat achieved a quorum from 2000 to March 2002. Almost three years were lost due to the low meeting frequency of the IACBGR, a problem that seriously hampered the processing of research applications.[24]

Resources for the formulation and implementation of EO 247 were extremely scarce all along. Nevertheless, the process developed acceptably up to 1998, but deteriorated markedly thereafter. Obviously, it was the political 'clout' among the leaders that drove the process up to 1998, and the lack of such a 'clout' that later turned the development in a negative direction.

The participative capacity of the involved government agencies was strong in the periods leading up to the EO and the DAO as compared to the resources available for the purpose, particularly with regard to the advocacy coalition of groups and organizations from civil society and other government entities. There was a spirit of inclusion of the consulted parties, particularly in the period prior to the adoption of the EO. The weakness was that business actors and actors with diverging views were included only at the end, when the basic principles were already in place. When controversies became evident it was not possible to account sufficiently for these, if indeed there was any political will at all. Participative capacity was basically confined to the advocacy coalition.

According to interviews with stakeholders in the EO 247 process,[25] the multi-stakeholder approach embodied in the IACBGR is generally seen as a positive

24 Most stakeholders and observers of the IACBGR process interviewed by Liebig et al. (2002) and by myself in March 2002 consider this to be the case.

25 The interviews were carried out by Klaus Liebig and his team from the German Development Institute from February to April 2002 (documented in Liebig et al., 2002).

feature of EO 247 implementation, since the intention was to ensure participatory and transparent decision-making on applications for research agreements. As we have seen, however, after 1998 it failed to live up to those intentions, due to low meeting frequency and low attendance. These are indicators of low integrative capacity in terms of the capacity to integrate and co-ordinate other government agencies and external networks for implementation.[26] For PAWB, these problems made the IACBGR somewhat of a burden.[27]

All in all, participative capacity with regard to the advocacy coalition was strong in the formulation phase, whereas integrative capacity for implementation was relatively weak. There were few signs of participation of external stakeholders in the process after the Implementing Rules and Regulations were adopted, except for the participation of some selected organizations/institutions in the IACBGR, and the initiative to formulate a manual on implementing EO 247, undertaken by *inter alia* the Foundation for the Philippine Environment (La Viña et al. 1997).

Most NGOs apparently saw their role as being that of agenda-setters and advocates for certain policies, with little ideas about what they could do in the implementation phase (with some outstanding exceptions such as SEARICE). Basically, they just let go. There are certainly reasons for that, not least with regard to their institutional capacity. However, the limited perspective that focused only on the adoption of a regulation – and not its sustained and successful implementation in a long-term perspective – is an important part of the explanation. If there had been a long-term perspective on goal achievement, we could have expected more action after the adoption of EO 247.

What lessons can we draw from applying the Institutional Capacity Approach? The experience from the formulation phase (where there were responsive state entities and leaders with political 'clout' but without sufficient participative capacity) and from the implementation phase (dependent on individuals and not sufficiently rooted in institutional structures, therefore vulnerable to political change) indicates institutional weaknesses that, ironically, facilitated the impact of the advocacy coalition on negotiation output, but limited the impact on implementation and on the actual political development concerning bio-prospecting. Combined with the generally short-sighted perspectives of the advocacy coalition, this meant there was no real future for EO 247.

No Philippine president, except Ferdinand Marcos, had ruled for more than one election period (6 years). Thus, it was widely expected that a presidential change would come in 1998, and that might remove central pillars in the advocacy coalition from their positions. This situation made the institutional structure for implementation of EO 247 highly fragile. Were preventive measures available? Probably the procedures of presidential changes are among the greatest challenges for continuity in Philippine politics, and many policies suffer the same destiny when

26 The EO 247 was not in conflict with other policies in the Philippines, so the integrative capacity of the relevant government agencies with regard to inter-policy co-ordination is not relevant in this context.

27 Based on interviews in PAWB in 2002 (see interview list).

broad segments of the leadership are exchanged and replaced so often. The problem lies in the system itself, with few possibilities for solving it within the bureaucracy.

These lessons raise wider questions as to the roles and tasks of advocacy coalitions in states with weak institutional government structures – even if they are responsive. How could the civil society counterparts of an advocacy coalition respond to the lack of participative capacity of state entities? It is not the anticipated role of civil society entities to take over the tasks of government entities to balance varying views and perspectives – but should they try to do so when this capacity is lacking on the government side? What can an advocacy coalition do when the core stakeholders in government entities disappear, and there are no adequate institutional structures to ensure compliance with an adopted regulation? Are there any possible roles or tasks that could support or pressure government officials to maintain implementation of the regulation? It all boils down to the time perspective and long-term commitment of the advocacy coalition. As long as the time perspective extends only until the adoption of a regulation, there is no further role for the coalition to play. In such a perspective, the coalition can be said to have been successful. If, however, the time perspective had been directed towards long-term compliance, then solutions would have had to be found to these questions.

Relative Effects of the CBD on EO 247 and Mechanisms of Influence

For analyzing the relative effects of the CBD on EO 247, five questions appear central:

1. How important was the CBD in influencing the positions of the actors involved in the formulation of EO 247 – as compared to other factors?
2. To what extent did the CBD influence the substantive contents of EO 247?
3. To what extent did it affect presidential approval?
4. To what extent did it affect implementation?
5. What were the mechanisms of influence?

The first question is crucial for identifying the pathways of influence, and the second for determining the contents of CBD influence. The third question is particularly relevant with regard to EO 247, since the initiative had come from civil society as a bottom–up initiative. Without the commitment of the President to the regulation, there would have been no Executive Order 247. The fourth question is central for understanding the conditions for implementation and whether the CBD has been useful in this regard, and the fifth sums up our findings with regard to mechanisms of influence.

Concerning the relative importance of the CBD for actors involved in formulating EO 247, we saw from the early history of EO 247 that four strains of motivation came together in the initiative to develop an executive order on bio-prospecting:

- The experience of People's Power from 1986 spurred a general sense of nationalism and the desire to regain control over Philippine resources.

- Philippine authorities were vested with new responsibilities concerning the management of bio-prospecting in the country, and needed guidelines.
- Philippine scientists demanded participation in the exploration of the biological resources of their country.
- With the adoption of the CBD, an international momentum was in place that gave the final push to developing a fully-fledged regulatory framework.

In other words, the need for a regulation on bio-prospecting emerged from the Philippines, independently of the CBD, as did the initiative to take action. As soon as La Vinã entered the scene, however, the CBD was brought into the context, and was used to frame the order as well as legitimize it. By then, the Philippines had ratified the Convention, and there was political will to pioneer its implementation, as we saw in Chapter 9.

Thus the CBD provided the key motivation to environmentalists in civil society, including those involved in PGRFA issues, and in government agencies, as well as to actors involved with indigenous peoples' issues, like La Viña. Although not an important motivation for the initiators of the regulation, it soon became accepted as a framework for the formulation, and it fit in well with the shared concepts that kept the advocacy coalition together.

The EO would have been a very different document without the influence of La Viña and the environmentalists – indeed, it might never have become an Executive Order at all. Therefore, we can conclude that the CBD was decisive for the development of EO 247, although it was not the sole factor that influenced all the actors involved.

What then of the relative influence of the CBD on the substantive content of what was to become the Executive Order? EO 247 derives its mandate partly from the CBD, and partly from the Philippine Constitution, in which ultimate responsibility for environmental conservation is vested. The elements introduced from the CBD can be summarized as follows:

- Benefit sharing from bio-prospecting was justified with the need for conservation.
- Prior informed consent to be obtained from indigenous peoples.[28]
- All biological and genetic resources, including PGRFA, were covered simultaneously.

As we can see from the EO 247 pre-history, PGRFA would most probably not have been included in Philippine regulations on bio-prospecting, and the responsibility for plant collection of PGRFA would have remained with the Bureau of Plant Industry, had it not been for the CBD. On this basis, we can say that the EO was highly influenced by the CBD, but not by the CBD as a whole. It was largely influenced by those parts of the CBD that address benefit sharing and its conditions. This can be explained on the basis of the advocacy coalitions and their priorities.

28 In the CBD, local communities are included, but after heated discussions these were left out.

Environmentalists and promoters of indigenous peoples' rights took the lead, and their motivation was mainly to provide for benefit sharing – for the benefit of the nation and the conservation of its natural resources. This was probably their perception of the relevance and importance of the CBD, so they used these parts of the Convention as a basis for framing the EO.

As to our third question, we have already noted that President Ramos was highly motivated to fulfill his country's obligations towards the CBD and to pioneer the implementation of the new Convention. As EO 247 was generally regarded as a pioneering initiative, we may say that this constellation was conducive to the approval granted by President Ramos.

Concerning the relative effect of the CBD for EO 247 implementation (question 4 above), it is important to realize that the extent to which developing countries can effectively implement the Convention depends critically on the transfer of financial and technological support from developed countries, as set out in Articles 18, 20 and 21. As noted above, the German GTZ intended to meet the financial requirements needed to solve the obstacles encountered in implementing EO 247, but became delayed for various reasons beyond its control. As the project was finally funded under the program 'Implementing the Biodiversity Convention' established at the GTZ, we can say that German support is directly related to implementing the CBD. This way the CBD affected EO 247 implementation in terms of financial and technical support. Had it not been for the difficulties on the Philippine side, support could have been provided at an earlier stage.

As to the fifth question, on mechanisms of influence, we have seen that the CBD was a core motivation for the 'advocacy coalition within the advocacy coalition', consisting primarily of environmentalists in civil society and 'progressive bureaucrats' in the DENR, and that this motivation spread in the coalition through a learning process. We have also seen that the coalition interpreted the CBD according to its own priorities, choosing to ignore counter-arguments that could have helped in correcting the direction. This shows how the CBD influenced the Philippine policy development through an advocacy coalition.

We have noted how institutional factors were decisive in this context, the most important being the 'political clout' of leaders in central government agencies and their integrative capacities. As the political leadership until 1998 was positive to the CBD, otherwise-scarce resources were pooled to frame the formulation of EO 247 and for its initial implementation. After the entry of the Estrada administration, the commitment to the CBD process was no longer evident – an important factor in explaining the failure of EO 247. We can thus see how the CBD influenced institutional factors. The CBD also influenced the institutional capacity for implementation through the support provided by the GTZ, which might have had impact had it been put to use earlier.

Relative Aggregate Effects of International Regimes Pertaining to EO 247

The TRIPS Agreement caused widespread anger among many NGOs and many of their allies in Government agencies in the Philippines – a substantial part of the EO advocacy coalition. This, combined with the belief that biotechnologies would open new possibilities for bio-prospectors to profit from the commercialization of biological and genetic material from the Philippines, fuelled their involvement in the EO 247 process. This motivation – which was in fact another side of the coin of the desire to raise funds for nature conservation – is probably important in explaining why these actors emphasized benefit sharing as compared to facilitated access to biological genetic resources.

The perceived absence of a separate international regulation on PGRFA management – although the International Undertaking was still in force – due to the negotiations leading to the ITPGRFA, left a vacuum with regard to the management of these resources in the Philippines. Many of my interviewees in central positions in various government bodies said that there was no will to take action on implementing the IU as long as a new international agreement on PGRFA was awaited. It was not considered feasible to take any steps in advance, because of scarce institutional capacity and the risk that policies would have to be changed.[29] Therefore no attention was paid to the IU during the formulation process of EO 247 – a situation hardly conducive to bringing in the particular management aspects of PGRFA. The general perception, according to my interviewees, was that since PGRFA were covered by the CBD, this was the basis for including them in EO 247 as well.

Confirming the Propositions and Summing Up

Our findings confirm the propositions set out in Chapter 3. We have seen that representatives of advocacy coalitions, slowly merging into one advocacy coalition, sought to exercise influence on the process that led to the adoption of the Executive Order on the prospecting of biological and genetic resources, confirming our fifth proposition (P_5). These representatives made reference to the CBD and interpreted it in line with their own ideas in seeking to influence the formulation of the EO, confirming our sixth proposition (P_6). We have also seen that EO 247 was largely formulated in accordance with the demands voiced by the advocacy coalition, whereas any diverging demands were less reflected – confirming our seventh proposition (P_7). Since there was no noticeable awareness of the special management requirements for PGRFA, and the IU was not followed up because it was under renegotiation, PGRFA were included in the EO according to the provisions of the CBD, without distinguishing them from other biological and genetic resources. Finally, the way in which advocacy coalitions influenced the formulation of the EO – by not taking counter-arguments seriously into account – negatively affected its implementation, since it resulted in an overly restrictive and bureaucratic regulation which could hardly function and which finally failed. This confirms our eighth proposition (P_8). The regulation proved to be negative to the conservation and development of PGRFA: and this is also the answer to our research question on how the process leading to the adoption of EO 247 affected its goal achievements with regard to PGRFA.

When it comes to institutional capacity, we have noted that there was little co-ordination between the ministries involved in PGRFA management, most notably

29 Based on interviews at the Department of Agriculture in May 2000 (see interview list).

between relevant agencies of the Department of Agriculture and DENR, confirming our ninth proposition (P_9). Moreover, the EO was largely formulated in line with the demands of the dominant influencing actors, the advocacy coalition – which confirms our tenth proposition (P_{10}). Finally, and confirming our eleventh proposition (P_{11}), we have seen that the advocacy coalitions needed responsive authorities and their institutional capacity to win through with their demands with regard to EO 247 and its legislation. When this capacity was decisively reduced in 1998, the prospects for implementation were reduced accordingly.

Advocacy coalitions tend to be short-lived, since they are normally geared towards the adoption of specific policies and regulations. The coalition had already crumbled by the time the trouble started in 1998, and there was no strategy for ensuring sustained implementation of the new regulation. Due to the weak institutional capacity of the government agencies and the removal of leaders with political clout, a continued strong advocacy coalition seems to have been the sole option for bringing the process back on track – if indeed that was possible. The major weakness of the advocacy coalition was its short-term perspective of getting the regulation adopted, which blinded it to important aspects that should have been taken into account in order to ensure sustainability.

The CBD provided an important rationale for core actors in the formulation of EO 247, and was the most decisive factor for the development of the document into a fully-fledged regulation, with presidential signature. It became the framework for the advocacy coalition and was also a driving force for central leaders in relevant government agencies who were part of the advocacy coalition. The motivation was mainly to provide for benefit sharing – for the benefit of the nation and the conservation of its natural resources, as set out in the CBD. This motivation was additionally spurred by resistance to the TRIPS agreement among core actors, who saw bio-prospecting and intellectual property rights as closely interlinked. On this background, benefit sharing was promoted to the detriment of access to genetic resources – which also reflects the distinction between core and secondary norms in the CBD, analyzed in Chapter 6. Since it was viewed as an implementing measure of the CBD, EO 247 received bilateral support, as part of Germany's commitment to support developing countries in their CBD implementation.

As we saw in Chapter 9, EO 247 was not conducive to PGRFA management. In this section we have seen how this situation came about. Few actors with specific responsibilities for PGRFA management participated in the decisive phases of the formulation of the new regulation. Most of them did not participate at all. In addition, they failed to realize the implications of the new regulation for PGRFA management. At that time, there was no awareness of the differences between the two categories of resources with regard to the management approaches needed to maintain them. This lack of awareness was related to the general inertia caused by the perception that the IU was not functioning, and that a new international treaty on PGRFA had to be awaited until new steps could be taken. The CBD was therefore the sole basis – in terms of international regimes – from which the EO was derived, and it did not signal any differences between the two categories of genetic resources. Therefore the CBD provided no help in avoiding the problem. Rather it encouraged it, by merging the two categories without providing a distinction between them with regard to management requirements. As the EO 247 pre-history indicates, PGRFA would probably not have been included in any bio-prospecting regulations in the Philippines, had it not been for the CBD. Plant collection of PGRFA would most likely have remained the responsibility of the Bureau of Plant Industry.

The Second Case: The Philippine Policy on Plant Breeders' Rights

As we saw in Chapter 9, the Plant Variety Protection Act (PVP Act) is not conducive to furthering genetic diversity in Philippine agriculture, but it may improve the availability of much needed high quality seeds for farmers who can afford to buy them. To the extent that the PVP Act leads to an increased use of improved seeds in agriculture, this can help to increase food production, provided that other requirements are fulfilled (other input factors and environmental conditions), but it would also further the replacement of traditional or locally bred varieties. Seed companies need financial and institutional capacity to apply for Certificates of Plant Variety Protection and to uphold these Certificates through annual fees, and thus small-scale Philippine breeders (including farmer-breeders) will hardly be in a position to make use of the system. For farmers, the new legislation represents a limitation of their practices in using PGRFA. In this section we ask how the legislation came about, and what role international regimes played, based on the theory framework in Chapter 3 and the method outlined in Chapter 4.

The Story of the Plant Variety Protection Act

The story of the PVP Act is the tale of how an initiative for a genuine *sui generis* system for plant varieties in the Philippines became transformed into an attempt to make the Philippines a member of UPOV'91, and how a foreign power proved to have a crucial influence on that process.[30]

The history of the PVP Act dates back to 1995, when Senator Orlando S. Mercado initiated the drafting of an 'Act Providing for the Registration and Protection of Plant Varieties and for Other Purposes'.[31] The draft was presented at the First Regular Session of the Tenth Congress of the Republic of the Philippines on 2 August 1995.

30 Philippine legislative processes follow two tracks: the Senate and the House of Representatives. Each of these has its own committees on political issue areas, and thus each has a Committee on Agriculture. The formal legislative processes start in these committees, where bills are drafted, discussed and disseminated for public hearings and consultations. Once a bill has been passed in the committees, it is presented to the Senate and the House respectively. After a bill has been passed in these bodies (normally in two readings) there is a bicameral procedure for harmonizing the two versions into one bill, which then goes to the Office of the President without public hearings, for presidential signature.

31 Senate of the Philippines (1995), the draft has the Senate reference number Senate S. No. 277/1995. The reference to 2 August as date for the presentation of the draft is provided in a letter from Deputy Secretary for Legislation Demaree J. B. Raval to the Chairperson of the Senate Committee on Agriculture and Food, Leticia Ramos Shahani, dated 2 August, reference number dj/25.

In fact, its contents were almost diametrically opposed to the contents of the Act that was finally approved in 2002.

In the introductory explanatory note, Senator Mercado refers to the NIPAS Act and to the necessity of protecting the rich biological diversity in the country. It is stated that this act would be sufficient for Philippine needs. However, since December 1994 the Philippines had been mandated by its acquiescence to the TRIPS Agreement to pass *sui generis* legislation for the protection of plant varieties in the country. The draft proposes the establishment of a national registry for plant varieties that are clearly distinguishable from known varieties, are stable in their essential characteristics, sufficiently homogeneous, and are being used by farm, rural, and tribal families and communities in the Philippines (Section 3). Any farm, rural and tribal family or community in the country could apply for registration – and also voluntary organizations or qualified local government units on behalf of such families or communities (Section 4). In addition, plant breeders of public-financed, private or corporate institutions could apply for registration on their own behalf or on the behalf of any farm, rural and tribal family or community in the Philippines.

Intellectual property rights over genetic information embodied in the variety should be owned by the state, for and on behalf of the farm, rural and tribal families and communities in the Philippines (Section 6). Breeders – whether public, private, national or foreign – wishing to introduce new varieties would have to produce an Environmental Impact Report with an assessment of the planned introduction of the variety on genetic erosion of locally used varieties. Filipino researchers should be allowed an 'absolute exemption' to use any variety covered under the Act for further breeding work, but could then not apply for any kind of proprietary right over the original materials or any products resulting from the use of the protected variety. The right of farmers to keep seeds for planting in their own farms should be 'absolute'.

A separate section (Section 8) was devoted to community intellectual property rights and farmers' rights.[32] The formulation has clear parallels to the FAO Conference Resolution 4/89 on farmers' rights, in that it recognizes the efforts made by the rural communities in the development of on-farm innovations and the enrichment and conservation of PGRFA. On this background it states that farmers and farming communities should have the right to ownership of biodiversity and genetic resources 'in perpetuity', including the right to select, modify and develop plant genetic material and to freely exchange and sell it without any authorization, except as brand names used by breeders and seed corporations. In addition, farmers should have the right to correct and complete information about seeds developed by breeders in both the public and the private sectors – as well as the right to fair compensation for the diversity developed by farmers and used without their permission by breeders. Violations against the Act would be punishable by fines, imprisonment or a combination of both.

32 As such, it appears that the draft Act has many parallels to India's Protection of Plant Varieties and Farmers' Rights Act of 2001. Some of my interviewees indicated that there were links between the two. This is possible, but I could not find any evidence of communication between the drafters from the two countries.

In October 1995, the Senate Committee on Agriculture and Food announced that public hearings would be held, and circulated the draft to stakeholders for comments.[33] And here end all traces of that bill. Probably, it was not pursued because it would clearly not meet the TRIPS requirements, and because it revealed insufficient understanding of the WTO system in general and intellectual property rights in particular.

As a reaction to the TRIPS Agreement, NGOs in the Philippines started to mobilize, with MASIPAG, SEARICE and GRAIN in the forefront. Flyers and brochures were produced and distributed, and meetings were held. Central arguments were that the TRIPS Agreement would pave the way for bio-piracy, 'the stealing of genetic material and knowledge from communities in the gene-rich developing countries',[34] and that it would breathe new life into the UPOV Convention as a *sui generis* system of intellectual property rights to plant varieties. Membership in UPOV would be detrimental to environmental and developmental concerns in the Philippines, for many reasons.[35] It would lead to genetic erosion because a few multinational breeding corporations would dominate the market[36] and replace local breeding firms that lacked the financial and institutional capacity to use the system. Farmers would lose their rights and control of production systems, and the benefits would go to multinational corporations in the North, which would also take control of agricultural production systems in the South. These developments would undermine the efforts of the FAO to establish an international regulation of PGRFA management and of the CBD with regard to PGRFA.

There followed a new, and genuine, *sui genesis* initiative when Nereus O. Acosta of the House of Representatives prepared a new draft Act for the First Regular Session of the 11th Congress in 1998. The draft was titled 'An Act to Provide Protection to New Plant Varieties, Establishing a National Plant Varieties Protection Board and for Other Purposes', and was presented as House Bill No. 1070. In his explanatory note to the Act, Acosta refers to the TRIPS Agreement and states that 'The UPOV, although not mentioned in the TRIPS Agreement, is considered by many to be a model of an effective *sui generis* system.' He goes on to discuss the differences between the 1978 and 1991 Acts of UPOV and concludes that the former is generally perceived as the most favorable for developing countries but that UPOV only allows membership in the latter. He discusses the reciprocity that follows from UPOV membership, and concludes that there will be 'more foreign breeders demanding protection from the government than local breeders who will correspondingly demand and enjoy foreign protection from UPOV member countries'. The conclusion is:

33 According to a letter from the Chairperson of the Committee on Agriculture and Food, Leticia Ramos Shahani, to President of the Philippine Seed Industry Association, Rene L. Mondragon, 17 October 1995.

34 MASIPAG et al. (1998).

35 GRAIN and the GAIA Foundation (1998).

36 (Ibid.) According to the authors, the world's top ten seed corporations controlled 40 per cent of the seed market in 1996, with Pioneer Hi-Bred as number one.

> It would appear, therefore, that the best option for the Philippines is to legislate its own plant variety protection act which would be close enough to the UPOV Act of 1978 for purposes of meeting the TRIPS requirements of giving effective protection for plant varieties, but which would not be burdened by the reciprocity and stringent requirements for the 1991 Act of the UPOV.

This draft is considerably more thoroughly prepared than the 1995 Senate draft and shows a deeper understanding of the WTO and the TRIPS Agreement. Also, the 1998 draft is more comprehensive in terms of objectives, definitions (an impressively long list), and operational provisions. On the other hand, some of the elements from the Senate draft were clearly taken into account, in revised form.

The draft proposes a long range of policy statements to establish the context of the Act, most of which are clearly derived from the CBD (Section 2): All biological resources shall be owned by the state, which is responsible for the conservation of biological diversity, the sustainable use of its components and the fair and equitable sharing of the benefits arising from the use of genetic resources. While recognizing intellectual property rights, this is to be done in a manner supportive of a healthy ecology, in accord with the 'rhythm and harmony of nature'. All access to genetic resources is to be subject to prior informed consent of the state, and, where relevant, the communities. In this, the state shall respect, preserve and maintain the knowledge, innovations and practices of indigenous and local communities and undertake measures to prevent or control risks associated with the use and release of genetically modified organisms. Against this backdrop, the draft states that it shall be the policy of the state to recognize the indispensable role of the private sector and to protect and secure the exclusive rights of scientists and other gifted citizens to their intellectual property and creations.

The Act would provide for protection of plant varieties that are distinct, uniform, stable and novel (Section 6). Breeders and representatives of indigenous cultural communities would be entitled to apply for plant variety protection (Sections 13 and 14). It should not be possible to apply for the protection of a plant variety if it stems from ancestral lands, except for by the indigenous people of that land, or if the variety was bred from Philippine materials outside the Philippines (Section 15).

The rights holder would be authorized with regard *inter alia* to offering for sale, selling and marketing, exporting, stocking and keeping inventories (Section 17). However, extensive exceptions were also included (Section 18 and 22): Harvested material would normally be exempted (except when acquired through unauthorized use of the protected variety), as would saving for the production of seeds or for use on own land holdings, the use for any purpose including the sale of crops and other material derived from the protected seeds, acts done for breeding purposes (except when this necessitates the repeated use of the protected variety). No reference is made to farmers' rights, but these seem to have been duly accounted for in the provisions mentioned.

The bill remained pending for approximately two years in the House of Representatives until it was replaced by a new draft. In the meantime there were hearings in 1999, to which we return later. Now we need to look at developments in the Senate, and at three bills in particular.

In June 1998 Senator Juan Flavier presented Senate Bill No. 108 on plant variety protection, with the same title as above. He was followed by Senator Teresa Aquino-Oreta in June 1999 with Senate Bill 1618 and by Senator Sergio Osmeña III shortly after, with Senate Bill No. 1912 (all with the same title).

The first two were very similar, but the third bill was substantially different. Whereas the first two drafts contained references to other Philippine Acts and concerns (access, benefit sharing, prior informed consent, rights of indigenous and local communities, environmental matters, and biosafety), the latter had no such references. It focused solely on the rights of plant breeders, and was almost a copy of the UPOV 1991 Act. Since the first two bills did not differ substantially from House Bill No. 1070 presented above, we will focus on Senate Bill 1912 here. It opens with the following statement as the suggested policy of the state:

> The State recognises that an effective intellectual property system in general and the development of new plant variety in particular are vital in attaining food security for the country. To this end, it shall protect and secure the exclusive rights of breeders with respect to their new plant variety particularly when beneficial to the people for such periods as provided for in this Act.

The addition of 'particularly when beneficial to the people' is interesting in this context, as it indirectly indicates that protection of the exclusive rights of breeders with respect to plant varieties is not necessarily beneficial to the people. However, in the regulatory rules, there is no follow-up to the option *not* to protect plant varieties if this is not beneficial to the people.

Senate Bill 1912 follows the provisions of the UPOV'91 Act in long passages, and the wording is largely identical. The criteria of novelty, distinctness, uniformity and stability are to constitute the sole basis for the granting of a Certificate of Plant Variety Protection (Title III). Those entitled to apply for such a Certificate would be breeders – no mention is made of farmers, indigenous peoples or communities (Section 17). There would be 'national treatment', meaning that any breeder from a country which by treaty convention or law offers the same privileges to Filipinos, would be entitled to the same treatment as Filipino applicants (Section 24).

The rights of the holders of Certificates of Plant Variety Protection would be identical to the rights prescribed in the UPOV 1991 Act, that is covering production, reproduction, conditioning for the purpose of propagation, offering for sale, selling or marketing, exporting, importing, and stocking for any of these purposes. There would be a breeders' exemption, but not if the varieties were essentially derived from protected varieties, or were not clearly distinct from them – or if production would require repeated use of the protected variety (Section 40).

Small-scale farmers would be allowed to use farm-saved seeds and harvested propagating material from protected varieties on their own landholdings and would be entitled to exchange such material (Section 43). However, the Board would determine the area of the landholdings subject to this exemption, taking into consideration the nature of the plant cultivated, grown or sown – and also the conditions for exchange.

All new grants for Certificates of Plant Variety Protection were to be made known in the *Plant Variety Gazette*, which would be critical for the possibility of opposition if the variety is not new, or for other reasons. The *Gazette* should be distributed to all concerned, particularly in the Senate and House Committees on Agriculture (Section 70). There is no mention of languages or how to get these announcements out to farming communities so that they might consider opposition.

Finally, the members of the new Board to be established for implementing the Act should come from various government agencies, the Institute of Plant Breeding, Philippine Seed Industry Association, Federation of Crop Science Society, and one 'outstanding farmer to be nominated by the Philippine Seed Industry Association and the Federation of Crop Science Society of the Philippines'. There would be no place for civil society organizations or other representatives of farmers.

Senate Bill 1912 stood in sharp contrast to all previous attempts to develop a genuine *sui generis* system adapted to the situation and needs of the Philippines. The Bill was drafted by a new actor in the process, AGILE,[37] the acronym for Accelerating Growth, Investment, and Liberalization with Equity – a program under the US Agency for International Development (USAID) to accelerate growth in all economically important sectors of the Philippines and, among other activities, promote competition in agricultural production and trade.[38] The main implementing institution of AGILE was the US-based consultancy firm Development Alternatives, Inc. (DAI), which was subcontracted by USAID, and which in turn established satellite offices in Philippine ministries, *inter alia* in the Department of Agriculture. As will be documented below, DAI was among other projects within the AGILE program tasked with introducing UPOV'91-compatible legislation in the Philippines on plant variety protection. The drafting of Senate Bill 1912 was the first step towards that end.

At about the same time as Senate Bill 1912 was presented to the Senate, UPOV, WIPO and the WTO jointly organized a regional workshop on the protection of plant varieties under Article 27.3(b) of the TRIPS Agreement. The workshop was organized in co-operation with the government of Thailand and the government of New Zealand and took place from 18–19 March 1999. My interviewees in the Philippines mentioned it frequently – particularly the lecture titled 'The 1978 and 1991 Acts of the UPOV Convention' by Barry Greengrass, UPOV Vice Secretary-General, copies of which were distributed in various networks in the Philippines. Greengrass's lecture explained the reasons for revising the Convention and its importance, and concluded:

> Intellectual property conventions are concerned with the protection and encouragement of innovation and creativity. They must be adjusted from time to time to respond to changes in the economic structure and value systems of society and to changes in technology. Each

37 This is documented in Bello (2003, 7) and various newspaper articles, and was confirmed in interviews at DA, at USAID and at AGILE itself (see list of interviews).

38 AGILE (2001): AGILE *Concise Work Plan 2001-03 Showing Linkages With The Medium Term Philippine Development Plan For 2001 – 2004 and the July 2001 State of The Nation Address*. Internal document obtained at USAID.

> of the important changes to the UPOV Convention that are described in this paper has been made to adjust the Convention to new realities.

The workshop provoked massive protests from NGOs throughout Asia, also in the Philippines, which saw it as an effort to force UPOV'91-compatible legislation on Asian countries.[39]

House Bill No. 1070, on plant variety protection, as proposed by Congressional representative Nereus Acosta, was discussed at a meeting in the Committee on Agriculture, Food and Fisheries of the House of Representatives 4 August 1999 – after comments had been compiled from concerned agencies. In addition to the Committee, the meeting was attended by representatives from the DA, IRRI, the Philippine Intellectual Property Office (IPO), the Ayala Corporation, Institute of Plant Breeding, and PSIA.[40]

The Chair, Angelito M. Sarmiento, presented the House Bill, and Congressman Acosta elaborated on the comments to the draft and how he intended to incorporate them. He also announced that he had changed his mind and would now advocate Philippine membership in UPOV 1991. All references to EO 247 were to be removed – likewise, references to indigenous peoples and the IPRA Act. The House Bill in its amended form was to cover the issue of plant breeders' rights only. The ensuing discussion revealed that exemption for farmers to save and use harvested protected material should refer solely to small-scale farmers, and harvested material from a protected variety could not be offered for sale without the authorization of the rights holder and the corresponding payment of royalty. Furthermore, it was proposed that membership in the Board which was to be established under the Act should be based not on representation, but on qualifications (previous knowledge and experience in the testing of the criteria of distinctness, uniformity, stability and novelty). Representatives from the seed industry expressed satisfaction with the development. A technical working group was established to prepare a final draft.[41] Undersecretary D. Panganiban from the DA urged rapid finalization of the Act, due to the WTO-TRIPS review of the Philippines, planned towards the end of 1999.

---

39 Joint press release (18 March 1999), from ActionAid India, Alternative Agriculture Network (Thailand), BIOTHAI (Thailand), Forum of the Poor (Thailand), GRAIN (international, Philippines), Kalpavriksh (India), MASIPAG (Philippines), Pesticides Action Network (Indonesia), Philippine Greens and Third World Network (international, Malaysia) titled 'WTO Pushing New Laws Against Asian Farmers – Protests Mount from India to the Philippines'.

40 According to the Minutes of the Meeting as certified by Committee Secretary Rosa D. Guzman. The representatives were Undersecretary Domingo Panganiban (DA), Deputy Director General for partnerships Dr William Padolina (IRRI), Atty. Josephine Santiago (IPO), Mr Laurel (Ayala Corporation), Director Dr Violeta Villegas (Institute of Plant Breeding), and President Mr Rene Mondragon (PSIA).

41 It would be composed of the following agencies: Philippine Seed Industry Association (PSIA), International Rice Research Institute (IRRI), Department of Agriculture (DA), Bureau of Plant Industry (BPI), Intellectual Property Office (IPO), Department of Environment and Natural Resources (DENR), Protected Areas and Wildlife Bureau (PAWB), Philippine Council for Agriculture, Forestry and Natural Resources and Development (PCARRD) and Technology Application and Promotion Institute (TAPI).

As we can see, this meeting represents the turn from a genuine Philippine *sui generis* legislation to a UPOV'91-friendly law in the House of Representatives, which was also in line with Senate Draft 1912 developed by consultants of AGILE.

The workshop on plant variety protection conducted at IRRI, Los Baños, Laguna, 16–18 February 2000 marked another milestone, according to several interviewees.[42] The workshop was titled 'Impact on research and development of *sui generis* approaches to plant variety protection of rice in developing countries' and was jointly organized by several institutions: IRRI, Council for Partnership on Rice Research in Asia (CORRA), Asia-Pacific Association of Agricultural Research Institutions (APAARI), the Global Forum on Agricultural Research (GFAR), the CGIAR Private Sector Committee (PSC), the US Agency for International Development (USAID), the Rockefeller Foundation, and the Ministry of Foreign Affairs of the Netherlands.

The background for the workshop[43] was the increased role of the private sector in rice breeding, and the obligations that most developing countries[44] had to implement TRIPS Article 27.3(b). The workshop was meant to provide a venue for discussion on current developments and issues affecting plant variety protection and the free flow of germplasm among rice producers in Asia.

Prominent representatives and scientists from, among others, UPOV, the World Bank, Monsanto, IRRI, and various universities in the US, Australia, UK and the Netherlands made presentations at the workshop. Among the papers presented was the later often-cited 'The Importance of the 1991 Act of the UPOV Convention for Developing Countries', by UPOV Vice-Secretary General, Barry Greengrass. The conclusion was that plant variety protection in rice may benefit research and breeding but would also restrict the sharing of new varieties of rice.[45] Various aspects of these effects were discussed and skills developed with regard to plant variety protection, licensing and material transfer agreements.

Senator Sergio Osmeña, supported by AGILE, now took the lead in the Senate process regarding the new act on plant variety protection. He chaired two public hearings on the three Senate Bills from 1998 and 1999,[46] which had been disseminated among stakeholders. The first hearing took place on 19 September 2000[47] and involved high-level representatives from the DA, PhilRice, IPO, Institute of Plant

42 Interviews conducted at IRRI, East-West Seed Company, Department of Agriculture and SEARICE in May 2000 and March 2002 (see interview list).

43 As described in an undated presentation brochure distributed prior to the workshop.

44 Those that were WTO members.

45 Summarized in the journal *Asian Seed & Planting Material* (2000), 'PVP Laws Benefit Rice R&D' (issue No. and author unknown), 9–10.

46 Senate of the Philippines (1998), Senate Bill 108; (1999a), Senate Bill 1618; (1999b) Senate Bill 1912.

47 Memorandum for the Honorable Senate President through the Secretary of the Senate, The Deputy Secretary for Legislation, the Executive Director for Legislative Services, the Director, Committee Affairs Bureau, the Service Chef, LCSS 'C' from Legislative Committee Secretary Arturo I. Mojica Jr. Subject: Report on the 1:P.M., 19 September 2000 Public Hearing of the Committee on Agriculture and Food, Chaired by Senator Sergio R. Osmeña (Re.: Senate Bill Nos. 108, 1618 & 1912 – Provision of Protection to New Plant Varieties), dated 6 November 2000.

Breeding, PCARRD, IRRI, PSIA, AGILE, and Pioneer Hi-Bred. The second hearing was held on 26 September 2000,[48] with high-level representatives from the DA, IPO, Institute of International Legal Studies, MASIPAG, SEARICE, IRRI and Flori Culture Industry of Davao.

Prior to these hearings, several of the invited institutions had prepared position papers, among them IRRI,[49] Philippine Seed Industry Association,[50] and a coalition of NGOs comprising a central farmers' union, other farmer' organizations, indigenous peoples' organizations, church organizations, and groups of scientists.[51] Participants in the hearings could be divided into three categories:[52]

- Those who explicitly supported Senate Bill 1912 (IRRI, PSIA, Department of Agriculture, Institute of Plant Breeding).
- Those who took Senate Bill 1912 as point of departure for discussion – among them, the representatives from PCARRD, PhilRice, and from the IPO. The latter, however, opined that the text of Senate Bill 1912 would place foreign applicants/nationals above Filipinos, because the latter would not enjoy the same benefits in any other country. They also had several other critical remarks to Senate Bill 1912, which could indicate that they regarded it as being too restrictive as it then stood.
- Those who were opposed to a bill on plant variety protection. Whereas SEARICE acknowledged the Senate Bills 108 and 1618 but rejected the idea of plant breeders' rights (Senate Bill 1912), the NGO coalition did not even express acknowledgement of the attempts within Senate Bills 108 or 1618 to

48 Memorandum for the Honorable Senate President through the Secretary of the Senate, the Deputy Secretary for Legislation, the Executive Director for Legislative Services, the Director, Committee Affairs Bureau, the Service Chef, LCSS 'C' from Legislative Committee Secretary Arturo I. Mojica Jr. Subject: Report on the 1:P.M., 26 September 2000 Public Hearing of the Committee on Agriculture and Food, Chaired by Senator Sergio R. Osmeña on Senate Bill Nos. 108, 1618 & 1912 (Provision of Protection to New Plant Varieties), dated 10 November 2000.

49 IRRI Position Paper (16 May 2000) on 'An Act to Provide Protection to New Plant Varieties, Establishing a National Plant Variety Protection Board and for Other Purposes'.

50 Statement of the Philippine Seed Industry Association (19 September 2000) to the Senate Committee on Agriculture and Food on Senate Bills 108, 1618 and 1912 'An Act to Provide Protection to New Plant varieties, Establishing a National Plant Variety Protection Board and for Other Purposes'.

51 Kilusang Magbubukid ng Pilipinas (KMP), Magsasaka at Siyentipiko Para sa Pagunlad ng Agham Pang-Agrickultura (MASIPAG), Kalipunan ng Katutubong Mamamayan sa Pilipinas (KAMP), Samahan ng Nagtataguyod ng Agham at Teknolohiya Para sa Sambayanan (AGHAM), Tunay na Alyansa ng Bayan Alay sa Katutubo (TABAK), Promotion of Church Peoples' Response (PCPR), Sibol ng Agham at Teknolohiya (SIBAT), Program Unit on Ecology and Environmental Protection (PUEEP of the National Council of Churches in the Philippines (NCCP) and TEBTEBBA Foundation, Inc (Indigenous Peoples' International Center for Policy Research and Education) (25 September 2001), Position Paper on Plant Variety Protection.

52 AGILE seems to have had an observer role in the meeting, or at least a low profile. It is not referred to in the minutes of the meeting, except for its being in attendance.

seek to incorporate the broader environmental and development context in the Philippines.

Proponents of Senate Bill 1912 argued against the other two bills that these dealt with issues that were better covered in existing legislation in the Philippines, and that the granting of a Certificate of Plant Variety Protection should not be confused with permission for the release of such a variety. In other words, concerns about environment, health and other issues could be addressed in legislation dealing with the release of these varieties, but should not be made a condition for the protection of an intellectual property. Moreover, such conditions for granting a Plant Variety Protection Certificate would be understood as discriminatory in a UPOV context. Finally, they noted the lack of provisions to ensure that essentially derived varieties would be covered by the rights of the Certificate holder; further, the rights of farmers should be limited to small-scale farmers and be more precise. IRRI suggested that the establishment of a community gene trust fund (as proposed in the other two bills) be included in the 1912 Bill, and that a representative each of indigenous cultural communities and consumers group be included on the Board. The PSIA, however, maintained that board membership should be based principally on expertise, training and experience relevant for plant variety protection.

The arguments of the organizations opposed to any legislation on plant breeders' rights went along two main lines. Whereas the NGO coalition vehemently opposed any legislation on plant breeders' rights, SEARICE opposed it, as long as the process did not include broad consultations with farmers and the text did not provide for a more radical integration of environmental and development concerns. SEARICE argued that the implications of such a system went beyond intellectual property rights, and must be understood in the broader context of biodiversity, food security, roles played by farming communities, etc. Such an act should particularly recognize farmers' rights to save, exchange, share, conserve, improve and sell plant genetic resources. The system, as it was proposed, would ignore the centuries of intellectual contribution of farming communities in developing varieties, and impede farmers' age-old practices of free exchange and sharing of seeds. SEARICE proposed that a broader and more thorough discussion with representatives of farmers' organizations/communities be conducted prior to any action on the bills discussed. The NGO coalition claimed that plant variety protection would be a violation of farmers' rights, a threat to biodiversity, sustainable agriculture and food security, would harm research and innovation and strengthen the control of multinational corporations over Philippine agriculture. They therefore demanded that the PVP bill be scrapped, and that instead a bill should be legislated that would protect farmers' rights over seeds and agrobiodiversity, prohibiting any type of intellectual property rights. At the 26 September public hearing, where some of the NGOs were represented, there were heated discussions on the topic. Senator Osmeña together with DA representatives argued strongly against the NGO representatives.

House Bill No. 1070 of 1998 had now been fully revised. The new version, titled 'An Act to Provide Protection to New Plant Varieties, Establishing a National Plant Varieties Protection Board and for Other Purposes', House Bill No. 10651, was proposed to replace the old bill. This new bill was sponsored by the Chair of the House Committee on Agriculture, Food and Fisheries, the Chair of the House Committee on Appropriations and representative Acosta, together with members of the former Committee. It was submitted to the House of Representatives by the Committee on Agriculture, Food and Fisheries and the Committee on Appropriation, which recommended its approval, without amendments. This new draft was almost identical to Senate Bill No. 1912 from Sergio Osmeña, which had been drafted by consultants of AGILE. In fact, most of the text, and all controversial parts, had been copied from the Senate Bill. Thus, now also the House had its first UPOV'91-compatible draft.

A seminar workshop titled 'Intellectual Property Rights and Technology Transfer' was conducted at PCARRD in Los Baños, Laguna, 20–22 September 2000. It was organized by DOST together with PCARRD and the Department of Trade and Industry together with the IPO. This time, the focus was on the Philippines and was particularly geared towards training. The Intellectual Property Code of the Philippines was presented, and training given in the development of patent applications and the drafting of patent claims. Furthermore the draft Philippine Plant Variety Protection Act was discussed along with plant variety registration and genetically modified organisms.

After the public hearings in 2000, Senators Sergio Osmeña III, Manuel B. Villar Jr. and Juan Flavier revised their drafts and sent the new versions out for comments. Senator Sergio Osmeña – supported by AGILE[53] – presented Senate Bill No. 170, Senator Manuel B. Villar presented Senate Bill No. 967 and Senator Juan Flavier presented Senate Bill No. 62. Again Senator Juan Flavier sought to incorporate environmental and developmental concerns into his draft, whereas Senator Sergio Osmeña followed up on the harmonization with the UPOV 1991 Act but incorporated some of the concerns raised in previous hearings. Senator Manual B. Villar Jr. took some steps in the direction of Senator Osmeña. However, Osmeña now received even broader support – also from those who had not clearly stated their positions in the previous hearings – due to his inclusion of some important concerns.

In their comments on the three revised bills, the Department of Agriculture[54] fully supported Senator Sergio Osmeña and argued against the other two drafts. Again they emphasized that it should not be necessary to include environmental and broader developmental concerns, nor would it be necessary to include a representative of indigenous peoples or from DENR on the Board. Here we should note that AGILE had a satellite office at the Department of Agriculture, located in the Policy and Planning Office – the office of the Assistant Secretary on Policy and Planning in the department, and as such a central provider of policy input to the DA Secretary.[55] The AGILE consultants held a central position, as shown by the fact that

53 According to interview with AGILE staff, March 2002.

54 Letter to the Chairman of the Senate Committee on Agriculture, Hon. Sen. Manuel B. Villar, Jr. from Secretary of the Department of Agriculture Leonardo Q. Montemayor dated 31 August 2001. Subject: Plant Variety Protection Bills (SB 62, 170 & 967).

55 According to interview with the DA Assistant Secretary on Policy and Planning and members of his staff in March 2002 (see interview list).

they represented the Department of Agriculture on various occasions,[56] and were actively engaged in processes concerning their mandate.[57] One AGILE consultant at the satellite office had also drafted the 1912 Bill for Senator Osmeña,[58] and it is likely that the consultants provided policy input to the DA Secretary in order to ensure support for this proposal. This may not prove influence by AGILE on the comments of the Secretary of Agriculture concerning the three revised bills, but it is surely a strong indication.

The PCARRD[59] supported Osmeña's draft and argued against Flavier's proposal, since these concerns would be covered by the IPRA Act and EO 247. By then, EO 247 had already been largely replaced by the new Wildlife Act (two months earlier), and implementation of the IPRA Act was still not functioning. These facts were, however, not taken into account. The Intellectual Property Office of the Department of Trade and Industry[60] supported Senator Sergio Osmeña as well, and suggested some technical improvements. IRRI[61] proposed that the drafts of Osmeña and Villar be re-worked into one harmonized draft and recommended the establishment of a community gene trust fund under the Act, to be administered by the Board. The NGOs continued to protest against legislation on plant variety protection as such.

After the hearings, Senators Flavier, Osmeña III and Villar Jr. re-worked their drafts into Senate Bill No. 1865, with the same title as above. The text was identical to House Bill 10651, except for several additions from Flavier and Villar – which had several parallels to the 1995 draft. AGILE consultants were central in formulating the draft.[62] According to Section 2, the Act was not to be implemented in a way inconsistent with the obligation to maintain a healthy ecology in accord with the rhythms and harmony of nature. All access to genetic resources was to be subject to prior informed consent; and the knowledge, innovations and practices of indigenous and local communities should be respected preserved and maintained by the state. Furthermore, the state should undertake all measures necessary to prevent or control risks associated with the use and release of living modified organisms. All these provisions were formulated as the policy of the state.

---

56 This is self-experienced. I met one of the consultants at a seminar on the International Treaty on Plant Genetic Resources for Food and Agriculture in Quezon City 22 March. She attended as a representative of the Department of Agriculture, although she was paid by AGILE and reported to DAI, and without any mention of this connection. Later I was told that she had represented DA on several occasions.

57 According to interview at AGILE March 2002 (see interview list).

58 See preceding note.

59 PCARRD-DOST Position and Inputs re: Senate Bills Nos. 170 and 62. Undated, but probably September 2001.

60 Letter to the Chairman of the Senate Committee on Agriculture, Hon. Sen. Manuel B. Villar, Jr. from Director General Emma C. Francisco, Intellectual Property Office of the Department of Trade and Industry, dated 19 September 2001. Subject: Senate Bill Nos. 62, 170, 967 '*Sui Generis* Protection of New Plant Varieties'.

61 IRRI Position Paper (4 September 2001), on Senate Committee on Agriculture and Food 'An Act to Provide Protection to New Plant Varieties, Establishing a National Plant Variety Protection Board and for Other Purposes'.

62 According to interview with AGILE staff in March 2002 (see interview list).

'Breeders' were defined to include farmers who breed new varieties (Section 3). The application for a Certificate of Plant Variety Protections should contain an affidavit stating that the applicant had secured prior permission from indigenous cultural communities for the use of varieties indigenous to them, as required under the IPRA, and had complied with all Philippine laws, including Section 24.6 of EO 247. Opposition to the grant of a Certificate could be accepted on the grounds that another person was entitled to a Certificate, that the variety could not be registered under the Act, that the grant of the Certificate would not be in the public interest, or that the variety might have adverse effects on the environment (Section 31).

Certain varieties were not to be registered if the Board believed that the prevention of commercial exploitation of such varieties would be necessary to protect human, animal or plant health, or to avoid serious harm to the environment (Section 60). In cases of emergency, the Board would have the authority to restrict the rights of the rights holder, if considered necessary and in the public interest (Section 36). The rights holder would then receive adequate compensation. The traditional rights of small-scale farmers to save, use, exchange, share or sell their farm produce of a variety protected under the Act should be exempted from the rights of the rights holder, except with sales for the purpose of reproduction under a commercial marketing agreement. The Board was to define the precise terms for this provision. Additionally, it was emphasized that the provision should extend to the exchange and sale of seeds among and between small-scale farmers.

A Community Gene Trust Fund was to be established for the benefit of *bona fide* organizations or institutions managing and operating an accredited community gene bank. This would be financed through an amount less than 20 per cent of the fees and charges paid to the Board for the Grant of Certificates (Section 71). In addition, an amount not less than 10 per cent of the income from fees and charges should be used for training small-scale farmers in the techniques of plant breeding. The Trust Fund, to be administered by the Board, should also be enabled to accept donations from national and international institutions interested in genetic conservation.

The Board should distribute copies of the *Plant Variety Gazette* at no cost to small-scale farmer groups and indigenous communities. This publication was to be written in dialects they understand, in order to enable them to check whether new varieties for which protection was being applied were known to them (Section 73). Finally, the Board should have one representative from a federation of small farmers' organizations, to be appointed by the Secretary of Agriculture (Section 66).

These concerns were included both as an attempt to meet the other drafts in an effort to harmonize the three, and to counter the arguments from the opposition. From the side of AGILE, however, there was a clear strategy behind their inclusion. They were not to be deleted – partly or fully – until the bicameral meeting, after which there would be no more public hearings.[63] This way, troublesome opponents would be calmed down and enactment could be achieved more quickly.

In the meantime, second thoughts seemed to be brewing in the House of Representatives, where several members had started to draft new versions of the plant variety protection act. Rep. J. R. Nereus Acosta wrote House Bill No. 202, Rep.

63 According to my interviewee at AGILE (see interview list).

Cynthia A. Villar House Bill No. 721, and Rep. Carlos O. Cojuangco House Bill No. 815, all with the same title as before.[64] This draft was again sponsored by the Chairs of the two committees, Rep. Alfredo G. Marañon (Committee on Agriculture, Food and Fisheries) and Rep. Rolando G. Andaya (Committee on Appropriations), in addition to the three representatives involved. The Bill was passed in the Committee and put to the House of Representatives. Comparison between this bill and the draft from 2000 (No. 10654) shows that the two are identical. We cannot know what moved the three representatives to write new drafts after the 2000 version, and unfortunately we do not know the contents of these drafts. What is known, however, is that these efforts were not successful, since the old version was approved in the Committee for the second time.

House Bill No. 4518, 'An Act to Provide Protection to New Plant Varieties, Establishing a National Plant Variety Protection Beard and for Other Purposes',[65] which was actually identical with House Bill No 10654 from 2000, was passed in the House of Representatives on 29 May 2002. After having been pending for about a year, the Senate Bill 1865 was finally passed in the Senate on 30 May 2002, shortly before the bicameral consolidation with the House Bill.

In the bicameral meeting, the Bills from the House and the Senate were worked together. The result was the deletion of most of the additions from the Senate version with reference to concerns involving human, animal and plant health, the environment, genetic erosion, farmers as breeders, indigenous and local communities, prior informed consent, genetically modified organisms, public interest, national emergency etc. As these were not merely normative provisions, but regulatory rules, this meant that the possibility of regulating the issue area in line with such concerns was removed.

Little remained of the additions from the old Senate version: one normative provision on consistency with a healthy ecology (Section 2.e); the regulatory provision on small-scale farmers' rights to save, use, exchange, share or sell their farm produce of protected varieties on conditions to be determined by the Board (Section 43.d); the establishment of the Gene Trust Fund, but without technical training for farmers (Section 71); the free distribution of the *Plant Variety Gazette* to farmers, in major local dialects; and representation on the Board of one person nominated by the Secretary of Agriculture to represent a federation of small farmers' organization (Section 66.g).

There were no public hearings or other forms of commenting on this final version, which was in line with procedures after bicameral consolidation of Senate and House Bills. However, AGILE was present at the bicameral meeting, where it

64 Unfortunately, I was not able to obtain these drafts during my field-work in the Philippines.

65 Based on three bills: Rep. J. R. Nereus Acosta wrote House Bill No. 202, Rep. Cynthia A. Villar House Bill No. 721, and Rep. Carlos O. Cojuangco House Bill No. 815, all with the same title as before.

could actively ensure that the deletions were made.[66] The version was enacted on 7 June 2002 and entered into force on 19 July in the same year.

The first task was now to develop the implementing rules and regulations. As will be shown below, AGILE supported this process actively. Administrative Order No. 07, Series of 2003, was signed on 20 February 2003 by the Secretary of Agriculture, who is also the Chair of the National Seed Industry Council, Luis P. Lorenzo, Jr. Since then, the institutions provided for in the Act have been established and training given to potential applicants. Provisional Certificates of Plant Variety Protection have been issued, pending the establishment of a complete system (Golez 2004, 23).

Only one day before the signing of AO No. 7, on 19 February 2003, Philippine newspapers got hold of the story of AGILE and how it had been influencing Philippine politics. The point of departure was AGILE's involvement in the enactment of the so-called Anti-Money Laundering Act, which had heated discussions in the Senate. In a Senate meeting on 18 February, Senator Ralph Recto had disclosed that AGILE maintained satellite offices in many central government agencies and lobbied the passage of several laws.[67] Senator Sergio Osmeña then voiced concern at the threat to national sovereignty, noting the powerful influence of AGILE's on the Arroyo administration. He claimed that AGILE was crafting various legislation that involved foreign interests and asked whether the Philippines was again becoming a US colony, Sen. Recto stated:[68]

> Philippine policy making is directed from Maryland, USA, where a US government funded lobby group is shaping Philippine laws to serve American interests.

Also other Senators found the news about AGILE alarming. Senator Vincente Sotto III declaimed:[69] 'they think they know what is best for us. I think we should let Filipinos themselves decide what is good for them'.

In a Senate Hearing on 19 February, Senator Manual Villar said that it was only now that he realized that all the bills that were given priority in the Legislative-Executive Development Advisory Council had been sponsored by AGILE. Now he understood why so little attention had been paid to bills on health, education and other topics that matter to the people.[70] He recommended banning AGILE from the Congress premises. (To this, Senate President Franklin Drilon responded that it would not be possible, since the committees are the ones to decide whom to invite.) Also Senator Osmeña claimed that he had not been aware of AGILE's strategy, and that he felt duped. He demanded investigations of the subversion and infiltration

66 According to my interview with AGILE staff, and confirmed on a general basis (stated by Senator Manual Villar at a Senate hearing 19 February 2003, cited in Rosario-Malonzo 2004, 2), AGILE used to take part in relevant bicameral meetings.

67 *Daily Tribune* (19 February 2003), 'US funded lobby group rapped for espionage', article by Angie M. Rosales.

68 Ibid.

69 Ibid.

70 *Daily Tribune* (20 February 2003), 'BAP chief's firm linked to AGILE-USAID-DAI funds', article by Angie M. Rosales.

of the bureaucracy, particularly as connected with policy formulation.[71] The Senate called AGILE officials to the hearing, but they claimed protection under an old law from 1952 that provided diplomatic immunity.

The US embassy posted a statement on AGILE on their website in February 2003, where they emphasized that AGILE was a joint project between the Philippine Government and USAID, jointly administered to assist the Philippines. The statement did not stop the criticism, however. Then, on 26 February, US ambassador Francis J. Ricciardone declared to the media that AGILE had rendered services to Philippine senators – including those now attacking the group – by providing technical assistance in their legislative work. He, however, declined to name any names.[72] This was considered a threat, and provoked harsh media criticism against the US ambassador.

Finally, a hearing could be conducted with Dr Ramon Clarete, AGILE Chief of Party, present, where facts about the work of AGILE were established and views exchanged. After this hearing, there was no more writing about it in the newspapers. Nothing was done concerning the acts that had been lobbied by AGILE. Since then, AGILE has ceased to exist and all references to the program have been deleted from the home page of the US embassy, except for the statement mentioned above. The DAI reports only two projects in the Philippines – one on micro-credits, and one on environmental issues. The Philippine DAI office is labeled an international office, in contrast to the DAI field offices elsewhere around the world where DAI is active. USAID continues, however, to 'assist' the Philippines in developing legislation and institutional frameworks for policy implementation in various fields of politics, including economic governance.

Structural Power and the Relative Influence of Actors Involved in PGRFA Management

There were powerful interests involved in the process leading to the adoption of the PVP Act and its initial implementation. The central actor was USAID, with its consultancy firm DAI and Philippine consultants at AGILE. At the other extreme were the NGOs, which followed a line of non-cooperation but were not united among themselves.[73] Between these two extremes were two groups: On one side were Congress representatives striving to develop a genuine *sui generis* system, but with little – if any – outside support. On the other side were the Congress representatives who presented the UPOV'91-compatible bills, together with supporters from the seed industry and the massive backing from USAID through AGILE.

There were no signs of advocacy coalitions in the process towards the PVP Act, except for the very limited coalition of the most radical NGOs. There were no

71 *Manila Times* Internet Edition (21 February 2003), 'Senators ignore USAID exec's defense of AGILE'.

72 *Daily Tribune* (27 February 2003), 'Senators rap Ricciardone as AGILE protector', article by Angie M. Rosales.

73 For example, SEARICE did not share the more radical stand of the NGO Coalition around MASIPAG.

alliances between Congress representatives who sought to develop a genuine *sui generis* system for the Philippines and NGOs that could have had the potential of an advocacy coalition. Thus we must conclude that the Advocacy Coalition Framework is not relevant for explaining the developments that led up to the PVP Act and its implementation.

What then of the theory of structural power, and its relevance? As shown in this chapter, the political economic structures of plant breeding and trading were decisively changed with the introduction of the PVP Act in the Philippines, thereby changing the conditions under which political institutions, economic enterprises and their professional people operate.

An important point to clarify is whether the power exercised in the Philippines was relational or structural – or a combination of both. It could be argued that the powerful role of USAID would indicate that this is more about relational than structural power, that it is a case of the United States influencing the Philippines. The US is certainly central in this regard. However, there were also other actors involved: transnational corporations like Pioneer Hi-Bred and Monsanto, as well as UPOV and the WTO. It is difficult to assess their roles and relative influence. We know only that they all pushed in the same direction. Nevertheless, it could be argued that the United States has worked through UPOV as well as the WTO for the same purposes as pursued in the Philippines – which is basically to secure US business interests in the Philippines, and elsewhere. When I discussed this with USAID staff (see interview list), they emphasized that the Philippines is a major trading partner for the United States and that there are substantial interests in the country. However, the Philippines also needed to develop its trade relations, so the assistance was really a win–win situation, they claimed. Also for political reasons, assistance is important, as the country is a valued ally in the 'fight against terrorism'. This has also been a key message in various documents issued by USAID.

In a power perspective, however, the interdependency between the two countries shows clear patterns of asymmetries (see Chapter 3), with the US far more powerful than the Philippines – a result not least of the colonial history and patterns established in connection with Philippine independence. Thus, the power relation between the two can be described as relational.

On the other hand, it is important to study very carefully where this relational power comes from, and how it works. In spearheading US business interests in the Philippines, transnational corporations are actually the main beneficiaries, since the seed companies that control most of the seed market are all transnational corporations – whether US-based or not. Also key arguments that the Philippines had to comply with its WTO commitments, and that this could best be done through UPOV'91-compatible legislation, indicate an international dimension. As we saw in Chapter 7, the initiative for provisions on intellectual property rights on plants in the TRIPS Agreement came from US business people, which shows that these multilateral instruments are used as leverage to push for changes in the global political economy to promote own interests. The initiative was soon supported by stakeholders in other Western countries, and eventually resulted in the inclusion of such provisions in the TRIPS Agreement. Clearly, then, there are larger structures through which key actors operate, and through which structural power is exercised.

In other words, and in line with our discussion of the Theory of Structural Power in Chapter 3, it makes sense to see relational and structural power as two sides of the same issue, in order to reach the fullest possible understanding of the situation. I have chosen to use the term 'structural power' to cover both approaches, since I see relational power as a component of structural power – it can be more, or less, dominant, but must nevertheless be understood in the broader context.

To have a precise discussion of structural power we need to identify the agents or actors that exercise this power. In a web of structural power, this is a challenging task, since power works through the structures. For example, the DAI consultants of AGILE would have had no power, had it not been for USAID – the powerful donor – with not only its financial resources but also its coercive force in terms of the extreme asymmetry between the two countries (see also Chapter 9). In turn, USAID would have had little power, had it not been for the US Congress, which grants the funds and draws up broad policy lines. However, the US Congress would have scant financial muscle if it were not for US business – and scant political legitimacy if it were not for the American voters.

If we seek to trace the flow of power 'downstream', we should note that the DAI consultants would not have managed to exercise their power, if the Senators and Representatives of the House who proposed the UPOV'91-compatible drafts had not responded to their efforts. Likewise, the Senators and House Representatives would not have not come far, had they not been able to convince the majority in Congress to abandon the genuine Philippine *sui generis* path and opt for a UPOV'91-compatible act. In this effort, they received argumentation help also from other international actors, particularly the UPOV Secretariat, through workshops arranged in Thailand and the Philippines.

All these actors – 'upstream' and 'downstream' – have in some way exercised power in this process. The patterns of interdependencies between them are perhaps the 'glue' of the structural power applied here. The most central agents of this structural power are nevertheless USAID, DAI and AGILE Chief of Party with the DAI-contracted consultants. What was their relative influence?

Relative influence of agents of structural power As noted, the discourse on the PVP Act was radically changed after AGILE entered the scene with their first draft (Senate Bill 1912), presented by Senator Osmeña. We have also seen that USAID and AGILE consultants followed the process closely, that AGILE consultants continued with drafting assistance, and that the PVP Act that was finally adopted was almost identical to the first version drafted by AGILE consultants, which indicates that AGILE had a decisive influence on the process. Let us now take a closer look at the nature of this influence to see whether it can be explained as a form of structural power.

AGILE was set up in 1997 to provide technical services to Philippine counterparts in training, consultants and production of information materials.[74] The overall goal was to revitalize the economy and transform governance to accelerate sustainable

74 Embassy of the United States in Manila (2003), U.S. Embassy Statement on Agile, at <http://manila.usembassy.gov/wwwhagil.html>.

growth. Towards this end various outcomes were defined, including 'commercial law strengthened and enforced', 'revenue administration strengthened', 'bank secrecy reduced and banking sector better regulated', 'competition increased in transportation services', 'competition increased in information and communication technology', 'and 'trade and agricultural competition increased'.[75] The latter is the outcome towards which AGILE's involvement in the PVP Act was directed.

DAI had established satellite offices in several government agencies in the Philippines, *inter alia* in the Central Bank, the Department of Finance, the Department of Trade and Industry, the Securities and Exchange Commission, the National Economic Development Authority, the Department of Transportation and Communications, the National Telecommunications Commission, and the Department of Agriculture.[76] For these offices, DAI subcontracted Dr Ramon Clarete as Philippine counterpart and Chief of Party, and several Philippine consultants.[77] The US Congress earmarked a total of USD 41,212,527 for AGILE between June 1998 and June 2003.[78]

USAID counts it to achievements that an Electronic Commerce Law was passed, a Procurement Watch Organization established, Road Users Charge Legislation signed, Securities Regulation Code revised, a Customs Valuation Bill ratified, and an Electricity Industry Reform Act was passed.[79] It engaged actively in initiating, drafting and finalizing the Intellectual Property Code (RA 8203) presented in Chapter 9, and the Electronic Commerce Act (RA 8792).[80] In addition, it 'helped' in enacting the General Banking Act, the Countervailing Measures Act, Anti-Dumping Act and Safeguard Measures Act.[81] According to central members of the Senate, AGILE was also crucial in the enactment of the Anti-Money Laundering Act.[82]

---

75 AGILE (2001), AGILE Concise Work Plan 2001-03 Showing Linkages With The Medium Term Philippine Development Plan For 2001 – 2004 and the July 2001 State of The Nation Address. Internal document obtained at USAID.

76 According to the brochure 'AGILE Accelerating Growth, Investment and Liberalization with Equity', produced by AGILE.

77 According to my interview at AGILE in March 2002, and confirmed in *The Manila Times* (28 February 2003), 'AGILE: The basic facts', by Rene Q. Bas, at <http://manilatimes.net/others/special/2003/feb/28/20030228spe1.html>.

78 Information from the DAI website at <www.dai.com/projects/text_only/asia_text_only/agile_text_only.htm> accessed 20 February 2003.

79 USAID (2001), USAID Achievements, at: <www.usaid-ph.gov/achievements_usaid.htm>.

80 Documented in Bello (2003, 7).

81 Information from the DAI website at <www.dai.com/projects/text_only/asia_text_only/agile_text_ only.htm> accessed 20 February 2003.

82 *Manila Bulletin* (19 February 2003), 'AMLA not yet officially rejected by FATF', by Mario B. Casayuran. Also cited in *Daily Tribune* (19 February 2003), 'US funded lobby group rapped for espionage', by Angie M. Rosales.

In the agricultural sector, the general policy outcome towards which AGILE worked was 'Trade and Agriculture Competition Increased'.[83] This involved three 'Result Packages':[84]

1. 'Grain Markets More Efficient and Equitable', including efforts to reduce grain prices and the liberalization of rice import, *inter alia* the drafting of various legislation and regulations.
2. 'Biotechnology Exploited Safely', including the development of guidelines and legislation for safe exploitation of biotechnology, to forestall the passage of an anti-biotechnology bill then pending in Congress, and to ensure that domestic biotechnology laws and regulation would be consistent with international treaty obligations.
3. 'Intellectual Property Rights Protected', including the reduction of piracy of optical disks and other electronic media, ensuring the protection of plant varieties developed through agricultural research, and meeting international obligations on intellectual property rights (WTO, TRIPS).

The latter is the result package under which the PVP work was based, primarily through the satellite office at the Department of Agriculture. As noted, the consultants of this satellite office worked as employees of the Department and represented it in various external contexts – but they were paid by DAI and reported to DAI. A central task was to ensure the adoption of a PVP Act.[85] According to the Philippine consultant with lead responsibility for this, her contract contained a deadline for the passage of the PVP Act: June 2002.[86] This is how DAI reports on its experiences in the Philippines:[87]

> In the Philippines, DAI (…) is working with the Department of Agriculture, redrafting PVP legislation to make it compliant with UPOV standards. DAI also is advocating for the PVP law in Congress, where it is pending. The key to DAI's strategy in the Philippines is to separate PVP from controversial issues of GMOs and the rights of indigenous communities. These issues are governed by separate legislation that requires environmental clearances before GMOs can be tested or marketed and that protects the rights of indigenous groups. DAI took key officials and congress persons to Argentina and the United States to learn about PVP programs and legislation. The Filipino Congress is expected to pass the PVP law within the next months. Anticipating the passage of the law, DAI staff have been preparing the ground for its implementation by helping the

83 This refers to page 44 in the DAI Statement of Work dated November 2001, an internal publication for AGILE in the Philippines, which I received at USAID.

84 DAI Statement of Work, November 2001, 44–51, an internal publication for AGILE in the Philippines.

85 According to the DAI Statement of Work for November 2001 (p. 50), a 'key expected accomplishment' of the work in the area of intellectual property rights was 'Plant Variety Protection (PVP) legislation enacted by February 2002'. The DAI Statement of Work is an internal publication for AGILE in the Philippines, which I received at USAID.

86 Interviewed 25 March 2002 (see interview list). The deadline had been extended from February to June.

87 Kent and Bash (2000).

Department of Agriculture develop rules and regulations, and establish the PVP board which is responsible for registration and enforcement of breeders' rights.

Whereas the US embassy holds that AGILE was a joint project between the US and the Philippines[88] and that the assistance was demand-driven,[89] the above quotes indicates a different situation, with DAI pursuing clearly defined objectives. These objectives had in turn been defined by their employer, USAID.

A central question with regard to whether USAID assistance through AGILE was demand-driven is whether the senators who received assistance from AGILE were aware of the background to this assistance – and particularly if Senator Osmeña, the key congressional member fronting the UPOV'91-compatible bill drafted by AGILE, knew about it. As we have seen, Senator Osmeña was among those who brought the role of AGILE to the surface and who raged over the 'manipulation', saying that he felt 'duped' on learning about AGILE's strategy and the larger context of US influence in the Philippines. Had he searched the Internet, or approached the relevant officials at USAID, the Senator could have acquired the information necessary to understand the situation. All information was accessible, as stated by the US Ambassador. However, it seems that not even the NGOs managed to find this information.[90] There is therefore reason to believe that Senator Osmeña was not informed about the full extent of and the intentions behind the assistance he received, and that his rage was therefore authentic.[91] This would further indicate that AGILE's services were not 'demand-driven'.

A factor that might be cited to explain why USAID could claim that the assistance was demand-driven was the intention to help the Philippines comply with its commitments under Article 27.3(b) of the TRIPS Agreement. As analyzed in Chapter 7, there are several possible ways of complying with these commitments, and not only the one that the US argued for (UPOV'91-compatible legislation). The latter is what USAID pursued in the Philippines, with argumentative support from the UPOV Secretariat, and they effectively ruled out any attempts at other *sui generis* solutions. Therefore, this argument is not valid. All in all, it appears that USAID did exercise power over the Philippines in the case of PVP legislation, and that this power can be analyzed in a structural power perspective.

Relative influence of proponents of a genuine sui generis system This tale also shows that there was apparent consensus in the Philippine Congress on the need for a genuine sui generis system for plant varieties protection – adapted to the needs and priorities of the country – until AGILE entered the scene in 1999. The AGILE draft (Senate Bill 1912) soon received considerable support from key stakeholders within

---

88 Embassy of the United States in Manila: U.S. Embassy Statement on AGILE, at <http://usembassy.state.gov/posts/rp1/wwwhagil.html>.

89 In the AGILE brochure 'AGILE Accelerating Growth, Investment and Liberalization with Equity', published by USAID and the Republic of the Philippines.

90 It took me about three weeks of intensive search to identify relevant actors and acquire the information necessary to document this story.

91 Aspects of this question are further discussed below, in connection with the political clout of the leadership.

and outside the Congress, several of whom had participated in a program arranged by AGILE that included travels to the US and to Argentina. The congressional representatives who continued to advocate a sui generis system had a challenging task. Without any noticeable support from groups or actors outside the Congress, they continued making proposals in the House and the Senate. Each time, the drafts had to be worked together with a UPOV'91 – compatible draft – and each time, the result was that the latter was used as foundation, with the occasional addition of aspects from the former. In the end, most of these additions were deleted in the bicameral session. This negotiation tactic from AGILE reduced to an absolute minimum the influence of the proponents of a genuine sui generis system.

Relative influence of NGOs NGOs were also involved in the PVP process – what was their relative influence? We have seen that they could not find a joint position, and that most NGOs rejected any proposal concerning PVP, and demanded that the PVP bill be scrapped. SEARICE initially took a slightly more communicative stance once the UPOV'91 conform bill was on the table. They opposed the bill and the process as long as it had not included broad and thorough discussions with farmers' representatives. None of these demands were heeded by their opponents, who argued strongly against them. As a result, they defined themselves out of the process in the hope that they could stop the new act – instead of participating in the process that was to define its terms and conditions. Thereby they lost the possibility to have any influence.

In retrospect it seems that the best chance the NGOs had of influencing the process in a direction that would have been more in line with their own goals would have been to join forces with those congressional members who were working for a genuine *sui generis* system. We will never know if such an approach would have made a difference, since the negotiation tactics of AGILE, and particularly the practice of ensuring the deletion of elements they opposed, may have rubbed out such joint efforts as well. Probably, it would at least have made it more difficult to delete proposals from a joint opposition to the AGILE draft. Still, all that we can say is that this potential avenue of collaboration – which might have resulted in an advocacy coalition – was not chosen, and in retrospect, this can be viewed as a lost opportunity for the NGOs.

Furthermore, the NGOs were not aware of the role of USAID and AGILE in the legislation process of the PVP Act.[92] True, they had noted that USAID was often present at meetings where draft PVP Acts were discussed, and they had heard about AGILE, but they were not aware of its influence in this particular process until shortly before the new act was adopted. Perhaps their strategy would have been different if they had known.[93]

---

92 According to interviews with representatives from central NGOs and with observers from academic circles in 2002 (see interview list).

93 When I had identified and analyzed the role of USAID and AGILE in the PVP process during my field work in the Philippines, I informed key stakeholders in the NGO community, academe and in relevant ministries about my findings. This was, however, shortly before the bicameral meeting in 2002. SEARICE produced an information flyer, but it was already too

Consequences for implementation of the PVP Act There were powerful forces behind the PVP Act. USAID began assisting in implementing the new act even before it was adopted, and continued to support its further implementation. According to the DAI Statement of Work of November 2001 (49–50)[94] AGILE was to accomplish the following in 2002 with regard to PVP legislation:

- 'Plant Variety Protection (PVP) legislation enacted by February 2002' (later extended to June).
- 'Support the DA in drafting Implementing Rules and Regulations (Administrative Order) for the PVP law.'
- 'The DA and other agencies responsible for implementing and enforcing the PVP law have staff with training and capacity to do their work by May 2002.'
- 'The Judiciary (...) is briefed on the intent and the legal and practical implications of the PVP law by May 2002.'

Their strategy was clearly not restricted to the adoption of the Act (as with the EO advocacy coalition) but aimed at the successful realization of its intentions in a longer time-perspective. It is still too early to assess the success of implementation, but there has been good progress in establishing the necessary institutions and in introducing the measures required.

Institutional Factors Pertaining to the PVP Act

Whereas the EO 247 process was initiated in civil society, was brought into relevant government agencies and then kept outside the congressional system, the PVP process unfolded within the congressional system itself. This marks an important difference, and one that is relevant for the influence of institutional factors. A further difference concerns implementation. The PVP Act is still so new that there is little empirical evidence, whereas there is ample evidence for analyzing EO 247 implementation.

The first drafts of a PVP legislation indicate that the congressional representatives who advocated the genuine *sui generis* path to PVP initially had little competence on such issues. However, more representatives became involved in the development of such drafts, and after a few years, the level of expertise seems to have improved, as indicated by the resulting drafts. There is no evidence of external support to these representatives in terms of personnel or financial resources.

For advocates of the AGILE draft, the situation was different indeed. They were offered secretarial and drafting assistance from AGILE, were invited on trips to the United States and Argentina to learn about PVP legislation, and were generally

late for the process. Also, there was fear in the NGO community, since several key NGO leaders had been exposed to anonymous threats and smear campaigns.

94 The DAI Statement of Work is an internal publication of AGILE in the Philippines, which I received at USAID.

supported and promoted. According to the DAI Statement of Work,[95] the Chief of Party, and Agriculture/Trade Advisor and a Trade Policy Task Manager were available for all AGILE activities concerning intellectual property rights, including PVP. The AGILE-PVP group included an intellectual property rights lawyer, an advocacy specialist, and an international expert with experience in implementing PVP law. All this support gave proponents of the AGILE draft clear advantages.

The differences between the two groups must be viewed in the light of structural power. At the outset the two had similar framework conditions. It was from the outside that the competence and support came, in an effort to influence the situation – in other words, to exercise power.

We have already touched on the political clout of the leadership as a factor. In this context, the 'leadership' means the representatives who took the lead in proposing bills of legislation. We find little evidence of leadership among the various representatives who proposed genuine *sui generis* systems of PVP: they were divided and could not manage to join forces – despite the marginal differences between their drafts. Instead, the only joint drafts to emerge were those that were prepared together with the AGILE draft as part of the required legislative process.

The representatives who proposed the AGILE draft in the Senate and the House showed more leadership, as they spearheaded the legislation. This leadership was, however, informed and controlled by AGILE, so there is reason to question its actual political clout. There are three possible ways of understanding the relationship between the proposing representatives and AGILE:

1. If the proposing representatives did not realize AGILE's influence until the day before the Implementing Rules and Regulations of the PVP Act were adopted, as they maintain, this means that they let themselves be manipulated. They themselves claimed that this was manipulation. If so, it means that AGILE was actually in charge of the leadership and that any political clout among the proposing congressional representatives was almost non-existent: they were merely instruments.
2. Another interpretation is that the representatives knew all along about AGILE and its strategy, and that they were persuaded to follow up due to the many incentives offered – consultants, personnel, study tours and the general power provided through the network of AGILE. In that case, they were aware of the situation and had better possibilities of intervening. However, AGILE was still in charge of the leadership, so the representatives had scant political clout also in this case.
3. Finally, it may be that the representatives who proposed the AGILE draft were convinced by the arguments and not by the incentives, and that they chose UPOV'91 for that reason. In that case they can be said to have shown more political clout, but without AGILE they would probably have had a hard time

95 DAI Statement of Work, November 2001, p. 49, an internal publication of AGILE in the Philippines.

arguing against the larger number of representatives who advocated a genuine *sui generis* path for the Philippines.

As we saw above, there is reason to believe that the first of these three options is closest to reality. In any case, however, it is reasonable to conclude that Philippine congressional representatives evinced little political clout in the PVP process, and that the real leadership of the process was with AGILE – which even had an internal deadline for when the legislative process was to be finalized.

The participative capacity of the proposing congressional representatives became manifest with the hearings they organized. There were also informal contacts between congressional representatives and several key institutions involved in the process.[96] These institutions were largely supportive of the AGILE draft, in line with their official positions as represented by their top management. However, this supportiveness was often weaker at lower institutional levels, where there was more criticism of the proposal.

There were few attempts to include actors with other perspectives, such as NGOs, in the process. Part of this can be explained by the low will of most NGOs to participate, as they were totally opposed to any PVP legislation whatsoever. The condition set by SEARICE, however, was not unrealistic and it was quite legitimate: for a broader hearing with affected farmers. Nevertheless, there was no attempt at meeting this condition, which in turn would indicate rather low participative capacity.

Important for the output is probably also the participative capacity of AGILE, which invited key stakeholders from the Philippine Congress and the various authorities on study tours to the US and Argentina to learn about PVP legislation. These trips were probably important for ensuring support for the AGILE draft.[97]

Integrative capacity in terms of the harmonization of various drafts and joint presentations may seem to have been high – at first glance. In practice, however, the only drafts that won through were the UPOV'91 drafts, whereas the concerns of the other representatives were never really attended to. Integrative capacity must therefore be said to be low. Moreover, the strategy of deleting such concerns in the bicameral meeting indicates that this low integrative capacity was intentional.

Participative and integrative capacities are necessary to ensure democratic legitimacy and ownership of political processes. When structural power is at work, however, it seems that these capacities are used for a different purpose – to exert influence. The purpose of participation then becomes more to ensure support than to pay heed to the insights and input of the institutions. The strategy chosen by AGILE may explain why there was a discrepancy between the top level of the most important key institutions, and the staff at lower institutional levels. It seems that democratic

96 Such as DA, PCARRD, PSIA, IPB, according to interviews at these institutions in May 2000 and March 2002 (see interview list).

97 The DAI Statement of Work from November 2001 (p. 49) (an internal publication of AGILE in the Philippines) also mentions 'funding and cooperating in media campaigns with industry groups and consumer groups' for the purpose of coalition building. However, I found no indications that such strategies were followed in practice.

legitimacy is less important for agents of structural power, but that ownership is still important for ensuring implementation. Ownership to political processes can obviously be achieved through various strategies, and AGILE's participative strategy seems to have been important in securing support for the draft legislation and the process leading up to its adoption among the stakeholders – in this case, the top management of central institutions.

As to the Institutional Capacity Approach, we find that it has less explanatory power concerning the PVP case than with EO 247. This is partly because it explains different processes – primarily an implementation process in the first case and a legislation process in the second case. The former clearly requires far more institutional capacity than the latter. It is, however, also because mechanisms of influence interact differently with institutional capacity. Broadly speaking, we can say that institutional capacity was an important reason why the EO advocacy coalition did not succeed, whereas in the PVP case it was clearly subordinate to structural power as a mechanism of influence. Structural power worked through the institutional structures and molded them for the purpose of the political goal. Therefore, structural power must be said to have had greater explanatory power – as compared to the Institutional Capacity Approach – in the PVP case than the Advocacy Coalition Framework had – as compared to the Institutional Capacity Approach – with EO 247. However, this does not mean that institutional capacity is less important in the context of structural power. To understand how structural power works, we need to know how it acts through relevant institutional structures, and how it molds and changes these structures for the purpose of political change, as seen above.

Relative Effects of the TRIPS Agreement Pertaining to the PVP Act

The TRIPS Agreement was decisive for the initiative to develop legislation on the intellectual property protection of plant varieties, as stated in all the explanatory notes following the various drafts, because the Philippines was expected to comply with Article 27.3(b) of the TRIPS Agreement by the year 2000 as part of its WTO obligations. Let us see whether and how the TRIPS Agreement affected the negotiation output.

For USAID, the TRIPS Agreement was an important vehicle for improving trade relations with the Philippines. The aim was to increase trade and agriculture competition in the Philippines. 'Meeting the international obligations on intellectual property protection (that is WTO, TRIPS)' was the last of three objectives under the 'result package' on the sub-goal 'Intellectual Property Rights Protected' (as referred to above, and documented in the DAI Statement of Work).[98] This gives a clear indication of priority – that assisting the Philippines in meeting its international obligations was seen as a means towards the higher goal of increased trade and agriculture competition.

98 DAI Statement of Work, November 2001, p. 49, an internal publication of AGILE in the Philippines.

As noted, Article 27.3(b) of the TRIPS Agreement does not provide any particular guidelines for the drafting of intellectual property rights to plant varieties, except for the requirement that they should be 'effective'. Therefore several options were available at the outset, as also reflected in the various draft bills. However, the result was a UPOV'91-compatible piece of legislation, with some formulations taken directly from the 1991 Act of the Convention.

The TRIPS Agreement did as such not influence the contents very much, but it influenced the direction, in that there was uncertainty as to the meaning of an 'effective' *sui generis* system. This was a situation that was used by powerful actors, like the US, to promote UPOV'91 as the solution. When UPOV'91 was used as the basis for the PVP Act in the Philippines, it was because of the US involvement and not because of any Philippine initiative. Thus, UPOV'91 influenced the process and the output decisively – through the work of AGILE/USAID.

It is still early to draw conclusions about the effects of these constellations for implementation of the PVP Act. However, implementation is well underway with support from USAID, and there is reason to believe that this support will continue in order to develop the full potentials of the new Act for US–Philippine trade relations, and thus meet the intentions with USAID support as set out in its overall objectives.

We have seen how the Structural Power Approach can provide a fruitful perspective for analyzing the legislation process leading to the PVP Act and its initial implementation – that structural power can serve as a central mechanism of influence in this regard. We have also seen that the Institutional Capacity Approach is a secondary mechanism of influence. Structural power worked through the institutional structures, molding them for the purpose of the political goals set for AGILE.

Relative Aggregate Effects of International Regimes Pertaining to the PVP Act

Before AGILE entered the scene, all efforts were focused on developing a PVP Act in full compliance with the CBD and the IU, pursuing the objectives of all three international agreements (the TRIPS Agreement, the CBD and the IU). AGILE's strategy was to separate issues that concerned the latter two agreements from those of the TRIPS Agreement, and to link implementation of the TRIPS Agreement to UPOV'91. Because of AGILE's success in this, the aggregate effects of the international regimes were largely limited to those of the TRIPS Agreement and the UPOV Convention.

The PVP Act was adopted half a year after the ITPGRFA. However, the implications of the Treaty for the new Act were not considered because the Philippines had already exceeded the deadline for compliance with TRIPS Article 27 3 (b), and the adoption of the PVP Act was considered urgent.[99]

99 According to interviews with DA staff.

Confirming Propositions and Summing Up

Our findings confirm the propositions set out in Chapter 3. We have seen how agents of structural power molded institutional structures and processes to win through with their demands concerning the PVP Act and its implementation. Agents of structural power, most notably USAID, sought to exercise influence on the process leading to the adoption of the PVP Act, confirming our first proposition (P_1). They used the TRIPS Agreement as a vehicle for pursuing US trade interests, as clearly stated in the overall goals for these activities, thereby confirming our second proposition (P_2). For this purpose, they interpreted the TRIPS agreement in line with their own interests, in line with UPOV'91, when influencing the legislation process. As a result, and as set out in the third proposition (P_3), the PVP Act was largely formulated in accordance with the demands of USAID, whereas any diverging demands were hardly reflected at all.

Finally, and in accordance with the fourth proposition (P_4), we found that the PVP Act was accompanied by measures to ensure a relatively strong institutional framework for implementation, as compared to other relevant Philippine legislation on the management of plant genetic resources. This can probably be ascribed to the level of detail of the UPOV Convention, combined with the support from USAID.

As to the explanatory power of the institutional factors, we did not find any evidence for lack of co-ordination between ministries, since that was largely irrelevant here. The Department of Agriculture was clearly in charge. The ninth proposition (P_9) therefore applies only to the first case (EO 247). However, we did find that the PVP Act was formulated in line with the demands of the dominant influencing actor, which confirms our tenth proposition (P_{10}). Finally, we have seen how agents of structural power managed to mold institutional structures and processes to win through with their demands concerning the PVP Act and its implementation. This confirms our twelfth proposition (P_{12}).

On this basis, we can conclude that the TRIPS Agreement worked through structural power as a mechanism of influence in the Philippines, and that relevant institutional structures and processes were shaped and used for this purpose. Without the comprehensive impact of structural power, the PVP Act of the Philippines would probably have become a considerably different piece of legislation.

PART 5
Conclusions, Relevance and Challenges

This book has been motivated by awareness of the enormous challenges facing the management of plant genetic resources for food and agriculture. These include:

- the rapid erosion of plant varieties from farmers' fields all over the world, varieties that have been developed over hundreds and thousands of years and are now forever extinct;
- the often poor domestic systems for gene bank conservation of plant varieties, resulting in great losses from these facilities as well as the loss of critical information about them;
- the emerging barriers to access to and use of genetic resources through legislative measures that particularly impinge on the livelihoods of farmers as well as small- and medium-scale breeders in developing countries;
- the lack of attention to the farmers who still safeguard these resources for present and future generations – the great majority of whom live in poverty without any reward for their contribution to the global genetic pool, and mostly without explicit rights to ensure their continued role as custodians and innovators of PGRFA.

Unless the course of these developments can be changed decisively, our generation risks depriving all future generations of vital means to ensure their food security – means which have been developed and handed over to us by previous generations of farmers. Also at stake are the livelihoods of more than one billion economically poor farmers in developing countries today, who depend on PGRFA diversity. Grasping the dynamics behind these developments represents one initial contribution to the ongoing work of improving the management of these vital resources – as well as to efforts to combat poverty. Since multilateral agreements are the most important instruments available to the international community for influencing the management of PGRFA, they were selected as a research focus here. An additional intention has been to contribute to the growing research body on regime formation and implementation.

Analyzing the effects of international regimes on the management of PGRFA in developing countries was to prove a challenging task – due to the high level of complexity in terms of the number of international regimes and their interaction, the many interrelated sub-issues of PGRFA management, the analysis of relative aggregate effects across scales from the international to the national level, and the identification of mechanisms of influence. In the two chapters of this final part, we draw conclusions from our findings, discuss their relevance for other developing countries, derive implications for PGRFA management and reflect on the contribution to regime studies.

Chapter 11

The Aggregate Effects of International Regimes on PGRFA Management in Developing Countries

Here we will draw conclusions from our findings as to the aggregate effects of international regimes on PGRFA in the Philippines and the relevance of these findings for other developing countries. We begin by explaining the regime constellations pertaining to each of the PGRFA management components identified in Chapter 3,[1] and the assumptions on aggregate effects derived for these components in Chapter 8. These assumptions are examined in light of the findings from the Philippines, before discussing their relevance for other developing countries. To substantiate these empirical generalizations, data from other developing countries will be used. In the second section we explain why. Taking the mechanisms of influence identified in Chapter 3, we offer conclusions as to their relevance in explaining the regime constellations pertaining to PGRFA management and the aggregate regime effects across scales to the Philippines, before discussing their relevance for other developing countries. To substantiate these analytical generalizations, data from other developing countries will be presented.

Aggregate Effects of International Regimes on PGRFA Management in Developing Countries

A key conclusion from this analysis is that the international regime constellations in question were largely negative to the management of PGRFA[2] in developing countries throughout the period from 1992 until 2004 – despite other intentions behind the individual agreements – due to regime interaction. After 2001, more positive prospects emerged with the adoption of the ITPGRFA. Whereas these signals were positive for PGRFA management, the Treaty only entered into force in 2004, and terms and conditions for implementation were not in place until 2006. Therefore, for the period under study here, we could find no signs of change related to the ITPGRFA with regard to the aggregate effects of international regimes pertaining to PGRFA management in developing countries.

1 Utilization of PGRFA will, for the sake of simplicity, be dealt with under the items of *in situ* management and access.

2 As defined in Table 3.1.

Aggregate Effects on In Situ Management of PGRFA

As a result of the interaction between all the regimes examined in this book, there were no clear international guidelines on *in situ* management of PGRFA from the entry into force of the CBD in 1993 until the ITPGRFA entered into force in 2004. At the initiative of the negotiating parties to the CBD, the IU was under renegotiation; in the meantime the CBD was the only legally binding agreement pertaining to *in situ* management of PGRFA.[3] Negotiations took time, due not least to the challenges posed by established regimes, particularly the TRIPS Agreement and the CBD. The Conference of the Parties to the CBD urged for, and awaited, the finalization of the negotiations leading to the ITPGRFA, but took no action in this regard until 2002 – in the form of a co-ordination plan. The 1996 FAO Global Plan of Action for the Conservation and Sustainable Use of PGRFA was an attempt to bring renewed attention to the challenge in the absence of international guidelines. However, the plan failed in this, since most international focus was directed towards finalizing the ITPGRFA negotiations.

On the other hand, a positive effect of the interaction was that the CBD negotiators initiated the renegotiation of the IU in their resolution adopting the CBD in 1992. This was probably decisive for the formation of the ITPGRFA as a legally binding international regime.

We expected the aggregate effects of these regime constellations for the period between 1992 and 2004 to be scant attention to *in situ* management of PGRFA in developing countries. In the Philippines, we found little attention to this issue, but only marginal lines of influence from international regimes. The argument of awaiting the negotiations that would lead to the ITPGRFA may have provided an excuse for granting so little attention to the issue, but the real reason was probably another set of political priorities, which was not conducive to *in situ* management of PGRFA.

Several studies have shown that *in situ* management of PGRFA receives little attention in developing countries.[4] We lack sufficient data to determine the influence of the international regime constellations on this situation. However, it seems clear that the regime constellations were not advantageous to the promotion of *in situ* management of PGRFA in the period in focus here.

A core intention behind the IU/ITPGRFA and the CBD has been to strengthen efforts for the conservation and sustainable use of PGRFA. However, interaction between these regimes did not improve this work during the period studied, and the erosion of PGRFA from the fields continued unabated.

Aggregate Effects on the Ex Situ Management of PGRFA

Regime interaction was also not particularly conducive to *ex situ* conservation of PGRFA. The CBD, as the only legally binding regime in this regard during the

3 The TRIPS Agreement and the UPOV Convention affected *in situ* management, but were not aimed at regulating it.

4 FAO (1998, 202); FAO/CGRFA (2002b), CGRFA-9/02/6; Andersen (2005b).

period, provided for domestic *ex situ* facilities in source countries of PGRFA, whereas the IU and the ITPGRFA called for international co-operation. Domestic *ex situ* facilities would provide greater control over the resources for source countries – as a response to the intellectual property rights regimes emerging through the TRIPS Agreement and the UPOV Convention. International co-operation would ensure better maintenance of acquired accessions, due to the general lack of financial and institutional resources at the national level. A basic assumption behind the CBD was that international support should ensure the financial and institutional resources required for developing countries to develop and maintain domestic gene banks. The risk was that germplasm collections could be seriously damaged if adequate funds failed to materialize – whether through lack of international support, or inadequate allocations in national budgets.

In the Philippines we found that both strategies were pursued – international co-operation (particularly for rice) and domestic gene banking. There were few lines of influence from the international regimes in question. With the non-rice gene banks, a critical point was the lack of financial resources to maintain storage facilities and develop their collections. For several years, three out of five cold storage facilities were out of order due to leakages. It seems that international support was not available, and national funds to repair the facilities were allocated only towards the end of the period under study here.

The Philippine case reflects the situation in many developing countries where gene bank collections are threatened by poor management, often due to scarce financial resources (FAO 1998; Global Crop Diversity Trust 2005). This situation highlights the risks of opting for domestic gene bank facilities without ensuring that necessary financial resources will be available. If national resources are not available (or prioritized), international co-operation is vital. But such co-operation has become more difficult, due to the current trend caused by regime interaction, whereby countries seek to ensure their control over the resources as a response to emerging intellectual property regimes and to ensure equitable benefit sharing. Again, the overall intentions behind the two regimes that directly seek to regulate *ex situ* conservation, the CBD and the IU/ITPGRFA, were the conservation of PGRFA – but the interaction was not conducive in this regard. Partly as a result of this, some – perhaps substantial amounts of – genetic resources are being lost from national gene banks.

Three major international achievements in terms of *ex situ* conservation of PGRFA during the period studied should, however, be highlighted:

- the establishment of the *International Network of Ex Situ Collections under the Auspices of the FAO*, which brought the collections of PGRFA in the gene banks of the CGIAR institutions under the auspices of the FAO, and increased their accessibility;

- the System-wide Information Network for Genetic Resources (SINGER) established by the CGIAR institutions as database for individual collections in the gene banks,[5] with all available characteristics and information;
- the recently established Global Crop Diversity Trust, which aims at providing permanent funding for *ex situ* crop diversity collections around the world.

These initiatives were the results of FAO and CGIAR initiatives, and thus cannot be traced back to any regime interaction.[6] Together, they provide promising entry points for future *ex situ* conservation.

Aggregate Effects on Access to PGRFA

As a result of the interaction among all the international regimes in question, access to PGRFA[7] was substantially reduced – in legal terms[8] – in several developing countries during the period of study. Since then, the number has only increased. Two developments brought about this change:

- *Access legislation:* The CBD provided for the establishment of access legislation on genetic resources, including PGRFA, among the contracting parties. An important aim was to ensure control over, and the equitable sharing of benefits from the use of, genetic resources – not least as a reaction to the emerging intellectual property rights regimes. As the CBD covers all biological diversity including PGRFA, no distinction was made between domesticated and non-domesticated plants. Since access to domesticated PGRFA is vital to the further maintenance of these resources, general regulation of access – without taking into account the specific management requirements for PGRFA – meant new problems for PGRFA management.
- *Intellectual property rights legislation:* Since 2000, developing country WTO members are to have established intellectual property rights legislation to plant varieties.[9] In these efforts, there has been massive pressure to follow UPOV, which generally limits access to and use of protected PGRFA, with some exemptions for farmers and breeders. Limitations on saving, exchanging and selling seeds and propagating material protected by intellectual property rights

5 SINGER covers the gene banks of the CGIAR system. However, the Science Council of the CGIAR has adopted priorities that call for the gene banks of the CGIAR to have a more global outlook and to partner more substantively with others. SINGER has started to follow up on this, and collaboration with national programs is underway – a development that would not have been possible without the ITPGRFA (according to Cary Fowler, e-mail correspondence in May 2006).

6 The Global Crop Diversity Trust is, however, closely linked to the ITPGRFA, and would not have been possible without it.

7 As defined in Table 3.1.

8 Whether access is reduced in practical terms depends on the implementation of laws and regulation, to which we return later in this section.

9 Least developed countries have a later deadline.

serve to reduce farmers' possibilities of maintaining their role as custodians of on-farm PGRFA.

The ITPGRFA seeks to balance these forces by ensuring continued facilitation of access to PGRFA of the crops included in the Multilateral System, and to ensure that they are not made subject to intellectual property rights in the form in which they are received. How this will affect access with regard to ITPGRFA remains to be seen, as the Treaty did not become operational in this regard until 2006.[10]

On this background, it was assumed that the aggregate effect of the international regimes in question would be a situation of increasingly limited access to PGRFA in developing countries in the period studied. In the Philippines these assumptions were supported by our findings, and we could trace the main lines of influence back to the TRIPS Agreement, UPOV, the CBD and their interaction. The first access legislation, which is considered among the strictest in the world, was derived largely from the CBD by actors who sought a strong counterweight to the emerging intellectual property rights, and therefore emphasized those parts of the CBD they considered useful for this purpose. Combined with weak institutional capacity for the implementation of regulations, the result was dramatically decreased legal access to PGRFA for foreign as well as Philippine breeders (although not all breeders were aware of the new legislation and continued to utilize traditional varieties at times). It is difficult to measure the impact of the legislation in practical terms, since data are scarce and we do not know how many breeders and conservers deliberately refrained from accessing PGRFA because of the legislation – as was the case with several pharmaceutical firms. For small-scale Philippine breeders the situation became particularly difficult after the introduction of plant breeders' rights, for which they generally lack the financial means and institutional capacity to apply, and which will strengthen the position of foreign companies in the Philippine seed market. For local farmers, the access situation remains unclear, since the new legislation on plant breeders' rights leaves substantial scope for interpretation with regard to terms and conditions for exemptions regarding farmers' practice of saving seeds and propagating material from protected varieties and to exchange and sell such material. All in all, we found that the influence of the international regime constellations on the facilitation of access to PGRFA in the Philippines has been substantial – and negative.

Similar features can be found in other developing countries as well. As of 2004, various forms of access and benefit regulations have been or were in the process of being adopted in 43 countries, as shown in Table 11.1 (based on ten Kate and Laird 1999, 16; GRAIN 2004). Whereas there are differences between them in terms of approach and restrictiveness, most of them cover genetic resources in general, including PGRFA.

10 With the Standard Material Transfer Agreement adopted at the First Session of the Governing Body in Madrid, June 2006.

Table 11.1 Developing countries that have, or are the in process of adopting, CBD-related regulation on access to genetic resources, as of 2004

Asia	Africa	Latin America	Pacific region
Bangladesh	Cameroon	Argentina	Fiji
India	Eritrea	Belize	Papua New Guinea
Indonesia	Ethiopia	Bolivia	Samoa
Korea, Rep. of	Gambia	Brazil	Solomon Islands
Laos	Ghana	Colombia	
Malaysia	Kenya	Costa Rica	
Pakistan	Lesotho	Ecuador	
Philippines	Malawi	Guatemala	
Thailand	Mozambique	Mexico	
Turkey	Namibia	Panama	
Vietnam	Nigeria	Peru	
Yemen	Seychelles	Venezuela	
	South Africa		
	Tanzania		
	Zimbabwe		

Sources: ten Kate and Laird (1999: 16) and GRAIN (2004).

Also, the number of countries that establish legislation on intellectual property rights in response to the TRIPS Agreement is steadily increasing, as Table 11.2 shows (based on GRAIN 2004; and UPOV[11]).

11 UPOV homepage on memberships available at <www.upov.int/en/about/members/pdf/pub423.pdf>.

Table 11.2 Developing countries that have, or are in the process of adopting, TRIPS-related intellectual property rights legislation pertaining to PGRFA, as of 2004

Asia	Africa	Latin America	Pacific region
Bangladesh	Algeria	Belize	
Bhutan	Egypt	Bolivia	
China	Kenya	Brazil	
India	Mauritius	Chile	
Iraq	Morocco	Colombia	
Jordan	South Africa	Costa Rica	
Korea, Rep. of	Tunisia	Ecuador	
Oman	Zimbabwe	Mexico	
Pakistan		Nicaragua	
Philippines		Panama	
Saudi Arabia		Paraguay	
Sri Lanka		Peru	
Taiwan		Trinidad and Tobago	
Thailand		Uruguay	
Vietnam		Venezuela	

Source: GRAIN (2004) and UPOV homepage on memberships available at <www.upov.int/en/about/members/pdf/pub423.pdf>.

The decisive point with regard to access is whether the systems facilitate expeditious access to PGRFA and whether they allow for breeders' and farmers' exemptions from the rights of the rights holders. Countries differ as to how strict their regulations are. A few seek to combine intellectual property rights with regulation of farmers' rights to local plant varieties. The general trend is, however, one of reduced access to PGRFA from various legal angles – and this is not conducive to the management of PGRFA and may seriously hamper the further maintenance of these resources, depending on forms of restrictions and their implementation.

Nevertheless, access to genetic resources is a central intention underlying the IU/ITPGRFA as well as the CBD. Even the OECD countries behind the provisions of the TRIPS Agreement on plants and plant varieties – with the United States in the lead – advocate free access to PGRFA that is not protected with intellectual property rights. Thus, in practice the regimes have worked to the contrary of core intentions of their proponents when it comes to access to PGRFA in the period under study – and this is due to regime interaction.

Aggregate Effects for Sharing Benefits from the Use of PGRFA

Thus far, the focus with regard to 'benefit sharing' has been mainly on the sharing of benefits between providers and users of genetic resources.[12] Until 2004, regime constellations were not conducive to such sharing of benefits from the use of PGRFA. First, the bilateral approach to benefit sharing established under the CBD – as a response to the emerging intellectual property rights systems pertaining to PGRFA – is hardly applicable for most PGRFA, because it takes as the point of departure the countries of origin of these resources. As genetic resources have been exchanged ever since the dawn of agriculture, it is generally impossible to identify the specific country of origin of a genetic resource as defined in the CBD. Second, benefit-sharing arrangements have been linked to the regulation of access to genetic resources; implementing such regulation requires administrative procedures and tends to be both bureaucratic and time-consuming. Thus, the prospects for achieving the most important of all benefits for the management of PGRFA – access to these vital resources – would be reduced in countries with such laws and regulations in force. Third, linking benefit sharing to access regulation is not a promising approach, because the demand for PGRFA is limited among the large multinational corporations that could be expected to have the greatest profits to share. These corporations already have their own well-stocked gene banks, reducing the demand for PGRFA from the 'countries of origin'. As benefit sharing between providers and users of genetic resources is linked solely to access regulation, we could therefore expect that there would be few cases of such sharing. For all these reasons, the regime constellations do not promote this direct form of benefit sharing between providers and users of PGRFA.

Again, the ITPGRFA seeks to deal with some of these challenges through the Multilateral System on access and benefit sharing, which is an approach intended to enable benefit sharing at the multilateral level and to distribute the benefits not only to providers of PGRFA but also to farmers who conserve and use PGRFA sustainably. However, also the Multilateral System takes access as the point of departure for benefit-sharing arrangements, so the central question is what can be expected in terms of benefit sharing from the Multilateral System. With the bilateral approach under the CBD, expectations for benefit sharing were immense – far out of proportion to the actual benefits achieved as result of the use of derivatives of genetic resources for pharmaceutical products. How the Multilateral System will work remains to be seen.

On this background, we assumed that, as an aggregate effect of the three regimes, the most important benefit with regard to PGRFA – access to these vital resources for food production – was reduced from 1992 and up to 2004. We also expected few, if any, examples of the sharing of benefits between providers and users. However,

12 This termed direct benefit sharing, *inter alia* in Andersen (2006, 5), as compared to indirect benefit sharing, which connotes the sharing of benefits between custodians of agrobiodiversity and society at large, in the North–South context particularly through development co-operation and through the Multilateral System of the ITPGRFA once it becomes operational (ibid., 5–8).

we assumed that we would find examples of indirect benefit sharing, in the form of development co-operation for the management of PGRFA. These assumptions were supported by the findings from the Philippines. Derived largely from parts of the CBD – in response to the TRIPS Agreement – Philippine Executive Order 247 was targeted at all genetic resources including PGRFA. As this did not result in any cases of benefit sharing based on the use of PGRFA, but instead restricted legal access to these resources, the effects must be said to have been negative. How negative they were is difficult to estimate, as we do not know how many breeders and conservers refrained from accessing PGRFA in the Philippines due to EO 247.

This situation is similar in other developing countries. I have searched for examples of direct benefit sharing based on bio-prospecting of PGRFA in the literature and through an international questionnaire survey (Andersen 2005b). Whereas there are a few cases of agreements on potential benefit sharing in cases of commercialization, I have not found one single case of direct benefit sharing from bio-prospecting of PGRFA in developing countries despite the increasing number of countries with bio-prospecting regulation. Even if such a case, or even a few, should be detected, the overall picture remains negative: thus far, bio-prospecting regulation has not resulted in the sharing of benefits between users and providers of PGRFA. Nevertheless, there exist various projects for the conservation and sustainable use of PGRFA and for farmer breeding and participatory plant breeding (ibid.). These are the most concrete manifestations of benefit sharing related to PGRFA management to date. Most of them are supported through development co-operation, and can as such be seen as indirect benefit sharing between the North and the South. Typically, however, these projects are not linked to access to genetic resources.

In terms of direct benefit sharing, it seems that the most important benefit of all with regard to PGRFA, and a crucial factor for food security – access to these vital resources – is increasingly being sacrificed for few if any benefits, as laws on access and benefit sharing are being adopted in more and more countries.[13] The interaction between the TRIPS Agreement and the CBD has brought about a situation that has worked against the intentions behind the CBD as well as the IU/ITPGRFA in this regard in the period under study.

Aggregate Effects on Farmers' Rights

The regime constellations pertaining to farmers' rights were multi-faceted. The FAO had adopted a resolution recognizing these rights in 1989,[14] but they were not defined. The CBD negotiators declared farmers' rights to be an outstanding issue when they adopted the Convention in 1992, an issue to be negotiated under the FAO in the process that led to the ITPGRFA. In practice, this meant that farmers' rights remained 'on hold' until the adoption of the ITPGRFA and its entry into force in

13 Among the latest examples is Ethiopia, with its Access to Genetic Resources and Community Knowledge, and Community Rights Proclamation (Proclamation No. 482/2006) adopted in 2006 (see *Federal Negarit Gazeta* 2006), based partly on the Philippine EO 247. For a presentation and analysis of this legislation, see Regassa (2006).

14 FAO (1989b), FAO C 1989/REP, Conference Resolution 5/89.

2004. In the meantime, two international regimes influenced the situation of farmers' rights decisively: the TRIPS Agreement (together with the UPOV Convention) and the CBD.

Although the TRIPS Agreement is not aimed at regulating farmers' rights, it stipulates that plant varieties are to be protected through some sort of intellectual property rights (Article 27.3 [b]). When farmers' rights came up as an issue in the FAO, it was as a reaction to the emerging systems of intellectual property rights to genetic resources as negotiated in the Uruguay Round of the WTO, and as a direct response to the demand that plant breeders' rights be recognized as compatible with the International Undertaking. It has been argued that, to be effective as required in the Article, implementation of TRIPS Article 27.3 (b) in terms of a *sui generis* system for the protection of plant varieties must follow the UPOV Convention. Following UPOV in this regard leads to restrictions on rights concerning farmers' traditional use and exchange of, plant varieties protected with intellectual property rights.

The provisions of the CBD can be conducive to farmers' rights, as well as conflicting with them, depending on how they are implemented. The provision on the recognition of traditional knowledge (Article 8j) deals with indigenous peoples and local communities – many of whom are farmers or farming communities. In the context of the CBD, this provision is closely interlinked with the provisions stipulating that access to genetic resources should be granted upon prior informed consent and on mutually agreed terms. It is often extremely difficult to identify which communities could legitimately decide on, for example, a plant variety, since plant varieties are typically developed through the exchange of genetic resources between farmers within and across farming communities over short and long distances.[15] Therefore this approach is not particularly applicable with regard to farmers' rights. Furthermore, it could be interpreted as a kind of exclusive ownership right and provide disincentives to sharing knowledge and propagating material among farming communities in cases where communities would expect benefits from these resources. As sharing is the precondition for access to genetic resources in traditional farming systems, such incentives could work to the contrary of farmers' rights in terms of access to PGRFA. Finally, and depending on implementation, it could make conservation efforts for the benefit of farmers difficult, because these efforts could also be attached to access procedures that include processes to ensure prior informed consent by farmers, as indicated above. Because of these problems, direct benefit sharing between farmers and receivers of PGRFA would be difficult. On the other hand, the CBD negotiators have contributed to an understanding of the need to recognize the rights of indigenous and local communities, and thereby also perhaps paved the way for an understanding of the legitimacy of farmers' rights. Furthermore, by referring farmers' rights to the FAO negotiations, the CBD negotiators helped to get these rights included in the ITPGRFA. And with the ITPGRFA, greater scope has emerged for the realization of farmers' rights.

On this background, we assumed that, as aggregate effects of the international regimes, little would have been done to realize farmers' rights in developing countries so far, but that more would have been achieved with regard to the enhancement

15 Increasingly also by including commercial varieties (Salazar et al. 2007).

and protection of traditional knowledge as set out in the CBD. Nevertheless, access legislation combined with regulations pertaining to the recognition of traditional knowledge, as derived from the CBD, could limit farmers' rights, depending on their design and implementation, as they could hinder conservation work for the benefit of farmers and provide disincentives to the sharing of seeds and traditional knowledge among farmers. The introduction of intellectual property rights to PGRFA could substantially limit these rights further by restricting customary practices of saving, using and exchanging seeds, particularly if introduced without legislation on the recognition of farmers' rights.

The findings in the Philippines largely support these assumptions and can generally be traced back to the regime constellations. We saw that little had been done to realize farmers' rights directly. The effect of the CBD was two-edged: on the one hand, EO 247 provides for prior informed consent for access to PGRFA, which indicates that farmers could have a say, perhaps also with regard to benefit sharing. On the other hand, it restricts access to PGRFA for breeders and conservers, which is an important precondition for the availability of adequate seeds to farmers. Furthermore, the approach of sharing benefits with the providers of PGRFA excludes from the scope of benefit sharing the large majority of farmers who participate in maintaining PGRFA for future generations. Even if implemented properly, EO 247 would never have reached more than a few farming communities. Whereas the regulation could have provided disincentives for farmers to share PGRFA, it did not develop this way – probably because farmers were simply not aware of the regulation or of its relevance to PGRFA.[16] Farmers did not experience any benefit-sharing arrangements based on bio-prospecting, but there have been several projects for conservation of PGRFA and participatory plant breeding, not linked to the EO 247, and funded through development co-operation that could be seen as benefit-sharing arrangements.

Until 2002, Philippine farmers were largely free to use, share and sell farm saved seeds and propagating material.[17] Then the new TRIPS-induced and UPOV-derived regulation introduced restrictions to these rights – with exemptions for small scale farmers – regarding their rights to save, use, exchange, share or sell their farm produce of varieties protected under the act. Farmers are not allowed to sell seeds or propagating material of protected varieties on a commercial basis. The terms and conditions for the exemptions for small scale farmers are to be determined by a Board which has only one member from a farmers' organization, appointed by the Minister of Agriculture. The 'novelty criterion' for plant varieties in the new regulation may enable intellectual property rights on farmers' varieties that have been 'discovered' and only marginally improved by the breeder, with the burden of proof as to whether the protected variety had previously existed resting with the farmers in question. This legal situation is clearly negative to farmers' rights.

---

16 Among many stakeholders it seemed to be associated with wild genetic resources, and its relevance for PGRFA was not always recognized.

17 Some instances of contracts between breeders and buyers of propagating material have, however, been reported, which restricts the free use of the material by farmers.

Participation in decision-making processes is among the components of farmers' rights. Some farmers' representatives were involved in several relevant decision processes in the Philippines up to around 2000, but then this started to decline, due to internal political developments. None of the developments pertaining to participation can be traced back to international regimes.

Interestingly, it seems that the IU/ITPGRFA had little effect on farmers' rights in the Philippines, except for its catalyzing effects for bottom–up initiatives to support farmers' *in situ* management of PGRFA, including participatory plant breeding. In this context also the CBD is relevant as a catalyst of donor funding. All in all, the international regime constellations were found to be important for the situation of farmers' rights in the Philippines, but we noted various other domestic lines of causation as well.

In general, the effects from international regimes balanced each other until 2002, but turned negative after the adoption of the PVP Act. Nevertheless, we should recall that the concept of 'farmers' rights' was included in the PVP Act. Moreover, the circumstance that the exemptions were more radical than provided for in UPOV'91 can partly be traced back to the negotiations on farmers' rights under the FAO, as shown in Chapter 10.

This overall picture has many similarities with the situation throughout much of the developing world.[18] Little has been done to promote explicit realization of farmers' rights on the national level in most countries, but there is one exception: India. The Indian Protection of Plant Varieties and Farmers' Rights Act of 2001, which aims to distribute rights equitably between breeders and farmers, is a novel and interesting approach to realizing farmers' rights.[19] It is also the most advanced example in terms of the realization of farmers' rights. A central driving force in the development of the farmers' rights provisions of this act was Prof. M.S. Swaminathan, who also played an important role in the work for recognition of farmers' rights in the FAO and the Keystone dialogues (Ramanna 2003). This indicates that there were close links between the Indian and the FAO processes.

In developing countries where strict access legislation has been introduced and linked with the protection of traditional knowledge, there are indications that the new incentive structures have limited the sharing of seeds and traditional knowledge and hampered conservation work for the benefit of farmers. This has been documented in, for example, Peru and Colombia (Brush 2004 and 2005; Andersen 2005b; Ruiz Muller 2006). Nevertheless, an international questionnaire survey on the state of farmers' rights (Andersen 2005b) shows that various processes are underway in many countries to register and support the maintenance of traditional knowledge on PGRFA, considered as protecting these resources not from bio-prospecting but from the specter of extinction.

---

18 This section is based largely on the findings from the *Farmers' Rights Project* at the Fridtjof Nansen Institute, initiated and led by the author (see <www.farmersrights.org>), and particularly on the findings from an international questionnaire survey carried out as part of the project (Andersen 2005b).

19 See Ramanna and Smale (2004) and Ramanna (2006) for in-depth discussions.

As expected and noted above, it has so far not been possible to trace direct benefit sharing between providers and users of PGRFA, despite legislation for this purpose in several developing countries. However, farmers generally participate, to varying degrees, in the indirect form of benefit sharing, particularly of *non-monetary* benefits. A questionnaire survey (Andersen 2005b) showed that the most often mentioned non-monetary benefits were access to seeds and propagating material, influence on breeding activities, and in some cases participatory plant breeding. Development co-operation was often involved. Nevertheless, farmers' access to and use of PGRFA is increasingly being restricted in many countries in Asia and Latin America and some countries in Africa due to intellectual property rights legislation, and in some cases due also to regulation of the protection of traditional knowledge combined with access legislation. These developments are closely related to the TRIPS Agreement and the CBD.

Participation in decision-making processes varies from country to country according to political systems and governance. These developments can rarely be traced back to the international agreements in question. Generally, such participation seems to be marginal in developing countries and is often confined to large-scale farmers who are normally not engaged in the maintenance of plant genetic diversity anyway (Andersen 2005b).

All in all, the situation with regard to farmers' rights has deteriorated as aggregate effects of the regime constellations in the period under study. This may impact on the livelihoods of traditional farmers as well as their ability to maintain PGRFA. In fact, the international agreements pertaining to the management of PGRFA have largely worked counter to these norms on farmers' rights. The ITPGRFA has the potential to change this situation – depending, however, on the political will of its contracting parties.

PGRFA and the Emerging Tragedy of the Anti-Commons

As we have seen, the greatest changes in the management of PGRFA in the period in focus affected access to these resources. The basis for access is, however, the availability of these resources, not least their conservation and sustainable use. Nevertheless, this fundamental issue has received little attention in policy development. At the international level, negotiations have focused on access and benefit-sharing regimes under the CBD as well as the ITPGRFA, and the Multilateral System of Access and Benefit Sharing is generally perceived the core of the International Treaty. As under the CBD, benefit sharing is linked to access also under the ITPGRFA.[20]

The ITPGRFA is designed so as to facilitate access to PGRFA more smoothly than the bilateral systems of access regulation. A fundamental question is what benefits this system will generate for sharing, due to the limited demand for PGRFA from the large, most profitable commercial breeders. As Brush (2005) points out with regard to the ITPGRFA, today's systems of benefit sharing linked to access to

20 Even if it applied only when PGRFA from the Multilateral System is made subject to intellectual property rights or other forms of commercialization, that is, was dependent on use.

genetic resources – whether derived from the CBD or the ITPGRFA – do not solve the classic 'tragedy of the commons' (Hardin 1968) that has beset the management of crop genetic resources, allowing breeders to benefit from the use of PGRFA without bearing the costs of maintaining them. Instead, it seems that various forms of development assistance remain the most likely source of funding for maintenance of PGRFA under current regimes. Ironically, this reverts to tools and principles that were established before the CBD, as Brush also notes (ibid.).

Nevertheless, all this focus on access in order to establish fair and equitable benefit-sharing systems failed to increase accessibility to PGRFA in the period studied in this book – what happened was rather the converse. Without access to PGRFA, no benefits can be generated, and thus there can be no benefits to share. Without access, also the conservation and sustainable use of PGRFA are hampered. Access to PGRFA is thus far the most important benefit from the use of these vital resources, a precondition for their maintenance – and for food security. It is therefore vital that benefit-sharing arrangements should not pose barriers to access to PGRFA. However, with the emergence of intellectual property rights systems, benefit-sharing arrangements seem necessary to ensure equity, and have led to an 'arms race' (Rosendal 2006). In turn, benefit-sharing arrangements and intellectual property rights erode access to these vital resources, ultimately leading to a 'tragedy of the anti-commons'.

The term 'anti-commons tragedy', in contrast to Hardin's classic 'tragedy of the commons' (1968), was first coined by Heller and Eisenberg (1998),[21] referring to the situation of biomedical research. Ramanna (2003) used the term to highlight how India's legislation on intellectual property rights to PGRFA and farmers' rights may pose problems through overlapping claims of ownership to genetic resources, resulting in an 'anti-commons tragedy' with a negative effect on agricultural development. This stand is developed further in Ramanna and Smale (2004). Brush (2004 and 2005) uses the concept to describe how the tension between intellectual property rights and access/benefit-sharing legislation leads to a situation where multiple owners have the right to exclude others from utilizing scarce resources and no one gets the effective privilege of use. He highlights the dangers of an access system based on market negotiations between purported 'owners' and 'users' of genetic resources, as this is likely to 'abuse the rights of people who have long been involved in the common pool of genetic resources but find themselves arbitrarily excluded in contracting' (Brush 2004, 255). The findings from this book underscore the gravity of this point and document how PGRFA management is emerging as a tragedy of the anti-commons in the South: a lose–lose situation for all, including future generations.

The ITPGRFA aims at facilitating access to genetic resources that are covered by the Multilateral System, attempting to ensure that no intellectual property rights will be claimed for resources from the system – in the form they are received. This is a promising approach with regard to Annex 1 crops. For all other crops than those

21 See also Heller (1998), who presented some aspects of the anti-commons tragedy in an earlier article, and Aoki (1998), who analyzed the contributions of Heller and Eisenberg in the context of intellectual property rights.

covered by the Multilateral System, access will have to be organized on a bilateral basis. This has proven problematic in cases where access legislation has been introduced – and such legislation is being introduced in more and more developing countries in response to intellectual property rights. Although the Multilateral System of the ITPGRFA is limited to the Annex 1 crops, it is nevertheless the most conducive approach today to international regulation of PGRFA management, and is a platform from which further strengthening of the international governance of PGRFA management can be developed.

Mechanisms of Influence of International Regimes

The basic assumption for this analysis was that response to international regimes, whether in terms of compliance with the regimes or as protest against them, takes place on the initiatives of actors. This makes it crucial to identify the actor constellations with regard to the policies in question, and to determine how various actors have used the international regimes – or been influenced by them – in their efforts to shape the policies. Another important point is to understand how they managed to achieve influence. In this study, the ways in which the norms and rules from the international regime constellations are transformed into domestic policies have been termed *mechanisms of influence*.

Whereas our focus has been on the effects of established international regimes in developing countries, it proved necessary to analyze regime formation and interaction as well, because this provides the basis for analyzing the effects of the regime constellations in developing countries. In analyzing regime formation, *mechanisms of influence* are central, showing how driving forces transform and channel their interests and/or ideas into international regimes. Thus, we have sought to analyze mechanisms of influence across scales from the national to the international level – and back to the national level.

Structural Power as a Mechanism of Influence

In the negotiations of the TRIPS Agreement with its Article 27.3(b) we see clear traces of structural power as a mechanism of influence. National and multinational corporations were the driving forces behind the group of industrialized countries that worked for the inclusion of an agreement on intellectual property rights in the WTO, more specifically with provisions on PGRFA. These forces had at their disposal various forms of power resources, particularly coercive force, which was involved when the developing countries finally accepted the TRIPS Agreement. The constellations of driving forces, their use of power resources and how this shaped and determined the structures of the global political economy within which developing countries, their political institutions, their economic enterprises and their professional people have to operate – all these indicate that structural power was at work, as the main mechanism of influence.

On this basis we expected that the TRIPS Agreement would work through structural power as a mechanism of influence in developing countries in general, and in the Philippines in particular. If so, we could expect to find that 'agents of structural power' had sought to exercise influence on the process leading to the adoption of the PVP Act; that they would refer to the TRIPS Agreement and interpret it in line with their own interests; and that the legislation would largely be formulated in line with their demands. Finally, we expected that the enforcement rules, including the institutional framework for implementation, of the new act would be stricter than for other relevant acts of Philippine legislation not influenced by agents of structural power. These propositions were all supported by our study, and we found clear pattern matching between the development in the Philippines and the elements of structural power as defined in Chapter 3. We thus concluded that structural power was a central mechanism of influence with regard to the PVP Act in the Philippines.

In the Philippines, the political economic structures of plant breeding and trading were decisively changed with the introduction of the PVP Act, which altered the conditions under which political institutions, economic enterprises and their professional people operate in the country. Powerful foreign interests were active in the process that led to the adoption of the PVP Act and its initial implementation. Their power resources involved great wealth and coercive force, and the methods were a combination of infiltration and the offering of incentives. These actors effectively weeded out all references to prior informed consent, benefit sharing, and other provisions that had been proposed by Philippine congress-members to ensure an equity dimension in the new Act (referring particularly the Convention on Biological Diversity). Thereby they used the TRIPS Agreement as leverage, arguing that UPOV-compatible legislation would be the only safe way to ensure that the TRIPS Council would accept the Act as Philippine compliance with Article 27.3(b). Without the comprehensive impact of structural power, the PVP Act of the Philippines would probably have looked quite different indeed.

The experiences from the Philippines are unique in certain ways, due to the historical ties between the United States and the Philippines, but also have features that are comparable across countries. The central element of these features is pressure. Considerable work remains to investigate how such pressure is exercised in various countries, but a common denominator is often that of the 'TRIPS-plus' agreements. Industrialized countries and groups of such countries negotiate bilateral and regional trade and investment agreements with individual or groups of developing countries. Under such constellations, the developing countries have a considerably weaker bargaining position. Increasingly, bilateral and regional agreements between developed and developing countries have come to include provisions on intellectual property rights. In general, developing countries have to introduce intellectual property rights at higher standards than those provided for under the TRIPS Agreement and they are often obliged to join UPOV 1991 by a certain deadline – thus, these agreements are frequently called 'TRIPS-plus' agreements (Vivas-Eugui 2003). The US, the European Free Trade Association and the European Union are active proponents and initiators of bilateral and regional agreements, and have established such agreements with many developing countries, as shown in Table 11.3.

Table 11.3 Developing countries[22] that are in the process of entering into, or have entered into, 'TRIPS-plus' agreements imposing intellectual property rights on biodiversity[23]

Status	Asia/Pacific	Africa	Latin America[a]
Agreement signed and/or entered into force	Bahrain (US)[b] Bangladesh (EU) Cambodia (US) Jordan (EFTA; EU; US) Korea (EFTA; EU; US) Laos (US) Lebanon (EFTA; EU) Mongolia (US) Oman (US) Palestinian Auth. (EFTA; EU) Singapore (US) Sri Lanka (EU; US) Syria (EU) Vietnam (US; Switzerland)	Algeria (EU) Egypt (EU) Morocco (EFTA; EU; US) South Africa (EU) Tunisia (EFTA; EU)	Bolivia (US) Chile (EFTA; US) Colombia (US) Dominican Republic (US) Ecuador (US) Mexico (EFTA; EU) Nicaragua (US) Peru (US) Trinidad & Tobago (US) Venezuela (US)
Ongoing negotiations as of September 2005	Bahrain (EFTA) China (EFTA) Iran (EU) Kuwait (EFTA) Oman (EFTA) Qatar (EFTA) Saudi Arabia (EFTA) Thailand (EFTA; US) United Arab Emirates (EFTA; US)	Algeria (EFTA) Botswana (EFTA) Egypt (EFTA) Lesotho (EFTA) Namibia (EFTA) South Africa (EFTA) Swaziland (EFTA)	Argentina (EU) Bolivia (EU) Brazil (EU) Caribbean Basin (USA) Colombia (EU) Ecuador (EU) Panama (US) Paraguay (EU) Peru (EU) Uruguay (EU) Venezuela (EU)

[a] The Free Trade Area of the Americas currently under negotiation is not covered in this overview. The draft contains strict provisions on intellectual property rights on plant genetic resources.
[b] Counterparts given in parentheses.
Source: Based on GRAIN (2005).

22 Some of these countries have entered into the agreement as groups, see GRAIN (2005).

23 Contents and documentation of the agreements are provided in GRAIN (2005).

In these cases, there is reason to assume that structural power may represent a central mechanism of influence that can help to explain why and how these countries comply with their obligations towards Article 27.3(b) of the TRIPS Agreement.

Whereas these data indicate where structural power could have been at work, the question of exactly how structural power would work in these countries has to do with analytical generalization. The effects of structural power in the introduction of intellectual property rights to PGRFA is assumed to be highest in developing countries where institutional capacity is weak (see below), and where powerful foreign or multinational stakeholders have an interest in better access to seed markets.

The decisive question with regard to structural power is how to deal with it in the formation and implementation of international regimes pertaining to PGRFA, so as to ensure policies conducive to the management of these resources. For this purpose we need to find out as exactly as possible how structural power works in specific cases, in order to identify entry points of action. In the Philippines, the 'agents of structural power' were not known to the vast majority of actors involved in the PVP process. Awareness of their role and methods might possibly have led to other strategies by the actors involved, and thus resulted in another act of legislation – possibly one more in line with the initial visions of ensuring equity and legitimacy. Further research is necessary to trace the methods of influence in other countries.

Learning and Norm Diffusion as Mechanisms of Influence

NGOs were decisive in initiating and pushing for the negotiations towards the CBD – they were the initial driving forces as such. When formal negotiations started, some coalitions had already been formed between NGOs and responsible government entities, particularly around the issues of conservation (largely Northern countries) and benefit sharing (largely Southern countries). It is clear that policy core beliefs about the value of conservation, or the justice of benefit sharing, formed the glue that kept these coalitions together. Some country delegations acted as mediators – for example the Nordic countries – which indicates that the CBD negotiations had patterns of advocacy coalitions. The negotiation focus changed from a conservationist approach to a more holistic one that included sustainable development and equity aspects, largely through a process of learning and norm diffusion. Thus, learning and norm diffusion are seen as the main mechanisms of influence.

On this basis we expected to find that learning and norm diffusion would work through the CBD as a mechanism of influence in developing countries in general and in the Philippines in particular. If so, we could expect to find that representatives of advocacy coalitions would have sought to exercise influence on the process that led to the adoption of EO 247, that they would refer to the CBD in this context and interpret it in line with their own ideas when seeking influence over the legislation process, and that the regulation would be formulated largely in accordance with the demands of these coalitions. Finally, we expected to find that the resulting institutional framework for implementation would be rather weak as compared to other relevant Philippine acts of legislation that have been influenced by agents of structural power. These propositions were all supported by our findings, and we noted clear pattern matching between the development in the Philippines and the elements of the Advocacy Coalition Framework as an ideational (learning and norm diffusion) approach.[24] We thus concluded that learning and norm diffusion were the central mechanisms of influence with regard to EO 247 in the Philippines.

The CBD provided an important rationale for core actors in the formulation of EO 247, and was the most decisive factor for the development of the document into a fully fledged regulation. It became the framework for the advocacy coalition and was also a driving force for central leaders in relevant government agencies who were part of the advocacy coalition. The main motivation was to provide for benefit sharing – to the benefit of the nation and the conservation of its natural resources, as set out in the CBD. This motivation was further spurred by resistance to the TRIPS agreement among core actors who saw bio-prospecting and intellectual property rights as closely interlinked. On this background, benefit sharing was promoted, to the detriment of access to genetic resources. This also reflects the distinction between core and secondary norms in the CBD, as analyzed in Chapter 6.

As we have seen, EO 247 was not conducive to the management of PGRFA. Few actors with specific responsibilities for PGRFA management were involved in the decisive phases of the formulation of the new regulation. Most of them did not participate at all. In addition, actors involved in, for example, the management of PGRFA did not realize the potential implications of the new regulation for PGRFA management. At the time, there was no awareness of the particular requirements for domesticated as compared to non-domesticated plant genetic resources. This was related to the general perception that the IU was not in function, and that one would have to await a new international treaty on PGRFA before taking any new steps. The CBD was therefore the basis – in terms of international regimes – from which EO 247 was derived, and it did not signal any differences between the two categories of genetic resources. Therefore the CBD did not help in avoiding the problem. Rather it encouraged it, because the two categories had been merged without providing for a distinction between them with regard to management requirements. Until then, the two categories had been kept separate in institutional terms in the Philippines.

The whole process leading to the adoption of EO 247 can be described as a learning process towards consensus building, starting out with a little group of scientists and evolving to include NGOs and core state entities, all sharing the basic norms behind the idea. New elements introduced by various parts of the coalition were openly received and considered, resulting in input to or revisions of the text – partly or fully in line with the 'lessons' learned. However, the coalition resisted any information that might produce conflicting inferences in that, for example, core beliefs were found invalid or unattainable. In the process leading up to EO 247, information that might challenge the basic beliefs was not formally sought until towards the very end, when business actors were finally invited for consultations. Despite harsh criticism after the adoption of EO 247, the coalition resisted it, also when formulating its implementing rules and regulations. Recognition of the actual effect of EO 247 as a barrier to bio-prospecting due to heavy bureaucratic burdens and long processing times can be described as a kind of slowly evolving shock. Particularly government agencies in the Department of the Environment and Natural

24 As defined in Chapter 3.

Resources took this experience seriously, and as a response to that, they included new and simpler rules on bio-prospecting in the new Wildlife Act – thereby overruling EO 247 with regard to wild biological and genetic resources. However, this was to have unforeseen consequences for access to PGRFA.

The advocacy coalition was geared towards the adoption of policies and regulations, and less towards sustained implementation. For various reasons the coalition fell apart soon after the regulation had been adopted. There was no shared strategy for its sustained implementation. The major weakness of the advocacy coalition was its short perspective in getting the regulation adopted, which blinded it to important aspects that should have been taken into account to ensure sustained accomplishments.

Again, the experiences from the Philippines are unique in many ways, but also have features that are comparable across countries. Data from Colombia, Costa Rica, Mexico, Chile and Malaysia (in Carrizosa et al. 2004) indicate that the norms of the CBD on access and benefit sharing fit well with national claims to gain control over genetic resources to ensure equitable sharing of benefits derived from their use in countries that have introduced such regulations. This may have represented an important glue and shared motivation for groups of stakeholders involved in the process. The data further indicate that learning and norm diffusion have been central in these contexts, but that the ways in which these processes have developed differ from country to country. In some countries NGOs played an important role, in other countries academic institutions were the driving forces, in yet others government entities were the central actors. Coalitions varied in form among the stakeholders of these categories, but we lack sufficient data to say whether these can be labeled advocacy coalitions according to the criteria of the Advocacy Coalition Framework (ACF).

Generally, we found that the prospects for learning and norm diffusion as a mechanism of influence in other countries are greatest when the core norms of the international regime – in this case the CBD – correspond with belief systems among groups of stakeholders in the country. Furthermore, a responsive state is required.

The ACF is one of several approaches to studying learning and norm diffusion. Further research is necessary to identify how learning and norm diffusion evolve in policy processes pertaining to PGRFA management in order to ensure conduciveness according to the principles defined in Chapter 3. The lessons from the Philippines show that the policy process led by the advocacy coalition could not ensure sustained implementation. Awareness of the roles of the various participants in the advocacy coalition – combined with a stronger sense of strategic thinking with regard to sustained accomplishments – might have ensured a better regulatory framework and more efficient implementation. It would probably also have attracted funding at an earlier stage. Analysis of ideationally based PGRFA policy processes in other countries might make it possible to identify entry points for improving the prospects for conduciveness as well as effectiveness.

Institutional Capacity and Mechanisms of Influence

Institutional capacity has been found to be a kind of intermediate mechanism of influence, not a separate mechanism in its own right: The uses of institutional capacity depend on the actors in charge of this capacity, so a central question is how these are influenced by structural power and/or learning and norm diffusion. Institutional capacity alone cannot be called a mechanism of influence of an international regime – only in relation to other mechanisms of influence. In the developments that led to the IU and the ITPGRFA, several mechanisms were at work. Economically strong Northern countries were hesitant to the IU at the beginning and demanded the acceptance of intellectual property rights before they would join the Undertaking, which indicates elements of coercive force and structural power. There were also elements of learning and norm diffusion – for example with the *State of the PGRFA* report (FAO 1998) – which helped to create an understanding of the critical situation of PGRFA. International implementation of the IU, as well as the negotiations pertaining to the ITPGRFA – and the follow-up after its adoption – struggled due to scarce institutional capacity in terms of funds as well as changing delegations and general lack of underlying political clout. This lack of institutional capacity can be understood as a result of low interest on the part of agents of structural power, as well as the lack of pressure from advocacy coalitions and other ideational actors: they were critical to decisions regarding the recognition of intellectual property rights in 1989, as well as how farmers' rights were finally recognized in the ITPGRFA.[25]

With the emphasis on institutional capacity as an intermediate mechanism of influence, our focus has been on how the other two mechanisms of influence work through institutional capacity. Basically we assumed that if institutional capacity is low in a country, the influence of agents of structural power and of advocacy coalitions will tend to be high. We thus expected to find little co-ordination between the branches of the ministries involved in PGRFA management, and that acts of legislation pertaining to these resources would generally be formulated in line with the demands of the dominant influencing actors. Advocacy coalition would depend on responsive authorities winning through with their demands, it was assumed, whereas agents of structural power would mold institutional structures and processes in order to win through with their demands. We found support for these propositions in the findings of the cases from the Philippines, and the Institutional Capacity Approach proved fruitful for identifying developments important to understanding the influence of international regimes. On this basis we concluded that agents of structural power worked through institutional structures and processes in the Philippines by molding them for the purpose of TRIPS implementation according to their demands, whereas advocates of learning and norm diffusion required a responsive institutional structure

25 According to the NGO representatives present at the First Session of the Governing Body of the ITPGRFA in Madrid 12–16 June 2006, most NGOs regard the ITPGRFA as a lost battle due to the way farmers' rights were recognized; this was why so few NGOs were present at the session (source: personal communication with representatives from various NGOs at the First Session).

to win through with their demands for implementation of access and benefit-sharing arrangements under the CBD.

In the case of EO 247, the political clout of the leaders in the involved government agencies – and after 1998, the lack of such clout – provides a central explanation for the success of the formulation process as well as the lack of implementation. The lack of integrative as well as participative capacity after 1998 and financial constraints are important additional factors. Together with the heritage from a fading advocacy coalition in terms of a regulation that was hardly feasible, the contours of a picture have emerged that can explain why EO 247 could not work. Experiences from the formulation phase (responsive state entities and leaders with political clout but without sufficient participative capacity) and from implementation (dependent on individuals and not sufficiently rooted in institutional structures, therefore vulnerable to political change) indicate institutional weaknesses that, ironically, facilitated the impact of the advocacy coalition on negotiation output (EO 247) but limited the impact on implementation and on the actual political development pertaining to bio-prospecting. Combined with the generally short-sighted perspectives of the advocacy coalition this meant there could be no real future for EO 247.

These lessons raise important questions on the roles and tasks of advocacy coalitions in states with weak institutional government structures – even if they are responsive. How could the civil society counterparts of an advocacy coalition respond to the lack of participative capacity of state entities? Have they any possible roles or tasks that could support or pressure the bureaucrats to maintain the implementation of a regulation? It all boils down to the time perspective and long-term commitment of the advocacy coalition. If the time perspective extends only until the adoption of a regulation, there is no further role for the coalition to play.[26] In contrast, had the time perspective been directed towards long-term compliance, it would also have been necessary to find solutions to these questions.

The Institutional Capacity Approach was found to have less explanatory power in the case of the PVP Act than in the case of EO 247. This is partly because it explains different processes – primarily an implementation process in the first case and a legislation process in the second case. The former obviously requires far more institutional capacity than the latter. This is, however, also because mechanisms of influence interact differently with institutional capacity. In broad lines, we can say that institutional capacity was an important reason why the advocacy coalition described in the case of EO 247 did not succeed, whereas institutional capacity in the PVP case was clearly subordinate to structural power as a mechanism of influence, as agents of structural power took the control over the legislation process as such. Structural power worked through the institutional structures and molded them for the purposes of the political goal. Therefore, we may say that in the case of the PVP

26 This finding is also relevant at the international level. As many NGOs lost interest in the ITPGRFA after its adoption, partly due to its provisions on farmers' rights, they have not been engaged in seeking to develop the options for implementation of these provisions. There is considerable scope for implementation, a potential that could be utilized by NGOs if they adopt a more far-sighted perspective to their work for farmers' rights than simply achieving the adoption of the respective provisions of the Treaty.

Act structural power had more explanatory power – as compared to the Institutional Capacity Approach – than the Advocacy Coalition Framework had – again, as compared to the Institutional Capacity Approach – in connection with EO 247. This indicates that institutional capacity is less important in the context of structural power. To understand how structural power works – and to enable intervention when required – it is necessary to grasp how it works through relevant institutional structures, molding them for the purpose of political change.

We have seen that analyzing how mechanisms of structural power and of learning and norm diffusion work through institutional capacity is particularly relevant where institutional capacity is weak, as is often the case in developing countries.

Summing Up

A key conclusion is that, despite the original intentions, the international constellations of norms and rules that resulted from the overlap and interaction between the regimes in question were largely negative to the management of PGRFA in developing countries throughout the period studied. Moreover, this was due to regime interaction. The result of these developments is an emerging anti-commons tragedy: a situation where multiple actors have the possibilities to exclude each other from the use of these vital resources.

The driving forces that produced this regime interaction transform and channel their interests and ideas into these developments along various patterns – mechanisms of influence – which can be traced across scales from the international to the national level: (1) structural power, that is the power to shape political-economic structures, (2) learning and norm diffusion, and (3) institutional capacity. The latter is not a separate mechanism in its own right, but one through which other mechanisms work and which also determines their influence. This study has shown how the driving forces shaped the regime interaction pertaining to PGRFA management through these mechanisms of influence. It has also shown how the identification of mechanisms of influence and their patterns of work can provide entry points for actors to shape governance in agrobiodiversity at the international as well as national levels.

After the entry into force of the ITPGRFA in 2004, new options have emerged. These might change developments for the crops included in the Multilateral System under the Treaty, in a direction more favorable for PGRFA management. What the effects of this new development will be for the management of these vital resources in developing countries remains to be seen. We begin the next, and final, chapter, by discussing the implications of our findings for the future governance of PGRFA management.

Chapter 12

Implications of the Findings and Challenges Ahead

In this final chapter we examine some of the most pressing implications of the findings presented in this book with regard to PGRFA management as well as regime studies, and we also discuss some of the central challenges ahead.

Governing PGRFA Management: Implications of the Findings

As this study has focused on the aggregate effects of a set of international regimes with fully or partly overlapping functional scopes, we will now return to these regimes and derive implications of our findings for their further implementation. A central concern is to identify potential ways of breaking out of the emerging anti-commons tragedy of PGRFA management.

Implementing the ITPGRFA

As a legally binding international regime, the ITPGRFA represents a considerable step in a direction conducive to PGRFA management. However, many questions remain. Central in this context is how the Multilateral System on Access and Benefit Sharing is to be implemented in practice. It is crucial for this system to succeed in ensuring expeditious access to PGRFA – but it is also important to have realistic expectations as to the shared benefits it may generate, and to realize that these will probably not be sufficient for global conservation needs. The funding strategy currently under development is an urgent concern, and additional funds will probably be required.

As the Multilateral System pertains to 35 food crops and 29 fodder plants in the public domain and under the control of the parties, a central question is what the future will be for all the other crops and fodder plants. Based on the findings in this book, the maintenance of these plant genetic resources would clearly have better prospects if they were included in the Multilateral System. One possibility is to propose re-opening the text of Annex 1 with the list of crops covered by the Multilateral System for renewed negotiations, with a view to including further crops. On the other hand, speedy implementation of the ITPGRFA is also vital, and embarking on renegotiations of Annex 1 might well delay that process. Thus, there are arguments for and against re-opening the list for negotiations. Another possibility is to consider how these issues can best be dealt with in an international regime on access and benefit sharing, as currently under negotiation in the CBD-COP. We will return to this question in the next section.

Moreover, there is the question of how to deal with *ex situ* accessions that were collected prior to the entry into force of the CBD and that are not covered by the Multilateral System – or indeed, basically not by any international regime. The ITPGRFA negotiators failed to find a solution to this challenge, which means that a new possibility emerges with the international regime on access and benefit sharing now being negotiated under the CBD. We return to this below.

If the tragedy of the anti-commons in PGRFA management is to be halted, a key challenge for the parties to the ITPGRFA will be to create and increase the 'legal space' (Andersen 2006) for farmers to maintain their role as custodians and innovators of agrobiodiversity. This is a basic requirement for continued *in situ* management of PGRFA. Creating and increasing legal space means mainstreaming legislation with a view to enabling farmers to continue their customary practices of conserving, developing, exchanging and selling PGRFA. These practices have been the foundation for the development of PGRFA diversity throughout history, and they will be a major precondition for the possibility to maintain these resources for present and future generations. Particular attention should be paid to seed laws, certification rules, and intellectual property legislation, as these regulations today represent the greatest barriers to such practices. Also access laws and regulations pertaining to traditional knowledge should be considered with a view to their effects on farmers' customary practices.

Here the concept of 'farmers' rights' assumes central importance. Creating and upholding farmers' legal space, and recognizing and rewarding farmers for their contribution to the global genetic pool, are basically what farmers' rights are about. The recognition and promotion of farmers' rights is a cornerstone in the implementation of the ITPGRFA. Moreover, it is vital to the livelihoods of many impoverished farmers in developing countries and thus to combating poverty and hunger. There exists no common understanding of how to follow up the provisions on farmers' rights in the ITPGRFA in practice – yet that is pivotal for making progress in the realization of farmers' rights and the maintenance of PGRFA. The resolution on farmers' rights adopted at the Second Session of the Governing Body, November 2007, is a first step.[1] It encourages Contracting Parties to report on their views and experiences, and requests the Secretariat to collect these views and experiences as a basis for an agenda item for consideration by the Governing Body at its Third Session to promote the realization of farmers' rights at the national level.

For many years, conservation and sustainable use of PGRFA have struggled with inadequate political attention, and thus also with lack of necessary funds. Promising approaches have been found to generate funds for, and improve, *ex situ* conservation of PGRFA (particularly with the Global Crop Diversity Trust), but *in situ* management remains a huge challenge. The Multilateral System may in future provide some, perhaps substantial, benefits to be shared among farmer custodians of PGRFA. For the time being, however, development co-operation remains the most important source of funds for this work. This highlights the importance of greater

1 The official reference for this resolution has not yet been announced, as the report from this session is still pending. The author of this book was, however, present at the session and has a copy of the resolution.

political attention to the issue of PGRFA management, as a central condition for development co-operation in this area.

When implementing the ITPGRFA, many will have to change their previously established policies and legislation. In the Philippines, there seems to be a certain degree of *regime fatigue* among those in charge in the various ministries, with less motivation for regime implementation than in the 1990s. A similar situation may also exist in other developing countries with scarce institutional capacity. Thus, for the ITPGRFA to become implemented, good support systems are needed.

Negotiating an International Regime for Access and Benefit Sharing under the CBD

We have seen that the current access and benefit sharing regime under the CBD has not worked out as envisaged and that it has not proven conducive to the management of PGRFA. Although many important food crops are now covered by the Multilateral System under the ITPGRFA, a substantial share remains under the international governance of the CBD. This means that the current negotiations of an international regime on access and benefit sharing under the CBD are highly relevant to the future management of these resources. How, then, might the sharing of benefits from the use of PGRFA be encouraged under such a regime, without contributing further to restricting access to these vital resources?

The ongoing negotiations under the CBD are based on the Bonn Guidelines on Access and Benefit Sharing and include proposed instruments, such as certificates of origin or legal provenance and disclosure of such information in applications for intellectual property rights. Also the use of existing customs procedures to regulate the import of genetic resources is being discussed. From the experiences presented in this book, it seems that this bilateral approach alone cannot suffice to provide effectively for access and benefit sharing for PGRFA, particularly not as long as the burden of proof remains with the providers of genetic resources. We need to ask whether it is possible to derive lessons from the Multilateral System on Access and Benefit Sharing under the ITPGRFA.

The Multilateral System applies to only a limited list of crops. Might it then be useful to develop an international regime on access and benefit sharing with different instruments for different types of genetic resources? For example, a model based on the Multilateral System could serve as a useful supplement to the bilateral approach discussed in the current negotiations – for cases where the country of origin or legal provenance cannot be established, or where the countries in question wish to apply such a simplified system.

A central provision of the Treaty on the Multilateral System is that access to genetic resources is to be provided expeditiously between the Parties to the Treaty – and without the need to trace individual accessions. These shall be provided free of charge, or if there is a fee, it is to cover only the minimum costs involved. The genetic resources in question are to be transferred with all available passport data. All this is possible because there exists a Standard Material Transfer Agreement that provides for all necessary arrangements regarding the use of the resources and benefit sharing. However, the Standard Material Transfer Agreement does not involve any procedures for prior informed consent by indigenous and local peoples, since it applies solely to material that is in the public domain – often in gene banks. The question of traditional knowledge is covered in separate provisions in the International Treaty on Farmers' Rights. Such a separation of topics might not be applicable to an international regime under the CBD, but it seems reasonable to evaluate this possibility in further detail, especially with a view to a possible multilateral component of the international regime, as indicated above.

An important prerequisite for getting access under the Multilateral System is that intellectual property rights are prohibited, in order to ensure that the material kept in the Multilateral System remains accessible. This provision applies to the genetic material from the Multilateral System in the form in which it was received – but *not* after it has been modified. Such a distinction may prove difficult in practice, as it is unclear just how much any given material must have been 'modified' for intellectual property rights to be allowed, and opinion differs on this point. For an international regime on access and benefit sharing, a similar provision – at least for access to PGRFA under such a regime – would probably be desirable, to ensure that the material provided remains available for use by others.

When it comes to benefit sharing, the arrangements are quite different from the Convention. Benefits are to be shared with the custodians of genetic resources – not with specific providers under the Multilateral System. This is an important difference between the Multilateral System and other models discussed under the CBD. Benefits do not flow back to the provider country of a specific resource, as provided for under the bilateral system envisaged in the CBD. Instead, they flow into a supranational system – the Multilateral System. Under the ITPGRFA, a so-called Third Party Beneficiary, acting on behalf of the Multilateral System, ensures that the benefits flow primarily to farmers, especially in developing countries and countries with economies in transition, who conserve and sustainably use genetic resources. This recognizes that *all* countries are net receivers of genetic resources, and thus there is no need for a complicated system of transfers of benefits between recipients and providers in the narrow sense. It should also make the transfer of the genetic resources much easier. Moreover, it ensures that the benefits contribute directly to the maintenance and further availability of these resources for food and agriculture. A similar system under an international regime on access and benefit sharing would certainly simplify matters, but it might not be feasible under a bilateral approach. If, however, the system involved a multilateral component, this model of benefit sharing would probably be a precondition.

The Standard Material Transfer Agreement (SMTA) is an agreement between the provider of a genetic resource (for example, a gene bank) and the receiver (for example a crop-breeding firm), stipulating the conditions for transfer of the material. Because it is standardized, it can be used equally by relevant actors in all countries and international organizations that are Parties to the ITPGRFA. Thus, there is no need for negotiation, which makes the facilitation of access much easier than if the terms and conditions must be negotiated each time. It also lowers the transaction costs considerably. This is an absolute asset of the multilateral system – but can it be translated to an international regime on access and benefit sharing? A precondition would be that the Contracting Parties establish a multilateral system, which supersedes the bilateral system that is practiced today, for example by adopting an agreed interpretation of the CBD.[2] But what might it look like in detail? One possibility is to extend the Certificates of Origin to a standardized Material Transfer Agreement. Admittedly, it would be a challenge to develop such an Agreement that would cover all genetic resources, related traditional knowledge, and that would account for benefit sharing with indigenous and local peoples. On the other hand, if complete standardization is not possible, a semi-standardized solution might be found, whereby the main part of the Agreement is standardized, while annexes are to be filled out by the Parties to the Agreement. Such annexes could include specifications on issues like the respect and recognition of traditional knowledge – and participation in benefit sharing. Much work would still be required to design such a standard agreement. A further possibility is to develop a standard material transfer agreement for a multilateral component of the international regime on access and benefit sharing under the CBD – based on the Multilateral System under the ITPGRFA.

Under the SMTA, the rates for the sharing of monetary benefits from commercialization are fixed. If the receivers of a genetic resource commercialize the resultant product, they are encouraged to make a payment to a so-called Third Party Beneficiary. If the product has been patented, or if other rights restrict its availability, then the receivers are obliged to pay a share from the sales. This share is fixed in the SMTA. This means that benefit sharing is actually linked to use, rather than to access as such. But is it possible to standardize the benefits from the many different uses of such different genetic resources and related knowledge as in an international regime under the CBD? Perhaps a set of rates could be developed for different uses and different resources, or at least for the material covered by a multilateral component of the international regime.

It should be noted, however, that this approach is solely monetary. Under the ITPGRFA, other forms of benefit sharing, which we can label indirect benefit sharing and which often involve non-monetary measures in the form of for example development co-operation, are covered by various provisions. This separation of direct and indirect mechanisms pertaining to benefit sharing should be studied in greater detail with a view to an international regime.

As stated in the SMTA, the Third Party Beneficiary is also in charge of monitoring compliance with the Material Transfer Agreements, on behalf of the Governing Body of the Treaty and its Multilateral System. It is to keep account of material transfers and payments, basically by keeping track of patents and the obligatory payments related to them. Such a Third Party Beneficiary could also be a relevant model for an international regime under the CBD, particularly if it can contain a multilateral component, as indicated above.

Under the Multilateral System, any of the three parties – the provider, the recipient, the Third Party Beneficiary – may initiate dispute settlement procedures, whether as amicable negotiations, as mediation or – if nothing else can solve the

---

2 This way of modifying international agreements was used by the FAO Conference in 1989 with FAO-Resolutions 4 and 5 related to the International Undertaking on Plant Genetic Resources.

conflict – as arbitration. The result of such arbitration is binding. Again, this model may prove useful for the design of an international regime.

As shown here, many questions will need to be clarified to find out whether this is a viable avenue – or whether elements of it can be helpful in designing an international regime on access and benefit sharing of genetic resources under the CBD. In any case, it is important that negotiators recognize the special needs related to the management of PGRFA when designing an international regime on access and benefit sharing covering such resources.

The TRIPS Agreement and the Doha 'Development Round'

The latest trade negotiation round under the WTO, the so-called Doha round, is also known as the 'Development Round' because it is meant to take into account the special needs and concerns of developing countries. In connection with PGRFA management, there are two concerns of particular relevance for developing countries:

- Many developing countries have been under pressure to develop intellectual property rights to plants and plant varieties that by far exceed the minimum standards set by the TRIPS Agreement, often in line with the 1991 Convention of UPOV. There seems to be a widespread perception that this is required for compliance with Article 27.3(b) of the TRIPS Agreement. That is, however, not the case. Norway is an example of a country that is still a member of UPOV according to the 1978 Convention, and which – in order to balance breeders' and farmers' rights – has decided *not* to accede to UPOV 1991. In this way, Norway fulfils its obligations towards the TRIPS Agreement with regard to the protection of plant varieties. A greater focus on the variety of possibilities for implementing Article 27.3(b) and support to countries seeking genuine *sui generis* systems could help developing countries in working out intellectual property systems for PGRFA that are more appropriate for their own needs and priorities.
- Benefit sharing has largely been a concern in the CBD-COP and the Governing Body of the ITPGRFA. There has been great resistance in the TRIPS Council to proposals addressing such concerns within the framework of intellectual property rights. However, only when user countries adapt measures to ensure benefit sharing will such a system become effective. Moreover, if the emerging tragedy of the anti-commons is to be halted, it is vital to get the focus of attention shifted to the use of genetic resources and to the users. If benefit sharing arrangements can be linked to the use of genetic resources rather than to access, much would be gained in terms of facilitated access to these resources and relieving provider countries – often developing countries with scarce institutional capacities – of burdensome procedures. Thus, proposals regarding disclosure of origin or legal provenance and other measures to ensure benefit sharing from the use of genetic resources related to intellectual property rights would need careful consideration in the TRIPS Council.

If the Doha Round is to meet the expectations expressed with the 'Development Round' – and based on the findings in this book – these two concerns would be of particular importance for PGRFA management in developing countries.

Improving Policy Coherence across Scales

When the functional scopes of different international regimes overlap on an issue-area, and these international regimes have different scopes and rationales, a central challenge is to provide for policy coherence. As we have seen in the case of PGRFA management, there has been a substantial lack of policy coherence across scales from the international to the national level. The crucial question is how policy coherence can be improved across scales.

This book has proposed ways and means to analyze regime overlap and interaction and to identify their potential aggregate effects on a specific issue-area. It has delineated how regime constellations can be analyzed as a basis for predictions on aggregate regime effects. On the basis of an historical analysis of regime development, core and secondary norms and rules were derived for each of the international agreements studied here. Against this background, overlaps of functional scopes were identified and their potentials for synergies and conflicts determined. This formed the foundation for our analysis of regime interaction pertaining to the identified issues of overlap, which enabled us to draw conclusions as to the overall regime constellations for the period under study. By analyzing these regime constellations in light of the identified mechanisms of influence, we were able to derive conclusions as to the assumed aggregate regime effects for developing countries.

The case-study findings supported these assumptions. Moreover, there is considerable evidence that similar effects can be found in many other developing countries as well. This indicates that this approach may be applicable also for predicting aggregate regime effects in other issue-areas as well – as a basis for improving policy coherence.

Contributions to Regime Studies

Whereas the approach described above may be useful for analyzing and improving policy coherence, the research strategy for predicting issue-specific aggregate regime effects has also been intended as a contribution to regime studies. In this final section, the contributions of this book to regime studies are summarized and some thoughts on future research challenges are offered.

Applicability of the Six Cs Model in Measuring Aggregate Regime Effects

In order to analyze the effects of regime constellations, good methods are crucial. While many regime studies have focused on the effectiveness of international regimes – with effects measured against the norms and rules of the regimes themselves – we sought to measure aggregate effects of international regimes against a set of criteria developed independently of these regimes. These criteria were formulated on the basis of a problem analysis, and a research strategy for predicting issue-specific aggregate regime effects was developed, as indicated above.

The basically qualitative Six Cs Model on the Measurement of Aggregate Regime Effects provided an approach that proved fruitful in the case studies. The first part of this model was devoted to the measurement of the dependent variable – the state of policy decisions and goal achievements, within a certain period, in a given country and with regard to the issue areas under scrutiny. We used three parameters for this purpose: *coverage*, *compatibility*, and *conduciveness*. Analyzing coverage was necessary to show whether the policy decisions or goal achievements pertained to the whole issue area or only to parts of it. Analyzing compatibility was necessary to indicate whether the policy decisions or goal achievements pertaining to the issue area were pulling in the same direction or in different ones, and whether policies pulling in different directions were conflicting. And finally, analyzing conduciveness was necessary to show whether the policy decisions or goal achievements were conducive to the management of the various components of PGRFA.

This background prepared us to trace the relative effects of international regimes on this state of the art, establishing causal links between the explanatory and dependent variables. The second part of the model introduced the last three parameters for analysis: *causation*, *correspondence*, and *conduciveness*. Here, causation refers to the degree to which policy decisions or goal achievements can be ascribed to an international agreement. Correspondence refers to the degree to which the identified effects were in accordance with norms and/or rules of an international regime. Conduciveness refers to the degree to which the effects of the international regimes could be described as positive for the management of the particular issue, determined on the basis of the indications of conduciveness defined previously. Whereas the second part of the Six Cs Model involved a rather brief assessment of all effects according to documentation of attribution, it could not enable us to trace lines of influence from international regimes in depth. For this purpose, embedded case studies were carried out in selected issue-areas deemed central to the overall issue. These case studies explained the identified regime effects in depth, using inductive process tracing combined with pattern matching concerning the proposed mechanisms of influence as methods.

By linking to institutional capacity the mechanisms of structural power, and of learning and norm diffusion, we could see how these mechanisms work under conditions of weak institutional capacity. Such knowledge is particularly relevant for identifying the challenges of implementing international regimes in developing countries.

Finally, we could draw conclusions across scales from the international to the national level with regard to the effects of international regime constellations on the overall issue-area in question and discuss the relevance of these findings for other countries – partly on the basis of empirical and partly on the basis of analytical generalization.

The model was shown to be fruitful for our purposes, and is expected to have applicability for the analysis of aggregate regime effects in other issue-areas as well.

Future Research Challenges

The research body on regime interaction is growing steadily. Nevertheless, little work has been done to reveal the aggregate effects of international regimes at the national level, even though this is vital for improving international governance. More research is necessary to fill this gap in order to provide correctives and guidance for the further development of international regimes.

A particular challenge is to study these effects across scales and down to the local level. Especially in the case of PGRFA management, such an approach would be important for precise identification of the effects and thus also entry points for possible changes. This would be highly challenging from a methodological point of view. Developing a research strategy for the purpose would probably benefit from including aspects of social anthropology, and would provide a good opportunity for further refining the inductive process tracing method.

Finally, further work is needed on the study of mechanisms of influence of international regimes. This is crucial for improving governance because it can also indicate entry points for change:

- This book has shown how regimes influenced by structural power have affected the management of PGRFA in the Philippines. If these processes had been understood in greater detail at an earlier stage, central actors could have intervened and might have achieved legislation more conducive to PGRFA management.
- We have also seen that regimes influenced by learning and norm diffusion tend to be weaker, in terms of implementation, than regimes based on structural power. A central question is thus how ideational regimes could become stronger. Here it seems fruitful to study the ideational elements of these regimes more systematically, so as to identify the reasons for their weakness as well as elements of strength. The fast-growing body of research on civil society may yield useful perspectives. Targeted research in this area could contribute to strengthening ideational regimes.
- In showing how agents of structural power and advocacy coalitions work through weak institutional capacity in a developing country, we have noted options for, as well as limitations to, policy influence. In the Philippines, we saw that such agents virtually took over policy formulation – with success, as seen from their perspective. However, the policies produced under such circumstances do not seem sustainable in the long run – depending most likely on the continued influence of the agents in question. Thus, a central question is how developing countries can improve their institutional capacity to keep the relevant authorities in the drivers' seat and to balance needs and demands of different stakeholder groups with a view to developing policies on a democratic basis, policies that can be sustained in a longer-term perspective. This is a broad field of research where much work remains to be done.

Whereas these research challenges concern the effects of international regimes in developing countries more in general, it is particularly important to follow them up with regard to agrobiodiversity. The management of plant genetic resources for food and agriculture is facing enormous challenges. Further research is urgently needed to strengthen international governance for the maintenance of these resources so vital to global food security and the eradication of poverty.

Bibliography

Literature

Aiken, W. and LaFollette, H. (eds) (1996), *World Hunger and Morality*, 2nd Edition (Upper Saddle River, NJ: Prentice-Hall).

Aiken, W. and LaFollette, H. (eds) (1997), *World Hunger and Moral Obligation* (Englewood Cliffs, NJ: Prentice-Hall).

Almond, G.A. and Coleman, J.S. (eds) (1960), *The Politics of the Developing Areas* (Princeton, NJ: Princeton University Press).

Andersen, R. (2001), *Conceptualising the Convention on Biological Diversity: Why is it Difficult to Determine the 'Country of Origin' for Agricultural Plant Varieties?* FNI Report 7/2001 (Lysaker, Norway: Fridtjof Nansen Institute).

Andersen, R. (2002), 'The Time Dimension in International Regime Interplay', *Global Environmental Politics* 2:3, 98–117.

Andersen, R. (2003), 'FAO and the Management of Genetic Resources', pp. 43–53 in Stokke and Thommessen (eds).

Andersen, R. (2005a), *The History of Farmers' Rights: A Guide to Central Documents and Literature*, Background Study 1 from the Farmers' Rights Project, FNI Report 8/2005 (Lysaker, Norway: Fridtjof Nansen Institute).

Andersen, R. (2005b), *Results from an International Stakeholder Survey on Farmers' Rights*, Background Study 2 from the Farmers' Rights Project, FNI Report 9/2005 (Lysaker, Norway: Fridtjof Nansen Institute).

Andersen, R. (2005c), 'Historien om poteten Linda', *Nationen*, 11 August, 3.

Andersen, R. (2006), *Realising Farmers' Rights under the International Treaty on Plant Genetic Resources for Food and Agriculture*, Summary of Findings from the Farmers' Rights Project (Phase 1), FNI Report 11/2006 (Lysaker, Norway: Fridtjof Nansen Institute).

Andersen, R. (2007), *Governing Agrobiodiversity: International Regimes, Plant Genetics and Developing Countries.* Dissertation for the Degree of Dr. Polit. (Oslo: Department of Political Science, Faculty of Social Sciences, University of Oslo).

Andersen, R. and Tvedt, M.W. (2002), 'Historien om patent på gule bønner i USA: Kan man patentere en farge?' ('The Story about Patents on Yellow Beans in the USA: Is it Possible to Patent a Colour?'). *Gen-i-alt* 4, 12–13.

Andersen, S.S. (1997), *Case-studies og generaliseringer. Forskningsdesign og strategi* (Bergen-Sandvigen: Fagbokforlaget).

Aoki, K. (1998), 'Neocolonialism, Anticommons Property, and Biopiracy in the (Not-So-Brave) New World Order of Intellectual Property Protection', *Indiana Journal of Global Legal Studies* 6, 11–58.

Arts, B. and Verschuren, P. (1999), 'Assessing Political Influence in Complex Decision-making: An Instrument Based on Triangulation', *International Political Science Review* 2:4, 411–24.

Balakrishnan, R. and Narayan, U. (1996), 'Combining Justice with Development: Rethinking Rights and Responsibilities in the Context of World Hunger and Poverty', pp. 231–48 in Aiken and LaFollette (eds).

Ball, T. (2001), 'New Ethics for Old? Or, How (Not) to Think About Future Generations', *Environmental Politics* 10:1, 89–110.

Bankoff, G. and Weekley, K. (2002), *Post-Colonial National Identity in the Philippines. Celebrating the Centennial of Independence* (Aldershot: Ashgate).

Bao-Rong, L. (1998), *A Report On IRRI-Philrice Cooperative Collecting for Oryza meyeriana in Palawan, Philippines* (Los Baños: IRRI).

Barber, C.V. and La Viña, A.G.M. (1997), 'Regulating Access to Genetic Resources: The Philippines Experience', pp. 115–41 in Mugabe et al. (eds).

Batta Bjørnstad, S-I. (2004), *Breakthrough for 'the South'? An Analysis of the Recognition of Farmers' Rights in the International Treaty on Plant Genetic Resources for Food and Agriculture*, FNI Report 13/2004 (Lysaker: Fridtjof Nansen Institute).

Bello, W. (1987), *Creating the Third Force: U.S. Sponsored Low Intensity Conflict in the Philippines*, Food First Development Report Series (San Francisco, CA: Institute for Food and Development Policy).

Bello, W. (2003), *Multilateral Punishment – The Philippines in the WTO, 1995–2003* (Bangkok, Thailand: Stop the New Round Coalition, Focus on the Global South) <www.apl.org.ph/ps/multilateral-punishment.pdf> accessed 22 November 2007.

Benavidez II, P.J. (2004), 'Philippines: Evolving Access and Benefit Sharing Regulations', pp. 153–76 in Carrizosa et al. (eds).

Boroña, M.L.J. (2002a), 'Higher Yielding Hybrid Rice Sows the Seeds of Opportunity in RP', Special Report in *Manila Times*, 3 September.

Boroña, M.L.J. (2002b), 'Farmers Wary of Hybrid Rice Despite Hyped Advantages', Special Report in *Manila Times*, 4 September.

Bourdieu, P. (1986), 'The Forms of Capital', in Richardson (ed.).

Brush, S.B. (ed.) (2000), *Genes in the Field. On Farm Conservation of Crop Diversity* (Rome: International Plant Genetic Resources Institute; Ottawa: International Development Research Center; Boca Raton, FL: Lewis).

Brush, S.B. (2004), *Farmers' Bounty: Locating Crop Diversity in the Contemporary World* (New Haven, CT & London: Yale University Press).

Brush, S.B. (2005), 'Protecting Traditional Agricultural Knowledge', *Washington University Journal of Law and Policy* 17, 59–109.

Cabanilla, L.S. and Paunlagui, M.M. (eds) (1999), *Food Security in the Philippines* (Manila: Institute of Strategic Planning and Policy Studies in cooperation with UP Center for Integrative and Development Studies).

de Candolle, A. (1882), *L'origine des plantes cultivées*, Paris. English version (1959), Origin of Cultivated Plants (New York: Hafner).

Carew-Reid, J. et al. (1994), *Strategies for National Sustainable Development: A Handbook for Their Planning and Implementation* (London: Earthscan).

Carew-Reid, J. (ed.) (2002), *Biodiversity Planning in Asia: A Review of National Biodiversity Strategies and Action Plans (NBSAPs)* (Cambridge: IUCN).

Carrizosa, S. et al. (eds) (2004), *Accessing Biodiversity and Sharing the Benefits: Lessons from Implementing the Convention on Biological Diversity*, IUCN Environmental Policy and Law Paper No. 54 (Gland, Switzerland and Cambridge, UK: IUCN).

Castro, N.T. (2000), 'Three Years of the Indigenous Peoples Rights Act: Its Impact on Indigenous Communities', *KASARINLAN – A Philippine Quarterly of Third World Studies* 15:2, 35–54.

Chapman, W. (1987), *Inside the Philippine Revolution: The New People's Army and Its Struggle for Power* (New York: W.W. Norton).

Checkel, J.T. (1997), 'International Norms and Domestic Politics: Bridging the Rationalist–Constructivist Divide', *European Journal of International Relations* 3:4, 473–95.

Checkel, J.T. (1999), *Why Comply? Constructivism, Social Norms and the Study of International Institutions*, ARENA Working Papers 99/24 (Oslo: ARENA).

Checkel, J.T. (2001), 'Why Comply? Social Learning and European Identity Change', *International Organization* 55:3, 553–88.

CIPR (Commission on Intellectual Property Rights) (2002), *Integrating Intellectual Property Rights and Development Policy* (London: CIPR).

Cohen, J.L. and Arato, A. (1992), *Civil Society and Political Theory* (Cambridge, MA: MIT Press).

Coleman, J.S. (1990), *Foundations of Social Theory* (Boston, MA: Harvard University Press).

Constantino, R. (1974), *Identity and Consciousness: The Philippine Experience* (Quezon City: Renato Constantino). First presented at the World Sociology Congress, 20 August 1974, Toronto, Canada.

Constantino, R. (1982), *The Miseducation of the Filipino* (Quezon City: Foundation for Nationalist Studies). First published in *Weekly Graphic*, 8 June 1966.

Cooper, H.D. et al. (eds) (2001), *Broadening the Genetic Bases of Crop Production* (New York: CABI, in collaboration with FAO and IPGRI).

Correa, C.M. (1998), *Implementing the TRIPs Agreement. General Context and Implications for Developing Countries* (Penang: Third World Network).

Correa, C. (2003), *Establishing a Disclosure of Origin Obligation in the TRIPS Agreement* (Geneva: Quaker United Nations Office).

Correa, C.M. (2004), 'The WIPO Draft Substantive Patent Law Treaty: A Review of Selected Provisions', *Trade-Related Agenda, Development and Equity Working Papers 17* <www.southcentre.org/publications/publist_category_WorkingPapers_index.htm>, accessed 23 November 2007.

Correa, C.M. and Yusuf, A.A. (eds) (1998), *Intellectual Property and International Trade. The TRIPS Agreement* (Dordrecht and London: Kluwer Law International).

Cramb, R.A. et al. (2000), 'Conservation Farming Projects in the Philippine Uplands: Rhetoric and Reality', *World Development* 28:5, 911–27.

Crocker, D.A. and Linden, T. (eds) (1999), *Ethics of Consumption* (Totowa, NJ: Rowman and Littlefield).

Crucible II Group (2000), *Seedling Solutions: Volume 1. Policy Options for Genetic Resources (*People, Plants, and Patents *Revisited)* (Ottawa, Rome, Uppsala: International Development Research Centre, International Plant Genetic Resources Institute and Dag Hammarskjöld Foundation).

Dahrendorf, R. (1964), *Homo Sociologicus. Ein Versuch zur Geschichte, Bedeutung und Kritik der Kategorie der Sozialen Rolle* (Köln, Opladen: Westdeutscher Verlag).

Dar, W.D. (1997), *The Philippine National Agricultural Research System. A Catalyst for Economic Growth and Sustainable Development* (Los Baños, Laguna: Philippine Council for Agriculture, Forestry and Natural Resources Research and Development, Department of Science and Technology).

Daroy, P.B. (1988), 'On the Eve of Dictatorship and Revolution', pp. 1–25 in Javate-de Dios et al. (eds).

Darwin, C. (1859), *On the Origin of Species* (Orig. edition London: John Murray; various reprintings and re-issues).

Darwin, C. (1868/1998), *The Variation of Animals and Plants under Domestication*, Vols 1 and 2 (Baltimore, MD: Johns Hopkins University Press).

Dawkins, R. (1982), *The Extended Phenotype: The Long Reach of the Gene* (Oxford: Oxford University Press).

De Klemm, C. (1982), 'Conservation of Species: The Need for a New Approach', *Environmental Policy and Law* 9:4, 118–28.

DENR (Department of Environment and Natural Resources, Philippines) (1988), Environment Management Bureau: National Workshop on the Formulation of the Philippine Strategy for Sustainable Development, 23–24 May 1988 (Workshop paper with draft strategy).

DENR and UNEP (United Nations Environment Program) (1997), *Philippine Biodiversity: An Assessment and Action Plan* (Makati City, Metro Manila: Bookmark).

Department of Agriculture (1995), *Philippines: Country Report to the FAO International Technical Conference on Plant Genetic Resources (Leipzig, 1996)* (Quezon City: Department of Agriculture of the Philippines).

Department of Agriculture (1999), *The Empowerment of The Filipino Farmer. Kasaganaan ng Sakahan at Kalikasan. The Philippine National IPM Program* (Quezon City: Department of Agriculture of the Philippines).

Diamond, J. (1999), *Guns, Germs, and Steel: The Fates of Human Societies* (New York: Norton & Company).

Diani, M. (2001), 'Social Capital as Social Movement Outcome', pp. 207–20 in Edwards et al.

Diokno, M.S.I. (1988), 'Unity and Struggle', pp. 132–175 in Javate-de Dios et al. (eds).

Doremus, P.N. (1996), 'The Externalization of Domestic Regulation: Intellectual Property Rights Reform in a Global Era', *Global Legal Studies Journal* 3:2.

Doronila, A. (ed.) (2001), *Between Fires: Fifteen Perspectives on the Estrada Crisis* (Metro Manila: Anvil Publishing and Philippine Daily Inquirer).

Dusik, R. (1986), *Philippinen Reise-Handbuch* (Köln: DuMont Buchverlag).

Dutfield, G. (2000), *Intellectual Property Rights, Trade and Biodiversity: Seed and Plant Varieties* (London: Earthscan).

Eckstein, H. (1975), 'Case Study and Theory in Political Science', in Greenstein and Polsby (eds).

Edwards, B. et al. (eds) (2001), *Beyond Tocqueville. Civil Society and the Social Capital Debate in Comparative Perspective* (Hanover, NH: University Press of New England).

Edwards, M. (2004), *Civil Society* (Cambridge: Polity Press).

Esquinas-Alcázar, J.T. (1993), 'Plant Genetic Resources', pp. 33–51 in Hayward et al. (eds).

Evans, P. and Walsh, J. (1994), *The EIU Guide to the New GATT* (London: The Economist Intelligence Unit).

Evenson, R. et al. (eds) (1998), *Agricultural Values of Plant Genetic Resources* (Wallingford: FAO/CEIS/CABI Publishing).

Federal Negarit Gazeta of the Federal Democratic Republic of Ethiopia (2006), Access to Genetic Resources and Community Knowledge, and Community Rights Proclamation (Proclamation No. 482/2006), 13:13, 27 February, 3353–73.

Ferrari, M.F. (2002), 'Philippines: Indigenous Peoples and the Convention on Biological Diversity', *World Rainforest Movement Bulletin* 62, <www.wrm.org.uy/bulletin/62/Philippines.html>.

Finnemore, M. (1996), *National Interests in International Society* (Ithaca, NY: Cornell University Press).

Finnemore, M. and Sikkink, K. (1998), 'International Norm Dynamics and Political Change', *International Organization* 52:4, 887–917.

Fowler, C. (1994), *Unnatural Selection: Technology, Politics, and Plant Evolution* (Yverdon, Switzerland: Gordon and Breach).

Fowler, C. (2001), 'Protecting Farmer Innovation: The Convention on Biological Diversity and the Question of Origin', *Jurimetrics* 41, 477–88.

Fowler, C. (2004), 'Assessing Genetic Resources: International Law Establishes Multilateral System', *Genetic Resources and Crop Evolution* 51, 609–20.

Fowler, C. and Hodgkin, T. (2004), 'Plant Genetic Resources for Food and Agriculture: Assessing Global Availability', *Annual Review of Environment and Resources* 29, 143–79.

Fowler, C. and Mooney, P. (1990), *Shattering: Food, Politics and the Loss of Genetic Diversity* (Tucson AZ: University of Arizona Press).

Fowler, C., Smale, M. and Gaiji, S. (2001), 'Unequal Exchange? Recent Transfers of Agricultural Resources and their Implications for Developing Countries', *Development Policy Review* 19:2, 181–204.

Frankfort-Nachmias, C. and Nachmias, D. (1992), *Research Methods in the Social Sciences*, 4th Edition (London: Edward Arnold).

Funston, J. (ed.) (2001), *Government and Politics in Southeast Asia* (London: Zed Books).

Galtung, J. (1971), 'A Structural Theory of Imperialism', *Journal of Peace Research* 8:2, 81–117.

Garcia, E. (1988), *The Filipino Quest: A Just and Lasting Peace* (Quezon City: Claretian Publications).

Gehring, T. and Oberthür, S. (2004), 'Exploring Regime Interaction: A Framework for Analysis', in Underdal and Young (eds).

George, A.L. and McKeown, T.J. (1985), 'Case Studies and Theories of Organizational Decision Making', *Advances in Information Processing in Organizations* 2, 21–58.

Giellespie, A. (2002), 'Forum Shopping in International Environmental Law: The IWC, CITES and the Management of Cetaceans', *Ocean Development and International Law* 33, 17–56.

Global Crop Diversity Trust (2005), *What is the Trust?* (Rome: The Global Crop Diversity Trust), <www.startwithaseed.org/items/trust.php>.

Golez, H. (2004), 'Philippine Plant Variety Protection Act of 2002 and its IRR, Status of Implementation' pp. 20–23 in *Proceedings of the Seminar-Workshop on Intellectual Property in Agriculture and Fisheries Sector*, 5–6 February 2004 (Science City of Muñoz, Nueva Ecija: PhilRice).

Gonzales, J.L. III (2001), 'Philippines. Continuing People Power', pp. 252–90 in Funston (ed.).

Goodin, R.E. (1985), *Protecting the Vulnerable: A Reanalysis of Our Social Responsibilities* (Chicago, IL: University of Chicago Press).

GRAIN (1997), 'Towards Our "Sui Generis" Rights', *Seedling, the Quarterly Newsletter of Genetic Resources Action International* 14:4.

GRAIN (1998), *Signposts to 'Sui Generis' Rights: Resource Materials from the International Seminar on 'Sui Generis' Rights*, Bangkok, 1–6 December 1997 (Bangkok, Thailand and Los Baños, Philippines: BIOTHAI and GRAIN), <www.grain.org/publications/signposts-en.cfm>.

GRAIN (2000), 'For a full review of TRIPS 27.3(b). An Update on Where Developing Countries Stand with the Push to Patent Life at WTO', published online at <www.grain.org/publications/reports/TRIPSfeb00.htm>.

GRAIN (2004), Biodiversity Rights Legislation (Los Baños, Philippines: GRAIN), website available at <www.grain.org/brl/>.

GRAIN (2005), 'Bilateral Agreements Imposing TRIPS-Plus Intellectual Property Rights on Biodiversity in Developing Countries', available online at <www.grain.org/rights/tripsplus.cfm?id=68>.

GRAIN and the GAIA Foundation (1998), 'Ten Reasons Not to Join UPOV', *Global Trade and Biodiversity in Conflict* 2:May, 1–10.

Greenstein, F.I. and Polsby, N.W. (1975), *Strategies of Inquiry. Handbook of Political Science,* Vol. 7 (Reading: Addison-Wesley).

Gupta, J. (1997), *The Climate Change Convention and Developing Countries. From Conflict to Consensus?* (Dordrecht: Kluwer Academic Publishers).

Haas, P.M. (ed.) (1992), 'Knowledge, Power and International Policy Coordination', *International Organization*, Special Issue 46:1.

Haas, P.M. et al. (1993), *Institutions for the Earth* (Cambridge: MIT).

Hanf, K. and Underdal, A. (1998), 'Domesticating International Commitments: Linking National and International Decision-Making', pp. 149–70 in Underdal (ed.).

Hardin, G. (1968), 'The Tragedy of the Commons', *Science* 162, 1243–8.

Harlan, J.R. (1975), 'Geographic Patterns of Variation in Some Cultivated Plants', *Journal of Heredity* 66, 182–91.

Harlan, J.R. (1992), *Crops and Man*, 2nd Edition (Madison, WI: American Society of Agronomy/Crop Science Society of America).

Harlan, J.R. (1995), *The Living Fields: Our Agricultural Heritage* (Cambridge: Cambridge University Press).

Hasenclever, A. et al. (1996), 'Interests, Power, Knowledge: The Study of International Regimes', *Mershon International Studies Review*, supplement to *International Studies Quarterly* 40:Suppl. 2, 177–228.

Hasenclever, A. et al. (1997), *Theories of International Regimes* (Cambridge: Cambridge University Press).

Hayward, M.D. et al. (eds) (1993), *Plant Breeding: Principles and Prospects* (London: Chapman & Hall).

Heclo, H. (1974), *Social Policy in Britain and Sweden* (New Haven, CT: Yale University Press).

Helfer, L.R. (1998), 'Adjudicating Copyright Claims Under the TRIPS Agreement: The Case for a European Human Rights Analogy', *Harvard International Law Journal*, 39:2, 357–98.

Helfer, L.R. (2002), 'Intellectual Property Rights in Plant Varieties: An Overview with Options for National Governments', FAO Legal Papers Online #31 July 2002 (Rome: FAO), available at <www.fao.org/Legal/prs-ol/years/2002/list02.htm>.

Helfer, L.R. (2004), 'Regime Shifting: The TRIPS Agreement and New Dynamics of International Intellectual Property Lawmaking', *Yale Journal of International Law* 29:1, 23–47.

Heller, M.A. (1998), 'The Tragedy of the Anticommons: Property in the Transition from Marx to Markets', *Harvard Law Review* 11, 621–88.

Heller, M.A. and Eisenberg, R.S. (1998), 'Can Patents Deter Innovation? Anticommons in Biomedical Research', *Science* 280:5364, 698–701.

Henecka, H.P. (1985), *Grundkurs Soziologie* (Opladen: Leske Verlag und Budrich).

Henne, G. (1998), *Genetische Vielfalt als Ressource. Die Regelung ihrer Nutzung.* (Baden-Baden: Nomos).

Héritier, A. (ed.) (1993), *Policy-Analyse. Kritik und Neuorientierung, Politisches Vierteljahresschrift*, Vol. 34, Sonderheft 24 (Opladen: Westdeutscher Verlag).

Hodgson, G.M. (1994), 'Cultural and Institutional Influences on Cognition', pp. 58–62 in Hodgson et al. (eds.).

Hodgson, G.M. et al. (eds.) (1994), *The Elgar Companion to Institutional and Evolutionary Economics* (Cheltenham, UK: Edward Elgar).

Höfer, H. (ed.) (1985), *Philippinen*, Apa Guides (Munich: Nelles).

Hornby, A.S. (1974), *Oxford Advanced Learner's Dictionary of Current English* (London: Oxford University Press).

IFAD (2001), *Rural Poverty Report 2001 – The Challenge of Ending Rural Poverty* (Rome: International Fund for Agricultural Development).

Ileto, R.C. (1979), *Pasyon and Revolution: Popular Movements in the Philippines, 1840–1910* (Quezon City: Ateneo de Manila University Press).

Jänicke, M. (1995), *The Political Systems Capacity for Environmental Policy.* FFU Report 1995/6 (Berlin: Forschungsstelle für Umweltpolitik, Freie Universität Berlin).

Javate-de Dios, A. et al. (eds) (1988), *Dictatorship and Revolution. Roots of People's Power* (Metro Manila: Conspectus).

Jenkins-Smith, H.C. and Sabatier, P.A. (1994), 'Evaluating the Advocacy Coalition Approach', *Journal of Public Policy* 14:2, 175–203.

Jenkins-Smith, H.C. and Sabatier, P.A. (eds) (1993), *Policy Change and Learning: An Advocacy Coalition Approach* (Boulder, CO: Westview Press).

Jocano, F.L. (1997), *Filipino Value System: A Cultural Definition. Anthropology of the Filipino People IV* (Metro Manila: PUNLAD Research House).

Johannsen, W. (1903), *Ueber Erblichkeit in Populationen und in reinen Linien. Ein Beitrag zur Beleuchtung schwebender Selektionsfragen* (Jena: Verlag Gustav Fischer).

ten Kate, K. and Laird, S.A. (1999), *The Commercial Use of Biodiversity: Access to Genetic Resources and Benefit-Sharing* (London: Earthscan).

Katzenstein, P.J. (1996), *The Culture of National Security: Norms and Identity in World Politics* (New York: Columbia University Press).

Kellman, M. and Tackaberry, R. (1997), *Tropical Environments: The Functioning and Management of Tropical Ecosystems* (London: Routledge).

Kellow, A. and Zito, A.R. (2002), 'Steering Through Complexity: EU Environmental Regulation in the International Context', *Political Studies* 50:1, 43–60.

Kent, L. and Bash, K. (2000), 'DAI a Top Banana in Securing Plant Breeders' Rights', *Developments – A Quarterly Newsletter of Development Alternatives, Inc.*, Spring, 9–10.

Keohane, R.O. (1989), *International Institutions and State Power* (Boulder, CO: Westview Press).

Keohane, R.O. (2000), 'Foreword', pp. ix–xviii in Lawton et al. (eds).

Keohane, R.O. and Levy, M.A. (eds) (1996), *Institutions for Environmental Aid. Pitfalls and Promise* (Cambridge, MA: MIT Press).

Keystone Center (1991), *Oslo Plenary Session. Final Consensus Report of the Keystone International Dialogue Series on Plant Genetic Resources: Global Initiative for the Security and Sustainable Use of Plant Genetic Resources*. Third Plenary Session, 31 May – 4 June 1991, Oslo, Norway (Keystone, CO: Keystone Center).

King, G. et al. (1994), *Designing Social Inquiry: Scientific Inference in Qualitative Research* (Princeton, NJ: Princeton University Press).

Kloppenburg, J.R. Jr. (1988), *First the Seed: The Political Economy of Plant Biotechnology 1492–2000* (Cambridge: Cambridge University Press).

Knoepfel, P. (1993), *Bedingungen einer wirksamen Umsetzung umweltpolitischer Programme. Erfahrungen aus Westeuropäischen Staaten*, Cahiers de l''IDHEAP no. 108 (Lausanne: Institut des Hautes Etudes en Administration Publique).

Kotte, H. (1988), *Das Parlament der Straße. Berichte von den Philippinen 1983–1987* (Frankfurt am Main: Fischer Taschenbuch Verlag).

Krasner, S. (1982), 'Structural Causes and Regime Consequences: Regimes as Intervening Variables', *International Organization* 36:2, 185–205.

Kuyek, D. (2000), 'Hybrid Rice in Asia: An Unfolding Threat', paper published jointly by Biothai; GRAIN; KMP; MASIPAG; PAN; Philippine Greens; and UBING). Available online at < http://www.grain.org/briefings/?id=136> accessed 26 November 2007.

La Viña, A.G.M. et al. (eds) (1997), *Regulating Access to Biological and Genetic Resources in the Philippines: A Manual on the Implementation of the Executive Order No. 247* (Quezon City: Foundation for the Philippine Environment and World Resources Institute).

Lawton, T. et al. (eds) (2000), *Strange Power. Shaping the Parameters of International Relations and International Political Economy* (Aldershot: Ashgate).

LePrestre, P.G. (ed.) (2002), *Governing Global Biodiversity. The Evolution and Implementation of the Convention on Biological Diversity* (Aldershot: Ashgate).

Leskien, D. and Flitner, M. (1997), *Intellectual Property Rights and Plant Genetic Resources: Options for a Sui Generis System*, No. 6, Issues in Genetic Resources (Rome: International Plant Genetic Resources Institute).

Liebig, K. et al. (2002), *Governing Biodiversity. Access to Genetic Resources and Approaches to Obtaining Benefits from their Use: the Case of the Philippines*, Reports and Working Papers 5/2002 (Bonn: German Development Institute).

Mann, E. (ed.) (1963), *Thomas Mann Briefe 1937–1947* (Frankfurt am Main: Fischer).

MASIPAG et al. (1998), *Biopiracy, TRIPS and the Patenting of Asia's Rice Bowl – A Collective NGO Situationer on IPRs on Rice*, 2. (Published by the authoring organizations: MASIPAG, GRAIN, BIOTHAI and PAN-Indonesia).

Mayr, E. (1982), *The Growth of Biological Thought: Diversity, Evolution, and Inheritance* (Cambridge, MA: Belknap Press of Harvard University Press).

Mayr, E. (1997), *This is Biology: The Science of the Living World* (Cambridge, MA: Belknap Press of Harvard University Press).

Mayuga, S. (1985), 'Aus Wasser und Feuer Geboren. Entstehung und Frühgeschichte', pp. 19–24 in Höfer (ed.).

McGraw, D.M. (2002), 'The Story of the Biodiversity Convention: From Negotiation to Implementation', in LePrestre (ed.).

Meniado, A.P. et al. (2002), 'The Philippines', pp. 215–36 in Carew-Reid (ed.).

Miles, E.L. et al. (2002), *Environmental Regime Effectiveness – Confronting Theory with Evidence* (Cambridge: MIT).

Miller, K.R. and Lanou, S.M. (1995), *National Biodiversity Planning: Guidelines Based on Early Experiences Around the World* (Washington DC: World Resources Institute, UNEP and IUCN).

Mitchell, M. (2001), 'Revenge of the Estrada Cronies', *Far Eastern Economic Review*, 1 March, 26–7.

von Moltke, K. (1997), 'Institutional Interactions: The Structure of Regimes for Trade and the Environment', pp. 247–72 in Young (ed.).

Mooney, P.R. (1979), *Seeds of the Earth: Private or Public Resources?* (Ottawa: Canadian Council for International Co-operation and the International Coalition for Development Action).

Moore, G. and Tymowski, W. (2005), *Explanatory Guide to the International Treaty on Plant Genetic Resources for Food and Agriculture*, IUCN Environmental Policy and Law Paper No. 57 (Gland, Switzerland and Cambridge, UK: IUCN).

Mouffe, C. (ed.) (1992), *Dimensions of Radical Democracy: Pluralism, Citizenship and Community* (New York: Verso).

Mugabe, J. et al. (eds) (1997), *Access to Genetic Resources: Strategies for Sharing Benefits* (Nairobi, Kenya: ACTS Press).

Musungu, S.F. and Dutfield, G. (2003), *Multilateral Agreements and a TRIPS-plus World: The World Intellectual Property Organization (WIPO)*. TRIPS Issues Papers 3 (Geneva: Quaker United Nations Office; Ottawa: Quaker International Affairs Program).

Mytelka, L.K. (2000), 'Knowledge and Structural Power in the International Political Economy', in Lawton et al. (eds).

Nagel, S.S. (ed.) (2002), *Environmental Policy and Developing Nations* (Jefferson, NC: McFarland).

Narlikar, A. and Wilkinson, R. (2004), 'Collapse at the WTO: a Cancun Post-mortem', *Third World Quarterly* 25:3, 447–60.

Naudón, C. (1970), *Amérika '70. Servidumbre o independencia en la presente década?* (Santiago de Chile: Ediciones Nueva Universidad).

NEDA (National Economic and Development Authority of the Republic of the Philippines) (1995), *Social Development in the Philippines: Vision, Challenges and Imperatives* (Manila: Philippine Information Agency, Office of the Press Secretary).

Negre, E.R. (2004), 'What is a Patent?' pp. 27–30 in *Proceedings of the Seminar-Workshop on Intellectual Property in Agriculture and Fisheries Sector*, 5–6 February 2004 (Science City of Muñoz, Nueva Ecija: PhilRice).

Nemenzo, F. (1988), 'From Autocracy to Elite Democracy', pp. 221–68 in Javate-de Dios et al. (eds).

Oberthür, S. and Gehring, T. (eds) (2006), *Institutional Interaction: Enhancing Cooperation and Preventing Conflicts between International and European Environmental Institutions* (Cambridge, MA: MIT Press).

Odum, E.P. (1997), *Ecology: A Bridge Between Science and Society* (Sunderland, MA: Sinauer Associates).

O'Neill, O. (1993), 'Ending World Hunger', pp. 235–79 in Regan (ed.).

Palacios, X.F. (1998), *Contribution to the Estimation of Countries' Interdependence in the Area of Plant Genetic Resources*. Background Study Paper No. 7, Rev. 1 (Rome: Commission on Genetic Resources for Food and Agriculture).

PAWB (1998a), *Implementing the Convention on Biological Diversity in the Philippines. The First National Report to the Convention on Biological Diversity* (condensed version) (Quezon City: Protected Areas and Wildlife Bureau, Department of Environment and Natural Resources).

PAWB (1998b), *The First National Report to the Convention on Biological Diversity* (Quezon City: Protected Areas and Wildlife Bureau, Department of Environment and Natural Resources).

PAWB (2000), Support to the Implementation of Executive Order No. 247: 'Bioprospecting and its Implementing Rules and Regulations', at <www.pawb.gov.ph/programs/biopros.htm>.

PAWB (2002), *The Second National Report to the Convention on Biological Diversity* (Quezon City: Protected Areas and Wildlife Bureau, Department of Environment and Natural Resources).

PCSD (1997), *Philippine Agenda 21: A National Agenda for Sustainable Development* (Pasig City, Metro Manila: Philippine Council for Sustainable Development).

PCSD Coordinating Secretariat (1997), *Rio in Retrospect: The Philippines and Global Agenda 21 1992–1996* (Pasig City, Metro Manila: Philippine Council for Sustainable Development).

PCSD (1998), *Implementing Philippine Agenda 21. A Continuing Journey Towards Sustainable Development* (Pasig City, Metro Manila: Philippine Council for Sustainable Development).

PCSD (1999a), *SIAD Guidebook – A Framework for the Localization of Philippine Agenda 21* (Quezon City: Center for Alternative Development Initiatives).

PCSD (1999b), *Sustainable Development in the Philippines: Up, Down and Forward.* (Pasig City, Metro Manila: Philippine Council for Sustainable Development).

PCSD (2000), *Meeting the Challenges of Sustainable Development in the New Millennium* (Pasig City, Metro Manila: Philippine Council for Sustainable Development).

Peng, M.K.K. (1990), 'Transnational Corporations: The Interests Behind GATT', *The Ecologist* 20:6.

Perlas, N. (2000), *Shaping Globalization: Civil Society, Cultural Power and Threefolding* (Quezon City: Center for Alternative Development Initiatives).

Perlas, N. (2001), 'NEDA Blocks Sustainable Development', *Philippine Star* (9 November).

Philippine Bureau of Agricultural Research (2001), *Strategic Plan 2001–2005* (Quezon City: Bureau of Agricultural Research, Department of Agriculture).

PhilRice (1999), *A Profile of One of the Country's Premier Research and Development Organizations PhilRice. A Government-Owned and Controlled Corporation Attached to the Department of Agriculture* (Muñoz, Nueva Ecija: PhilRice).

Pistorius, R. (1997), *Scientists, Plants and Politics. A History of the Plant Genetic Resources Movement* (Rome: International Plant Genetic Resources Institute).

Pogge, T.W. (1999), 'A Global Resources Dividend', pp. 501–38 in Crocker and Linden (eds).

Pogge, T.W. (2001), 'Priorities of Global Justice', *Metaphilosophy* 32:1/2, 6–24.

Powell, W.W. and DiMaggio, P.J. (1991), *The New Institutionalism in Organizational Analysis* (Chicago, IL: University of Chicago Press).

Princen, T. and Finger, M. (1994), *Environmental NGOs in World Politics: Linking the Local and the Global* (London: Routledge).

Putnam, R.D. (1995), 'Bowling Alone: America's Declining Social Capital', *Journal of Democracy* 6:1, 65–78.

Putnam, R.D. et al. (1993), *Making Democracy Work: Civic Traditions in Modern Italy* (Princeton, NJ: Princeton University Press).

Ragin, C.C. and Becker, H.S. (eds) (1992), *What is a Case? Exploring the Foundations of Social Inquiry* (Cambridge: Cambridge University Press).

Ramanna, A. (2003), 'India's Plant Variety and Farmers' Rights Legislation: Potential Impact on Stakeholder Access to Genetic Resources', Environment and Production Technology Division Discussion Paper No. 96 (Washington, DC: International Food Policy Research Institute).

Ramanna, A. (2006), *Farmers' Rights in India – A Case Study, Background Study 4 from the Farmers' Rights Project*, FNI Report 6/2006 (Lysaker, Norway: Fridtjof Nansen Institute).

Ramanna, A. and Smale, M. (2004), 'Rights and Access to Plant Genetic Resources under India's New Law', *Development Policy Review* 22:4, 423–42.

Raustiala, K. and Victor, D.G. (1996), 'Biodiversity since Rio: The Future of the Convention on Biological Diversity', *Environment* 38:4, 16–20; 37–45.

Raustiala, K. and Victor, D.G. (2004), 'The Regime Complex for Plant Genetic Resources', *International Organization* 58:2, 277–309.

Regan, T. (ed.) (1993), *Matters of Life and Death: New Introductory Essays in Moral Philosophy*, 3rd Edition (New York: McGraw-Hill).

Regassa F. (2006), *Farmers' Rights in Ethiopia – A Case Study, Background Study 5 from the Farmers' Rights Project*, FNI Report 7/2006 (Lysaker, Norway: Fridtjof Nansen Institute).

Richardson, J.G. (ed.) (1986), *Handbook of Theory and Research for the Sociology of Education* (New York: Greenwood Press).

Risse, T. (2000), '"Let's Argue!" Communicative Action in World Politics', *International Organization* 54:1, 1–39.

Risse, T., Ropp, S.C. and Sikkink, K. (eds) (1999), *The Power of Human Rights: International Norms and Domestic Change* (Cambridge: Cambridge University Press).

Robinson, R.A. (1996), *Return to Resistance: Breeding Crops to Reduce Pesticide Dependence* (Davis, CA: agAccess, in association with the International Development Research Centre, Ottawa).

Rosario-Malonzo, J. (2004), 'Philippines: USAID's AGILE: Toying with Governance to Globalize the Philippines?' in *The Reality of Aid* (The Reality of Aid Project). <www.realityofaid.org/roareport.php?table=roa2004&id=65>.

Rosendal, G.K. (1999), *Implementing International Environmental Agreements in Developing Countries: The Creation and Impact of the Convention on Biological Diversity,* PhD Dissertation (Oslo: Faculty of Social Sciences, University of Oslo).

Rosendal, G.K. (2000), *The Convention on Biological Diversity and Developing Countries* (Dordrecht: Kluwer Academic).

Rosendal, G.K. (2001), 'Impacts of Overlapping International Regimes: The Case of Biodiversity', *Global Governance* 7, 95–117.

Rosendal, G.K. (2006) 'The Convention on Biological Diversity: Tensions with the WTO TRIPS Agreement over Access to Genetic Resources and the Sharing of Benefits', pp. 79–103 in Oberthür and Gehring (eds).

Rosenfield, I. and Ziff, E. (1983), *DNA for Beginners* (London: Writers and Readers Publishing Cooperative).

Rothwell, D.R. (2004), 'Building on the Strengths and Addressing the Challenges: The Role of Law of the Sea Institutions', *Ocean Development and International Law* 35:2, 131–56.

Rucht, D. (1992), 'Studying the Effects of Social Movements: Conceptualisations and Problems', paper for the European Consortium for Political Research Joint Sessions, Limerick, Ireland.

Rudolph, S.H. (2000), 'Civil Society and the Realm of Freedom', *Economic and Political Weekly* 13 May.

Ruiz Muller, M. (2006), *Farmers' Rights in Peru – A Case Study, Background Study 3 from the Farmers' Rights Project*, FNI Report 5/2006 (Lysaker, Norway: Fridtjof Nansen Institute).

Sabatier, P.A. (1993), 'Advocacy-Koalitionen, Policy-Wandel und Policy-Learnen: Eine Alternative zur Phasenheuristik', *Politische Vierteljahresschrift*, Sonderheft 24/1993: Policy-Analyse: Kritik und Neuorientierung, 116–148.

Salazar, R. et al. (2007), 'Protecting Farmers' New Varieties: New Approaches to Rights on Collective Innovations in Plant Genetic Resources', *World Development* 35:9, 1515–1528.

Sauvé, R. and Watts, J. (2003), 'An Analysis of IPGRI's Influence on the International Treaty on Plant Genetic Resources for Food and Agriculture', *Agricultural Systems* 78, 307–27.

Schirmer, D.B. and Shalom, S.R. (eds) (1987), *The Philippines Reader: A History of Colonialism, Neocolonialism, Dictatorship and Resistance* (Quezon City: KEN Inc.).

Secretariat of the Convention on Biological Diversity (2001), *Handbook of the Convention on Biological Diversity* (London: Earthscan).

Senghaas, D. (ed.) (1972), *Imperialismus und strukturelle Gewalt. Analysen über abhängige Reproduktion* (Frankfurt aM: Suhrkamp).

Senghaas, D. (ed.) (1974), *Peripherer Kapitalismus. Analysen über Abhängigkeit und Unterentwicklung* (Frankfurt aM: Suhrkamp).

Sheehan, D. (2001), 'More Power to the Powerful', *Far Eastern Economic Review*, 1 February, 16–20.

Shue, H. (1996), 'Solidarity among Strangers and the Right to Food', pp. 113–22 in Aiken and LaFollette (eds).

Singer, P. (1972), 'Famine, Affluence, and Morality', *Philosophy and Public Affairs* 1:1, 229–43.

Skjærseth, J.B. (1999), 'Designing Qualitative Case Studies in Cross-National Research: Principal Problems and Possible Solutions'. Trial lecture for the degree of Dr. Polit., University of Oslo; mimeo at the Fridtjof Nansen Institute, Lysaker, Norway.

de Soto, H. (2000), *The Mystery of Capital: Why Capitalism Triumphs in the West and Fails Everywhere Else* (London: Black Swan).

Sprinz, D.F. et al. (2004), 'Separating and Aggregating Regime Effects'. Paper presented at the 45th Annual Convention of the International Studies Association, 17–20 March 2004, Montreal.

Steinberg, P.F. (2001), *Environmental Leadership in Developing Countries* (Cambridge, MA: MIT Press).

Stokke, O.S. (2000), 'Managing Straddling Stocks: The Interplay of Global and Regional Regimes', *Ocean and Coastal Management* 43, 205–34.

Stokke, O.S. (2001a), *The Interplay of International Regimes: Putting Effectiveness Theory to Work*, FNI Report 14/2001 (Lysaker: Fridtjof Nansen Institute).

Stokke, O.S. (ed.) (2001b), *Governing High Seas Fisheries: The Interplay of Global and Regional Regimes* (Oxford: Oxford University Press).

Stokke, O.S. and Thommessen, Ø.B. (eds) (2003), *Yearbook of International Co-operation on Environment and Development 2003/2004* (London: Earthscan).

Strang, D. and Chang, P.M.Y. (1993), 'The International Labor Organization and the Welfare State: Institutional Effects on National Welfare Spending, 1960–80', *International Organization* 47:2, 235–62.

Strang, D. and Meyer, J.W. (1993), 'Institutional Conditions for Diffusion', *Theory and Society* 22, 487–511.

Strange, S. (1988), *States and Markets* (London: Pinter).

Strange, S. (1996), *The Retreat of the State: The Diffusion of Power in the World Economy* (Cambridge: Cambridge University Press).

Strange, S. (1998), *Mad Money: When Markets Outgrow States* (Ann Arbor, MI: University of Michigan Press).

Svarstad, H. and Dhillion, S. (eds) (2000), *Responding to Bioprospecting. From Biodiversity in the South to Medicines in the North* (Oslo: Spartacus).

Swanson, T. (1999), 'Why is there a Biodiversity Convention? The International Interest in Centralized Development Planning', *International Affairs* 75:2, 307–31.

Swanson, T. and Johnston, S. (1999), *Global Environmental Problems and International Agreements* (Cheltenham, UK: Edward Elgar).

Swiderska, K. et al. (2001), *Participation in the Formulation of the Philippines Executive Order No. 247 On Access to Genetic Resources* (London: IIED).

Teehankee, J.C. (2001), 'Emerging Dynasties in the Post-Marcos House of Representatives', *Philippine Political Science Journal* 22:45, 55–78.

Tiglao, R.D. (1988), 'The Consolidation of the Dictatorship', pp. 26–69 in Javate-de Dios et al. (eds).

Tvedt, M.W. (2005), 'How Will a Substantive Patent Law Treaty Affect the Public Domain for Genetic Resources and Biological Material?', *The Journal of World Intellectual Property* 8:3, 311–74.

Tvedt, M.W. (2007), 'The Path to One Universal Patent', *Environmental Policy and Law* 37:4, 297–305.

Underdal, A. (1997), 'Modelling the International Climate Change Negotiations: A Non-Technical Outline of Model Architecture', CICERO Working Papers 1997:8 (Oslo: CICERO).

Underdal, A. (1998), 'Explaining Compliance and Defection: Three Models', *European Journal of International Relations* 4:1, 5–30.

Underdal, A. (ed.) (1998), *The Politics of International Environmental Management* (Dordrecht: Kluwer Academic Publishers).

Underdal, A. and Young, O.R. (eds) (2004), *Regime Consequences: Methodological Challenges and Research Strategies* (Dordrecht: Kluwer Academic).

UNDP (2004), *Human Development Report 2004* (New York: Oxford University Press).

Uphoff, N. (1998): 'Understanding social capital: learning from the analysis and experience of participation'. In I. Seragelding and P. Dasgupta (eds): *Social Capital* (Oxford: Oxford University Press).

Van Wijk, J. et al. (1993), *Intellectual Property Rights for Agricultural Biotechnology: Options and Implications for Developing Countries*, ISNAR Research Report No. 3 (The Hague: International Centre for National Agricultural Research).

Vavilov, N.I. (1949/50), 'The Origin, Variation, Immunity, and Breeding of Cultivated Plants', *Chronica Botanica* 13, 1–366.

Vavilov, N.I. (1951), 'The Origin, Variation, Immunity and Breeding of Cultivated Plants', *Chronica Botanica* 13:1/6 (with selected writings of Vavilov).

Vavilov, N.I. (1992), *Origin and Geography of Cultivated Plants.* Articles and lectures of Vavilov 1924–1940, first collected and published in book form in Russian in 1987 (Cambridge: Cambridge University Press).

Vavilov, N.I. (1997), *Five Continents.* Descriptions by Vavilov of his expeditions between 1916 and 1940, from manuscripts saved during the Second World War and the rest of the Stalin era. (St. Petersburg: N.I. Vavilov Research Institute of Plant Industry and Rome: International Plant Genetic Resources Institute).

Victor, D.G. et al. (eds) (1998), *The Implementation and Effectiveness of International Environmental Commitments. Theory and Practice* (Cambridge: MIT).

Vivas-Eugui, D. (2003), *Regional and Bilateral Agreements and a TRIPS-plus World: The Free Trade Area of the Americas (FTAA)*, TRIPS Issues Papers 1 (Geneva/Ottawa: Quaker United Nations Office, Quaker International Affairs Program and International Center for Trade and Sustainable Development).

Walzer, M. (1992), 'The Civil Society Argument', pp. 89–107 in Mouffe (ed.).

Watson, R.A. (1997), 'Reason and Morality in a World of Limited Food', pp. 115–23 in Aiken and LaFollette (eds).

Weissman, R. (1991), 'Prelude to a New Colonialism: The Real Purpose of GATT', *The Nation* 18 March, 336.

Wettestad, J. (1999), *Designing Effective Environmental Regimes. The Key Conditions* (Cheltenham, UK: Edward Elgar).

Wilson, E.O. (1992), *The Diversity of Life* (Cambridge, MA: Belknap Press of Harvard University Press).

Wolfrum, R. and Stoll, P-T. (1996), *Der Zngang zu genetischen Ressourcen nach dem Übereinkommen über die biologische Vielfalt und dem deutschen Recht.* Forschungsvorhaben des Umweltbundesamtes Nr. 101 01 002, Endbericht.

World Bank (2001), *Philippines Poverty Assessment*, Vol. I and II (Manila: World Bank Office Manila).

World Commission on Environment and Development (1987), *Our Common Future* (Oxford: Oxford University Press).

Yin, R.K. (1994), *Case Study Research: Design and Methods.* (Thousand Oaks, CA: SAGE).

Young, O.R. (1986), 'International Regimes. Towards a New Theory of Institutions', *World Politics* 39:1, 104–22.

Young, O.R. (1991), 'Political Leadership and Regime Formation: On the Development of Institutions in International Society', *International Organization* 45:3, 281–308.

Young, O.R. (1994), *International Governance – Protecting the Environment in a Stateless Society* (Ithaca, NY: Cornell University Press).

Young, O.R. (1996), 'Institutional Linkages in International Society: Polar Perspectives', *Global Governance* 2:1, 1–24.

Young, O.R. (ed.) (1997), *Global Governance: Drawing Insights from the Environmental Experience* (Cambridge: MIT).

Yusuf, A.A. (1998), 'TRIPS: Background, Principles and General Provisions', pp. 3–20 in Correa and Yusuf (eds).

Zaide, S.M. (1999, with supplement from 2001), *Philippine History and Government* (Quezon City: All-Nations Publ.).

Zeven, A.C. (1998), 'Landraces: A Review of Definitions and Classifications', *Euphytica* 104:2, 127–39.

Zhang, Z.X. (1998), 'Greenhouse Gas Emissions and the World Trading System', *Journal of World Trade* 32:5, 219–39.

FAO, IU and ITPGRFA References

FAO (1983), Conference Resolution 8/83, International Undertaking on Plant Genetic Resources (IU), adopted at the Twenty-second Session of the FAO Conference, Rome, 23 November 1983 <www.fao.org/ag/cgrfa/iu.htm>.

FAO (1985a), CPGR/85/REP. Report of the First Session of the Commission on Plant Genetic Resources, 11–15 March 1985, Rome.

FAO (1985b), C 1985/REP. Report of the Conference of FAO, Twenty-third Session, 9–28 November 1985 <www.fao.org/docrep/x5562e/x5562e00.htm>.

FAO (1987a), CL 91/14. Report of the Second Session of the Commission on Plant Genetic Resources, 16–20 March 1987.

FAO (1987b), CL 91/REP. Report of the Council of FAO, Ninety-first Session, 17–26 June 1987.

FAO (1989a), FAO C 1989/REP. Conference resolution 4/89, Agreed Interpretation of the International Undertaking. Report of the Conference of FAO, Twenty-fifth Session, Rome, 11–29 November 1989, <ftp://ext-ftp.fao.org/ag/cgrfa/Res/C4-89E.pdf>.

FAO (1989b), FAO C 1989/REP. Conference resolution 5/89, Farmers' Rights. Report of the Conference of FAO, Twenty-fifth Session, Rome, 11–29 November 1989 <ftp://ext-ftp.fao.org/ag/cgrfa/Res/C5-89E.pdf>.

FAO (1991), FAO C 1991/REP. Conference resolution 3/91, (Untitled). Report of the Conference of FAO, Twenty-sixth Session, Rome, 9–27 November 1991 <ftp://ext-ftp.fao.org/ag/cgrfa/Res/C35-91E.pdf>.

FAO (1993), FAO Conference resolution 7/93, Revision of the International Undertaking of Plant Genetic Resources. Twenty-seventh Session of the FAO Conference, Rome, 6–24 November 1993 <ftp://ext-ftp.fao.org/ag/cgrfa/Res/C7-93E.pdf>.

FAO (1994), CPGR-Ex1/94/3. Revision of the International Undertaking: Mandate, Context, Background and Proposed Process. Commission on Plant Genetic Resources, First Extraordinary Session, Rome, 7–11 November 1994 <ftp://ext-ftp.fao.org/ag/cgrfa/Ex1/E1W3E.pdf>.

FAO (1995a), Conference Resolution 3/95. Twenty-eighth Session of the FAO Conference, Rome.

FAO (1995b), CL 110/REP, paragraphs 13–14. Council Resolution 1/110, Adoption of the Statutes of the Commission on Genetic Resources for Food and Agriculture, 2–3 November 1995.

FAO (1996a), Global Plan of Action for the Conservation and Sustainable Utilization of Plant Genetic Resources for Food and Agriculture, adopted in 1996 by the International Technical Conference on Plant Genetic Resources, Leipzig, 17–23 June 1996 <www.fao.org/ag/AGP/AGPS/GpaEN/gpatoc.htm>.

FAO (1996b), Leipzig Declaration on Conservation and Sustainable Utilization on Plant Genetic Resources for Food and Agriculture, adopted in 1996 by the International Technical Conference on Plant Genetic Resources, Leipzig, 17–23 June 1996 <www.fao.org/focus/e/96/06/more/declar-e.htm>.

FAO (1998), *State of the World's Plant Genetic Resources for Food and Agriculture* (Rome: Food and Agriculture Organization of the United Nations).

FAO (2001), Conference resolution 3/2001, ITPGRFA, International Treaty on Plant Genetic Resources for Food and Agriculture, adopted at the Thirty-first Session of the FAO Conference, Rome, 2–13 November 2001 <www.fao.org/ag/cgrfa/compliance.htm#text>.

FAO/CGRFA (2002a), CGRFA/MIC-1/02/6. Terms of Reference for the Expert Group on the Terms of the Standard Material Transfer Agreement. Document prepared for the First Meeting of the CGRFA acting as the Interim Committee for the International Treaty on Plant Genetic Resources for Food and Agriculture, 9–11 October 2002 <www.fao.org/ag/cgrfa/docsic1.htm> and <www.fao.org/ag/cgrfa/mic1/m1w6e.pdf>.

FAO/CGRFA (2002b), CGRFA-9/02/6. Country Progress Report on the Implementation of the Global Plan of Action for the Conservation and Sustainable Use of Plant Genetic Resources for Food and Agriculture. Report prepared for the Ninth Regular Session of the CGRFA, 14–18 October 2002 <www.fao.org/ag/cgrfa/docs9.htm>.

FAO (2007), IT/GB-2/07/Inf.11. Experience of the Centres of the Consultative Group on International Agricultural Research (CGIAR) with the Implementation of the Agreements with the Governing Body, with Particular Reference to the Standard Material Transfer Agreement. Item 10 of the Provisional Agenda, International Treaty on Plant Genetic Resources for Food and Agriculture, Second Session of the Governing Body, Rome, 29 October – 2 November 2007.

CBD and UNEP References

Secretariat to the Convention on Biological Diversity (2001), *Handbook of the Convention on Biological Diversity* (London: Earthscan). Includes the CBD and reports from all meetings of the Conference of the Parties and subsidiary bodies up to 2001.

UNEP (1987), Decision 14/26, Rationalization of International Conventions on Biological Diversity, 14th Governing Council Session <www.unep.org/Documents.Multilingual/Default.asp?DocumentID=100>.

UNEP (1989), Decision 15/34, Preparation of an International Legal Instrument on the Biological Diversity of the Planet. 15th Governing Council Session <www.unep.org/Documents.Multilingual/Default.asp?DocumentID=71>.

UNEP (1990), UNEP/GCSS.II/3. Report of the Governing Council on the Work of its Second Special Session, 1–3 August, Nairobi, Annex I.

UNEP (1991), Decision 16/42, Preparation of an International Legal Instrument on Biological Diversity, 16th Governing Council Session. <www.unep.org/Documents.Multilingual/Default.asp?DocumentID=55>.

UNEP (1992), Nairobi Final Act of the Conference for the Adoption of the Agreed Text of the Convention on Biological Diversity, 22 May <http://biodiv.org/doc/handbook/cbd-hb-09-en.pdf>.

UNEP (1997a), UNEP/CBD/COP/3/38. Report of the Third Meeting of the Conference of the Parties to the Convention on Biological Diversity, Buenos Aires, Argentina, 4–15 November 1996.

UNEP (1997b), UNEP/CBD/SBSTTA/3/INF/9. Compilation of National Contributions on Agricultural Biodiversity.

UNEP (1997c), UNEP/CBD/SBSTTA/3/INF/21, Report of the Joint CBD Secretariat, FAO Technical Consultations.

UNEP (1999), UNEP/CBD/SBSTTA/5/Inf. 10. Agricultural Biological Diversity: Assessment of Ongoing Activities and Instruments. Note by the Executive Secretary (November).

UNEP (2000), UNEP/CBD/COP/5/11. Agricultural Biological Diversity: Review of Phase I of the Program of Work and Adoption of a Multi-Year Work Program. Note by the Executive Secretary to the Conference of the Parties to the CBD, on item 16.2 of the Provisional Agenda, Fifth Meeting in Nairobi, 15–26 May 2000.

UNEP/CBD (2002), UNEP/CBD/COP/6/20. Report of the Sixth Meeting of the Parties to the Convention on Biological Diversity, The Hague, 7–19 April 2002 <www.biodiv.org/convention/cops.asp#>.

UNEP/CBD (2004), UNEP/CBD/COP/7/21. Report of the Seventh Meeting of the Parties to the Convention on Biological Diversity, Kuala Lumpur, 9–20 and 27 February 2004 <www.biodiv.org/convention/cops.asp#>.

UNEP (2004a), UNEP/CBD/COP/VII/Inf. 18. Implications of the International Treaty on Plant Genetic Resources for Food and Agriculture on the issues under Article 8(j) and related provisions. Note from the Executive Secretary of the CBD to the COP of the CBD, Seventh meeting, Kuala Lumpur, 9–20 and 27 February 2004.

UNEP (2004b), UNEP/CBD/COP Decision VII/19. Access and benefit-sharing as related to genetic resources, Paragraph D1.

UNEP/CBD (2006), UNEP/CBD/COP/8/31. Report of the Eight Meeting of the Parties to the Convention on Biological Diversity, Curitiba, 20–31 March 2006 <www.biodiv.org/convention/cops.asp#>.

GATT, TRIPS, UPOV, WIPO and WTO References

GATT (1986), General Agreement on Tariffs and Trade (GATT): Punta del Este Declaration, Ministerial Declaration of 20 September 1986 <www.sice.oas.org/trade/Punta_e.asp>.

GATT (1988), MTN.GHG/NGII/W/26. Group of Negotiations on Goods (GNG), Negotiating Group on Trade-Related Aspects of Intellectual Property Rights including Trade in Counterfeit Goods: Guidelines and Objectives Proposed by the European Community for the Negotiations of Trade-Related Aspects of Substantive Standards of Intellectual Property Rights, 7 July 1988 <http://aei.pitt.edu/5391/01/001870_1.pdf>.

GATT (1989), MTM.GNG/NGII/W/47. Group of Negotiations on Goods (GNG), Negotiating Group on Trade-Related Aspects of Intellectual Property Rights including Trade in Counterfeit Goods: Standards for Trade-Related Intellectual Property Rights. (Submission from Canada), 25 October 1989.

GATT (1990), MTN.GNG/NG11/W/70. Group of Negotiations on Goods (GNG), Negotiating Group on Trade-Related Aspects of Intellectual Property Rights including Trade in Counterfeit Goods. Draft Agreement on the Trade-Related Aspects of Intellectual Property Rights, Communication from the United States, 11 May 1990.

TRIPS (1994), Agreement on Trade-Related Aspects of Intellectual Property Rights, signed in Marrakesh, 15 April 1994 <www.wto.org/english/tratop_e/trips_e/t_agm0_e.htm>.

UPOV (1991), International Convention for the Protection of New Varieties of Plants, adopted 2 December 1961 in Paris. Revised three times in Geneva: 10 November 1972, 23 October 1978 and 19 March 1991 <www.upov.int/> and <www.upov.int/en/publications/conventions/1991/act1991.htm>.

WIPO (2000), WO/GA/26/6. Matters Concerning Intellectual Property and Genetic Resources, Traditional Knowledge and Folklore. Document prepared by the Secretariat dated 25 August 2000, for the 26th General Assembly (12th Extraordinary Session), Geneva, 25 September – 3 October 2000 <www.wipo.org/documents/en/document/govbody/wo_gb_ga/pdf/ga26_6.pdf>.

WIPO (2001), A/36/14. Agenda for Development of the International Patent System. Memorandum of the Director General dated 6 August 2001. Assemblies of the Member States of WIPO, 36th Series of Meetings, Geneva, 24 September–3 October 2001 <www.wipo.org/documents/en/document/govbody/wo_gb_ab/pdf/a36_14.pdf>.

WIPO (2003a), SCP/10/2 Draft Substantive Patent Law Treaty prepared by the International Bureau, 30 September 2003, for the Standing Committee on the Law of Patents (SPLT), Tenth Session, Geneva, 10–14 May 2004 <www.wipo.org/edocs/mdocs/scp/en/scp_10/scp_10_2.pdf>.

WIPO (2003b), Convention establishing the World Intellectual Property Organization, signed in Stockholm 14 July 1967; entered into force in 1970; amended in 1979 and 2003 <www.wipo.int/treaties/en/convention/trtdocs_wo029.html>.

WIPO (2004), SCP/10/4. Draft Substantive Patent Law Treaty [Clean text] prepared by the International Bureau, for SPLT, Tenth Session, Geneva, 10–14 May 2004 <www.wipo.org/edocs/mdocs/scp/en/scp_10/scp_10_4.doc>.

WTO (2000), IP/C/W/216. Council for Trade-Related Aspects of Intellectual Property Rights, *The Convention on Biological Diversity and the Agreement on Trade-Related Aspects of Intellectual Property Rights, Note by the Secretariat.*

WTO (2001), WT/MIN(01)/DEC/W/1. Doha Ministerial Declaration, Ministerial Conference, 4th Session, Doha, 9–14 November 2001.

WTO (2002a), IP/C/W/369. Council for Trade-Related Aspects of Intellectual Property Rights, Review of the Provisions of Article 27.3(b), Summary of Issues Raised and Points made 8 August 2002. Note by the Secretariat.

WTO (2002b), IP/C/W/383. Council for Trade-Related Aspects of Intellectual Property Rights, Review of Article 27.3(b) of the TRIPS Agreement, and the Relationship between the TRIPS Agreement and the CBD and the Protection of Traditional Knowledge and Folklore, 'A Concept Paper', 17 October 2002.

WTO (2003a), IP/C/W/400. Council for Trade-Related Aspects of Intellectual Property Rights, Article 27.3(b), the Relationship between the TRIPS Agreement and the Convention on Biological Diversity, and the Protection of Traditional Knowledge, 28 May 2003.

WTO (2003b), IP/C/W/403. Council for Trade-Related Aspects of Intellectual Property Rights, Relationship between the TRIPS Agreement and the Convention on Biological Diversity and the Protection of Traditional Knowledge, 24 June 2003.

WTO (2003c), IP/C/W/404. Council for Trade-Related Aspects of Intellectual Property Rights, Taking Forward the Review of Article 27.3(b) of the TRIPS Agreement, 26 June 2003.

WTO (2003d), IP/C/W/273/Rev.1. Council for Trade Related Aspects of Intellectual Property Rights, Review of the Provisions of Article 27.3(b), Illustrative List of Questions.

WTO (2003e), *Understanding the WTO* (Geneva: WTO).

Philippine Laws and Regulations

DENR (1989), Administrative Order No. 120: General Rules and Regulations on the Participation of NGOs in DENR Programs.

House of Representatives of the Philippines (1998), House Bill No. 1070, an Act to Provide Protection to New Plant Varieties, Establishing a National Plant Varieties Protection Board and for Other Purposes, First Regular Session of the 11th Congress.

House of Representatives of the Philippines (2000), House Bill No. 10651, an Act to Provide Protection to New Plant Varieties, Establishing a National Plant Varieties Protection Board and for Other Purposes.

House of Representatives of the Philippines (2002), House Bill 4518, an Act to Provide Protection to New Plant Varieties, Establishing a National Plant Variety Protection Board and for Other Purposes, approved 14 May 2002.

Joint DENR-DA-PCSD-NCIP (2005), Administrative Order No. 1, Series of 2005: Guidelines for Bio-prospecting Activities in the Philippines, 12 January 2005.

Manila Declaration Concerning the Ethical Utilization of Asian Biological Resources (1992), from the Seventh Asian Symposium on Medicinal Plants, Spices and Other Natural Products (ASOMPS VII), Manila, 2–7 February 1992.

National Commission on Indigenous Peoples (1998), Administrative Order No. 1, Series of 1998: IPRA, Rules and Regulations Implementing Republic Act No. 8371, Otherwise Known as '"Indigenous Peoples" Rights Act of 1997', 9 June 1998.

Philippine Council for Sustainable Development (PCSD) (1997), Philippine Agenda 21, a National Agenda for Sustainable Development, (Manila: Philippine Council for Sustainable Development) <http://pcsd.neda.gov.ph/pa21.htm>.

Republic of the Philippines (1973), Presidential Decree No. 232: Creating a Philippine Coconut Authority, 30 June <www.lawphil.net/statutes/presdecs/pd1973/pd_232_1973.html>.

Republic of the Philippines (1975), Presidential Decree No. 729, Establishing the Institute of Plant Breeding in the College of Agriculture, University of the Philippines at Los Baños, 5 June 1975 <www.lawphil.net/statutes/presdecs/pd1975/pd_729_1975.html>.

Republic of the Philippines (1976), Presidential Decree No. 1046-A, Amending Presidential Decree No. 729 by Providing for the Establishment of the National Plant Genetic Resources Laboratory in the Institute of Plant Breeding; Increasing the Area that the Institute Shall Acquire and Providing the Mode and Cost of its Acquisition; and Appropriating the Necessary Funds Therefore, 12 November 1976.

Republic of the Philippines (1985), Executive Order No. 1061, Creating the Philippine Rice Research Institute of PhilRice, 5 November 1985.

Republic of the Philippines (1986), Executive Order No. 18, Creation of Sugar Regulatory Administration, 28 May 1986 <www.sra.gov.ph/SRA%20EO%2018.html>.

Republic of the Philippines (1989), Republic Act No. 6766, An Act Providing for an Organic Act for the Cordillera Autonomous Region.

Republic of the Philippines Senate and Congress (1992), Republic Act 7308, Seed Industry Development Act, an Act to Promote and Develop the Seed Industry in the Philippines and Create a National Seed Industry Council and for Other Purposes, approved 27 March 1992 <www.chanrobles.com/republicactno7308.htm>.

Republic of the Philippines, Senate and House of Representatives (1992), Republic Act 7586, NIPAS, National Integrated Protected Areas System Act, an Act Providing for the Establishment and Management of National Integrated Protected Areas System, Defining its Scope and Coverage, and for other Purposes, 1 June 1992 <www.tanggol.org/environmental_laws/ra7586.html>.

Republic of the Philippines (1992a), Republic Act No. 7607, Magna Carta for Small Farmers. An Act Providing a Magna Carta for Small Farmers, 4 June 1992 <www.chanrobles.com/republicactno7607.htm>.

Republic of the Philippines (1992b), Administrative Order No. 25, Series of 1992: National Protected Areas System (NIPAS) Implementing Rules and Regulations, 29 June 1992, Department of Environment and Natural Resources (DENR) <www.tanggol.org/environmental_laws/dao25.html>.

Republic of the Philippines (1992c), Executive Order No. 15, Office of the President, Creating a Philippine Council for Sustainable Development (PCSD), 1 September 1992 <http://pcsd.neda.gov.ph/eo15.htm>.

Republic of the Philippines (1993), Memorandum Order No. 126, Implementing Kasagaanna ng Sakaban at Kalikasan (Kasakalikasan), National Integrated Pest Management Program, 3 May 1993.

Republic of the Philippines (1995a), Republic Act 7900, High-Value Crops Development Act, an Act to Promote the Production, Processing, Marketing and Distribution of High-Valued Crops, Providing Funds Therefore, and for Other Purposes, approved 23 February 1995 <www.chanrobles.com/republicactno7900.htm>.

Republic of the Philippines (1995b), Republic Act 8048, Coconut Preservation Act, an Act Providing for the Regulation of the Cutting of Coconut Trees, its Replenishment, Providing Penalties Therefore and for Other Purposes <www.chanrobles.com/republicactno8048.htm>.

Republic of the Philippines (1995c), Executive Order 247, Prescribing Guidelines and Establishing a Regulatory Framework for the Prospecting of Biological and Genetic Resources, Their By-Products and Derivatives, for Scientific and Commercial Purposes and For Other Purposes, 18 May 1995 <www.pctc.gov.ph/laws/s95EO247.htm>.

Republic of the Philippines (1995d), Memorandum Order No. 288, PCSD, Philippine Council for Sustainable Development, Directing the Formulation of the Philippines' Agenda 21 and Activating its Formulation Process, 5 July 1995 <http://pcsd.neda.gov.ph/mo288.htm>.

Republic of the Philippines (1995e), Memorandum Order 289, National Biodiversity Strategy and Action Plan, Directing the Integration of the Philippines' Strategy for Biological Diversity Conservation in the Sectoral Plans, Programs and Projects of the National Government Agencies and the Operationalization of the Objectives of Sustainable Biological Diversity Resources Management and Development, 5 July 1995.

Republic of the Philippines Senate and Congress (1996), Republic Act 8178, Agricultural Tariffication Act, an Act Replacing Quantitative Import Restrictions on Agricultural Products, except Rice, with Tariffs, Creating the Agricultural Competitiveness Enhancement Fund, and for Other Purposes, approved 28 March 1996 <www.marina.gov.ph/policy/ra/RA8178.htm>.

Republic of the Philippines (1996a), Department Administrative Order (DAO) No. 96-20, On the Prospecting of Biological and Genetic Resources (Implementing Rules and Regulations for EO 247), Department of Environment and Natural Resources, 21 June 1996 <http://66.249.93.104/search?q=cache:Gqm929Sv26cJ:www.sipo.

gov.cn/sipo2006/ztxx/yczyhctzsbh/zlk/gglf/200509/P020060403599027505930.pdf+philippines+DAO+96-20&hl=no&gl=no&ct=clnk&cd=2>.

Republic of the Philippines (1996b), Memorandum Order No. 399, 1996: PCSD, Philippine Council for Sustainable Development, Manila, Directing the Operationalization of the Philippine Agenda 21 and Monitoring its Implementation, 26 September 1996 <http://pcsd.neda.gov.ph/mo399.htm>.

Republic of the Philippines (1996c), Executive Order No. 370: PCSD, Strengthening the Philippine Council for Sustainable Development, 26 September 1996 <http://pcsd.neda.gov.ph/eo370.htm>.

Republic of the Philippines (1997a), Republic Act No. 8293, Intellectual Property Code of the Philippines, An Act Prescribing the Intellectual Property Code and Establishing the Intellectual Property Office, Providing for Its Powers and Functions, and for Other Purposes, 6 June 1997 <www.lawphil.net/statutes/repacts/ra1997/ra_8293_1997.html>.

Republic of the Philippines (1997b), Republic Act 8435, 1997: Agricultural and Fisheries Modernization Act (AFMA), An Act Prescribing Urgent Related Measures to Modernize the Agriculture and Fisheries Sector of the Country in Order to Enhance Their Profitability, and Prepare Said Sectors for the Challenges of Globalization through an Adequate, Focused and Rational Delivery of Necessary Support Services, Appropriating Funds Therefore and for Other Purposes, adopted by 10th Congress, Third Regular Session, 28 July 1997 <www.da.gov.ph/AFMA/ra8435a.html>.

Republic of the Philippines (1997c), Republic Act 8371: Indigenous Peoples' Rights Act (IPRA), an Act to Recognize, Protect and Promote the Rights of Indigenous Cultural Communities/Indigenous Peoples, Creating a National Commission on Indigenous Peoples, Establishing Implementing Mechanisms, Appropriating Funds Therefore, and for Other Purposes, adopted by 10th Congress, Third Regular Session, 29 October 1997 <www.iapad.org/pafid/docs/IPRA.pdf>.

Republic of the Philippines (1998a), Administrative Order No. 6, Implementing Rules and Regulations Pursuant to Republic Act 8435: Agriculture and Fisheries Modernization Act of 1997, Department of Agriculture, 10 July 1998 <www.da.gov.ph/agrilaws/ra/afma.pdf>.

Republic of the Philippines (1998b), First Philippine National Report to the Convention on Biological Diversity, Department of Environment and Natural Resources, Protected Areas and Wildlife Bureau (PAWB) <www.biodiv.org/doc/world/ph/ph-nr-01-en.pdf>.

Republic of the Philippines (2001a), Administrative Order No. 05, Series of 2001, Establishment of the Philippine Corn Research Center under the Department of Agriculture, Department of Agriculture, 11 January 2001 <www.da.gov.ph/agrilaws/AO2001/AO_05s01.html>.

Republic of the Philippines (2001b), Administrative Order No. 04, Series of 2001, Establishment of the Philippine Tropical Fruit Research Center (Philfruits) under the Department of Agriculture, Department of Agriculture, 11 January 2001 <www.da.gov.ph/agrilaws/AO2001/AO_04s01.html>.

Republic of the Philippines (2001c), Republic Act 9147: Wildlife Resources Conservation and Protection Act of 2001, an Act Providing for the Conservation and Protection of Wildlife Resources and Their Habitats, Appropriating Funds Therefore and for Other Purposes, 30 July 2001 <www.chanrobles.com/republicactno9147.html>.

Republic of the Philippines (2001d), Executive Order No. 62: PCSD, Further Strengthening the Philippine Council for Sustainable Development. 10 December 2001.

Republic of the Philippines (2001e), Administrative Order 25, 2001: Hybrid Rice Commercialization Program, 12 December 2001.

Republic of the Philippines (2002a), Executive Order 76: Providing for the Implementation of the Hybrid Rice Program Transferring the Philippine Rice Research Institute from the Department of Agriculture to the Office of the President and for Other Purposes, 4 March 2002.

Republic of the Philippines (2002b), Second Philippine National Report to the Convention on Biological Diversity, Department of Environment and Natural Resources, Protected Areas and Wildlife Bureau (PAWB) <www.biodiv.org/doc/world/ph/ph-nr-02-en.pdf>.

Republic of the Philippines (2002c), Republic Act No. 9168: The Plant Variety Protection Act (PVP Act), an Act to Provide Protection to New Plant Varieties, Establishing a National Plant Variety Protection Board and for Other Purposes, approved by President Gloria Macapagal-Arroyo on 7 June 2002 <www.ecap-project.org/fileadmin/ecapII/pdf/en/information/philippines/act_9168_plant_variety_protection.pdf>.

Republic of the Philippines (2003), Administrative Order No. 07, Series of 2003, Implementing Rules and Regulations of the Philippine Plant Variety Protection Act of 2002, Department of Agriculture, February 2003.

Republic of the Philippines (2004), Joint DENR-DA-PCSD Administrative Order No. 01, Series of 2004: Joint Implementing Rules and Regulations (IRR) Pursuant to Republic Act no. 9147: 'An Act Providing for the Conservation and Protection of Wildlife Resources and Their Habitats, Appropriating Funds Therefore and for other Purposes', Palawan Council for Sustainable Development, Puerto Princesa City, Palawan, 18 May 2004 <www.pcsd.ph/resolutions/resolutions/wildlife/jointao01.htm>.

Senate of the Philippines (1995), Senate S. No. 277/1995. A Senate initiative for a sui generic legislation, Act Providing for the Registration and Protection of Plant Varieties and for Other Purposes.

Senate of the Philippines (1998), Senate Bill 108. Committee on Agriculture: An Act to Provide Protection to New Plant Varieties, Establishing a National Plant Variety Protection Board and for Other Purposes.

Senate of the Philippines, (1999a). Senate Bill 1618. Committee on Agriculture: An Act to Provide Protection to New Plant Varieties, Establishing a National Plant Variety Protection Board and for Other Purposes.

Senate of the Philippines (1999b), Senate Bill 1912. Committee on Agriculture: An Act to Provide Protection to New Plant Varieties, Establishing a National Plant Variety Protection Board and for Other Purposes.

Senate of the Philippines (2001a), Senate Bill No. 62. Committee on Agriculture: An Act to Provide Protection to New Plant Varieties, Establishing a National Plant Variety Protection Board and for Other Purposes.

Senate of the Philippines (2001b), Senate Bill No. 170. Committee on Agriculture: An Act to Provide Protection to New Plant Varieties, Establishing a National Plant Variety Protection Board and for Other Purposes.

Senate of the Philippines (2001c), Senate Bill No. 967. Committee on Agriculture: An Act to Provide Protection to New Plant Varieties, Establishing a National Plant Variety Protection Board and for Other Purposes.

Senate of the Philippines (2002), Senate Bill No. 1865. Committees on Agriculture and Food: An Act to Provide Protection to New Plant Varieties, Establishing a National Plant Variety Protection Board and for Other Reasons, 11 March 2002.

Interviews

During fieldwork in the Philippines in May 2000 and March 2002, I conducted 60 interviews with 60 people altogether. Some of these were interviewed several times; others were interviewed in groups, this is indicated in the list below. Moreover, I have conducted interviews at FAO and with other relevant informants, 12 interviews altogether. In addition, I have used the material I collected in an international stakeholder questionnaire survey of the Farmers' Rights Project (see <www.farmersrights.org>), covering 60 respondents from 31 countries (22 developing countries) (see Andersen 2005b). The following are the lists of interviews directly related to this book.

Interviews in the Philippines, May 2000

1 May, Nicanor Perlas, President, Center for Alternative Development Initiatives (CADI), Quezon City.

2 May, Maria Lourdes M. Lagarde, Assistant Director, National Project Management Unit, Governance for PA 21 Project, Philippine Council for Sustainable Development (PCSD), National Economic Development Authority (NEDA), Pasig City.

3 May, Dr Michael Jackson, Head, Genetic Resources Center, International Rice Research Institute (IRRI), Los Baños.

3 May, Dr Oscar B. Zamora, Professor of Agronomy, Department of Agronomy, University of the Philippines, Los Baños (UPBL).

3 May, Rene Velvé, Head, Philippine Office, Genetic Resources Action International (GRAIN), Los Baños.

3 May, Dr Pamela G. Fernandez, Associate Professor, Seed Science and Technology Division (SSTD), Department of Agronomy, UPBL.

3 May, Réné Salazar, initiator and board member of SEARICE, for Chair of the CBDC, independent consultant, Quezon City.

4 May, Amparo C. Ampil, Officer-in-Charge, Policy Advocacy and Legislative Support Division (group interview), Department of Agriculture (DA), Quezon City.

4 May, Marga Domingo, Planning Officer, Policy Advocacy and Legislative Support Division (group interview), DA, Quezon City.

5 May, Dr Dolores A. Ramirez, University Professor, Institute of Plant Breeding (IPB), UPBL.

5 May, Dr Pamela G. Fernandez, (second interview, see 3 May), Department of Agronomy, UPBL.

5 May, Dr Agnes Rola, University Professor, Institute of Strategic Planning and Policy Studies, UPBL.

5 May, Dr Gelio T. Castillo, Rural Sociologist and National Researcher of the Philippines, IRRI, Los Baños.

5 May, Manny Yap, Director,Magsasaka at Siyentipiko Para sa Pag-unlad ng Agrikultura (MASIPAG), Los Baños.

6 May, Nicanor Perlas, (second interview, see 1 May), CADI, Quezon City.

Interviews in the Philippines, March 2002

4 March, Giacomo Rambaldi, Advisor, Information and Communication Specialist, ASEAN Regional Centre for Biodiversity Conservation (ARCBC), Quezon City.

4 March, Arne Erik Jensen, Co-director for the World Bank Technical Assistance to Biodiversity Conservation (TABC) in the Philippines, Denmark's International Environmental Assistance Advisory Committee, Nordic Agency for Development and Ecology (NORDECO), Quezon City.

4 March, Altea Lota, Assistant Director, Protected Areas and Wildlife Bureau (PAWB), Department Environment and Natural Resources (DENR), Quezon City.

6 March, Arne Erik Jensen (second interview, see 4 March), NORDECO, Quezon City.

6 March, Réné Salazar (second interview, see 3 May 2000), Independent consultant, Quezon City.

7 March, Francisco P. Fellizar, Deputy Director for Professional Services and Administration (group interview), Southeast Asian Ministers of Education Organization (SEAMEO), Regional Center for Graduate Study and Research in Agriculture (SEARCA), Los Baños.

7 March, Mariliza, V. Ticsay, Monitoring and Evaluation Specialist, Biodiversity Research Project, Natural Resources Management Program (group interview), SEAMEO/SEARCA, Los Baños.

7 March, Sonny P. Tababa, Network Administrator for the Biotechnology Information Center (group interview), SEAMEO/SEARCA, Los Baños.

7 March, Gil C. Sguiguit, Manager for Research and Development (group interview), SEAMEO/SEARCA, Los Baños.

7 March, John MacKinnon, Co-director, ASEAN Regional Center for Biodiversity Conservation, Los Baños.

7 March, Dr Pamela G. Fernandez (third interview, see 3 May 2000), Department of Agronomy, UPBL.

7 March, Dr Oscar B. Zamora (second interview, see 3 May 2000), Department of Agronomy, UPBL.

8 March, Nicanor Perlas (third interview, see 1 May 2000), CADI, Pasig City (moved from Quezon City since 2000).

8 March, Angelita P. Meniado, Supervisor and Officer in Charge of the Planning Staff, Protected Areas and Wildlife Bureau (PAWB) of the Department of Environment and Natural Resources (DENR), Quezon City.

8 March, Norma Molinyawe, Officer in Charge, Protected Areas Division, Protected Areas and Wildlife Bureau (PAWB) of the Department of Environment and Natural Resources (DENR), Quezon City.

12 March, Dr. Saturnina C. Halos, Senior Project Development Advisor (BAR); Consultant, DNA Analysis Laboratory, NCRI; and member of the Interagency Committee on Biological and Genetic Resources (IACBGR), Bureau of Agricultural Research (BAR) at Department of Agriculture (DA); and Natural Sciences Research Institute (NCRI), at the University of the Philippines, Diliman, Quezon City.

12 March, Zandro Racoma, Engineer III, Environmental Impact Assessment Division (IEA Division), Environmental Management Bureau (EMB) of the Department of Environment and Natural Resources (DENR), Quezon City.

13 March, Dr Eliseo R. Ponce, Director, Bureau of Agricultural Research, Department of Agriculture, Quezon City.

13 March, Dr Segfred R. Serrano, Assistant Secretary, Policy and Planning, Department of Agriculture (DA), Quezon City.

13 March, Wilbur Dee, Program Manager of NIPA for the CPPAP project, NIPA, Quezon City.

14 March, Dr Patricio S. Faylon, Director, PCCARD, Los Baños.

14 March, Richard M. Juanillo, Director of the Planning and Development Division, PCCARD, Los Baños.

14 March, Concepcion A.E. Magboo, Supervising Science Research Specialist in the Technology Outreach and Promotion Division (group interview), PCCARD, Los Baños.

14 March, Juanito T. Batalon, Country Officer of PROSEA (group interview), PCCARD, Los Baños.

14 March, Dr. Oscar B. Zamora, Professor of Agronomy, Department of Agronomy, University of the Philippines, Los Baños.

15 March, Diego G. Ramos, Branch Manager, PhilRice, Los Baños.

15 March, Nestor C. Altoveros, Assistant Director for Plant Genetic Resources (group interview), National Plant Genetic Resources Laboratory (NPGRL), Institute of Plant Breeding (IPB), Los Baños.

15 March, Felipe S. Dela Cruz, University Researcher and Head (group interview), National Plant Genetic Resources Laboratory (NPGRL), Institute of Plant Breeding (IPB), Los Baños.

15 March, Gregorio D. Reyes, Forester and Chief Science Research Specialist in the Upland Farms Ecosystm Research Division, Ecosystems and Research Development Bureau (ERDB) of the Department of Environment and Natural Resources (DENR), Los Baños.

15 March, Aida B. Lapis, Supervising Science Research Specialist in Rattan Taxonomy, Forest Ecology and Silvics at the Grasslands and Degraded Areas Research Division (group interview), Ecosystems and Research Development Bureau (ERDB) of the Department of Environment and Natural Resources (DENR), Los Baños.

15 March, Simplicia A. Pasicolan, Senior Science Research Specialist of the Coastal Zones and Freshwater Ecosystems Research Division (group interview), Ecosystems and Research Development Bureau (ERDB) of the Department of Environment and Natural Resources (DENR), Los Baños.

15 March, Lope A. Calanog, Technical Expert on Biodiversity Conservation, Protected Area Management and Community Based Natural Resources Management (group interview), Ecosystems and Research Development Bureau (ERDB) of the Department of Environment and Natural Resources (DENR), Los Baños.

15 March, Eliseo M. Baltazar, Supervising Science Research Specialist, Ecosystems and Research Development Bureau (ERDB) of the Department of Environment and Natural Resources (DENR), Los Baños.

18 March, Julio Galvez Tan, Executive Director, Foundation for the Philippine Environment (FPE), Quezon City.

18 March, Hector D. Soliman, Legal Advisor to the Secretary of Agriculture at the Office of the Secretary, Department of Agriculture (DA), Quezon City.

18 March, Neth Daño, Executive Director, SEARICE.

19 March, Filemon G. Romero, Oceanographer and Consultant (group interview), Kabang Kalikasan ng Pilipinas, which is the Philippine Branch of the Worldwide Fund for Nature, Quezon City.

19 March, Jose Angelito Palma, Director (group interview), Kabang Kalikasan ng Pilipinas, which is the Philippine Branch of the Worldwide Fund for Nature, Quezon City.

19 March, Jose Padilla, Senior Policy Officer (group interview), Kabang Kalikasan ng Pilipinas, which is the Philippine Branch of the Worldwide Fund for Nature, Quezon City.

19 March, Carmen L. Villaseñor, Conservation Officer, Kabang Kalikasan ng Pilipinas, which is the Philippine Branch of the Worldwide Fund for Nature, Quezon City.

19 March, Mayo Grace C. Amit, Supervising Economic Development Specialist (agricultural staff), Philippine Council for Sustainable Development (PCSD) at the National Economic and Development Authority (NEDA), Pasig City.

20 March, Dr. Doracie B. Zoleta-Nantes, Associate Professor, College of Social Sciences and Philosophy, Department of Geography, University of the Philippines, Diliman, Quezon City.

20 March, Maria Teresa S. Robielos, Economic Policy Assistance Manager, Office of Economic Development and Governance, US Agency for International Development (USAID), Pasay City.

20 March, Fatima S. Verzosa, Women in Development and Project Development Specialist, Program Resources Management Office, US Agency for International Development (USAID), Pasay City.

20 March, Nicanor Perlas, President, Center for Alternative Development Initiatives (CADI), Pasig City.

21 March, Marleen van Luijk, Senior Plant Breeder (group interview), East-West Seed Company, Inc., San Rafael, Bulacan.

21 March, Martin Hinlo, Senior Plant Breeder (group interview), East-West Seed Company, Inc., San Rafael, Bulacan.

21 March, Mary Constancy C. Barrameda, Faculty Member, Department of Social Anthropology, College of Social Sciences and Philosophy, University of the Philippines, Diliman, Quezon City.

21 March, Dr Klaus Liebig, Economist, German Institute of Development Politics, Bonn (interviewed together with his project staff in Quezon City).

22 March, Leo F. Solomon, Sales and Marketing Division Manager (EWS), and President of the PSIA, East-West Seed Company, Inc., San Rafael, Bulacan; and Philippines Seed Industry Association (short dialogue at seminar in Quezon City).

22 March, Jason Cainglet, Advocacy Officer, Integrated Rural Development Foundation of the Philippines (short dialogue at seminar in Quezon City).

22 March, Dr. Peter H. Foerester, Adviser/ Natural Resource Development for the Bondoc Development Program People Peace Progress, German Gesellschaft für Technische Zusammenarbeit (GTZ) (short dialogue at seminar in Quezon City).

22 March, Réné Salazar, initiator and board member of SEARICE, independent consultant, Quezon City.

22 March, Mary Constancy C. Barrameda, Faculty Member, Department of Social Anthropology, College of Social Sciences and Philosophy, University of the Philippines, Diliman, Quezon City.

25 March, Ma. Fe Esperanza V. Madamba, Advocacy Specialist, Accelerating Growth, Investment, and Liberalization with Equity (AGILE) and Consultant to the Department of Agriculture (DA), Development Alternatives Inc. (DAI), contracted for the implementation of AGILE (Pasig City) of the USAID.

25 March, Ernesto S. Guiang, Natural Resources Management Specialist in the Division for Rural Development and Natural Resources, World Bank, Philippines, Pasig City.

Interviews in Europe

November 2003, on several occasions. Grethe Helene Evjen, Senior Advisor and leader of Norwegian delegations to FAO (PGRFA related), Ministry of Agriculture and Food, Norway (interviews in Rome).

7 November 2003, Lim Eng Siang, Principal Assistant Secretary (leading several of the negotiation rounds for the ITPGRFA), Ministry of Agriculture, Malaysia (interviewed in Rome).

7 November 2003, Radha Ranganathan, Director, Technical Affairs, International Seed Federation (interviewed in Rome).

10 November 2003, Jose T. Esquinas-Alcazar, Secretary, CGRFA, FAO, Rome.

10 November 2003, Clive Stannard, Assistant Secretary of the CGRFA, FAO, Rome.

10 November 2003, Martin Eric Smith, Senior Officer, Secretary of CGRFA, FAO, Rome.

10 November 2003, Brad Fraleigh, Senior Officer, Seed and Plant Genetic Resources Policy, Seed and Plant Genetic Resources Services, and Plant Production and Protection Division, FAO, Rome.

11 November 2003, Jan Engels, Director, Genetic Resources Science and Technology Group, IPGRI, Italy.

11 November 2003, Michael Halewood, Scientist and Legal Specialist, IPGRI, Italy.

11 November 2003, Jamie Watts, Scientist, Impact Assessment and Evaluation Specialist, IPGRI, Italy.

On several occasions: Peter J. Schei, Director and leader of Norwegian delegations to the CBD-COP, The Fridtjof Nansen Institute, Norway.

On several occasions: Cary Fowler, then Professor, NORAGRIC, Norway.

Index